CARBENES

VOLUME I

REACTIVE INTERMEDIATES IN ORGANIC CHEMISTRY

Edited by GEORGE A. OLAH

Case Western Reserve University

A series of collective volumes and monographs on the chemistry of all the important species of organic reaction intermediates:

CARBONIUM IONS
Edited by George A. Olah of Case Western Reserve University and Paul v. R. Schleyer of Princeton University: Vol. I (1968), Vol. II (1970), Vol. III (1972), Vol. IV (1973), Vol. V (in preparation)

RADICAL IONS
Edited by E. T. Kaiser of the University of Chicago and L. Kevan of the University of Kansas (1968)

NITRENES
Edited by W. Lwowski of New Mexico State University (1970)

CARBENES
Edited by Maitland Jones, Jr., of Princeton University and Robert A. Moss of Rutgers University, The State University of New Jersey: Vol. I (1973), Vol. II (in preparation)

FREE RADICALS
Edited by J. K. Kochi of Indiana University: Vol. I (1973), Vol. II (1973)

Planned for the Series

CARBANIONS
By M. Szwarc of the State University of New York, College of Forestry, Syracuse

ARYNES
Edited by M. Stiles of the University of Michigan

CARBENES

Edited by

MAITLAND JONES, JR.

Department of Chemistry
Princeton University
Princeton, New Jersey

ROBERT A. MOSS

Wright Laboratory
School of Chemistry
Rutgers University
The State University of New Jersey,
New Brunswick, New Jersey

VOLUME I

A WILEY-INTERSCIENCE PUBLICATION

JOHN WILEY & SONS, New York · London · Sydney · Toronto

Library of Congress Cataloging in Publication Data:

Main entry under title:

Carbenes.
(Reactive intermediates in organic chemistry)

1. Carbenes. I. Jones, Maitland, 1937– ed.
II. Moss, Robert A., ed. III. Series.

QD305.H7C35 547′.412 72-6324

ISBN 0-471-44740-4

Printed in the United States of America

10-9 8 7 6 5 4 3 2 1

To Sandy and Susan

Authors of Volume I

WILLIAM J. BARON, *Department of Chemistry, Princeton University, Princeton, New Jersey*

N. R. BERTONIERE, *Department of Chemistry, Louisiana State University in New Orleans, New Orleans, Louisiana*

MARK R. DECAMP, *Department of Chemistry, Princeton University, Princeton, New Jersey*

GARY W. GRIFFIN, *Department of Chemistry, Louisiana State University in New Orleans, New Orleans, Louisiana*

MICHAEL E. HENDRICK, *Department of Chemistry, Princeton University, Princeton, New Jersey*

MAITLAND JONES, JR., *Department of Chemistry, Princeton University, Princeton, New Jersey*

RONALD H. LEVIN, *Department of Chemistry, Princeton University, Princeton, New Jersey*

ROBERT A. MOSS, *School of Chemistry, Rutgers University, The State University of New Jersey, New Brunswick, New Jersey*

MIRIAM B. SOHN, *Department of Chemistry, Princeton University, Princeton, New Jersey*

Introduction to the Series

Reactive intermediates have always occupied a place of importance in the spectrum of organic chemistry. They were, however, long considered only as transient species of short life-time. With the increase in chemical sophistication many reactive intermediates have been directly observed, characterized, and even isolated. While the importance of reactive intermediates has never been disputed, they are usually considered from other points of view, primarily relative to possible reaction mechanism pathways based on kinetic, stereochemical and synthetic chemical evidence. It was felt that it would be of value to initiate a series that would be primarily concerned with the reactive intermediates themselves and their impact and importance in organic chemistry. In each volume, critical, but not necessarily exhaustive coverage is anticipated. The reactive intermediates will be discussed from the points of view of: formation, isolation, physical characterization and reactions.

The aim, therefore, is to create a forum wherein all the resources at the disposal of experts in the field could be brought together to enable the reader to become acquainted with the reactive intermediates in organic chemistry and their importance.

As the need arises, it is anticipated that supplementary volumes will be published to present new data in this rapidly developing field.

Preface

The renaissance of carbene chemistry in recent years had its origins in the pioneering work of Meerwein, Hine, Doering, and Skell. By the late 1950s the ground had been prepared for rapid advances in both mechanistic understanding and synthetic expertise. Highlights of recent years have included the physical studies of Herzberg, Hutchison, Wasserman, Closs and their co-workers, the development of the carbenoid concept by Closs, Miller, Köbrich, and their collaborators, and the introduction of new synthetic techniques such as Seyferth's route to halocarbenes using organomercurials, and the Simmons-Smith, Bamford-Stevens, and Gaspar-Roth procedures.

The past decade has seen carbene chemistry change from a compendium of novelties to a discipline requiring close attention to the intricate details of mechanism. For example, careful identification of the electronic state and degree of complexation became important considerations. Carbenes began to assume a more varied role in synthesis as well. No longer were small rings the sole targets of carbene chemists; larger molecules such as presqualene alcohol, biologically active heterocyclic compounds, and modified steroids became important.

During the 1960s widely scattered reviews appeared, and the fine first-generation texts of Hine and Kirmse were published. Now second-generation texts are beginning to appear. It seems likely to us that individuals will find it difficult to cope with the entire field in a *critical* way. We hope that these critical surveys of topical areas will prove useful both to those in the midst of things and to those who wish to plunge in. The first volume of "Carbenes" is offered in this spirit.

MAITLAND JONES, JR.

ROBERT A. MOSS

Princeton, New Jersey
Cambridge, Massachusetts
May 1972

Contents

CHAPTER 1

Carbenes from Diazo Compounds

William J. Baron, Mark R. DeCamp,* Michael E. Hendrick,†
Maitland Jones, Jr., Ronald H. Levin,‡ and Miriam B. Sohn§
Department of Chemistry, Princeton University, Princeton, New Jersey 08540

* University Fellow, 1969–1970.
† NSF Predoctoral Fellow, 1967–1971.
‡ NDEA Fellow, 1967–1970.
§ NDEA Fellow, 1969–1970.

I. INTRODUCTION

Our subject is an enormous one, for carbene chemistry is useful and interesting to an amazingly broad spectrum of chemists. There is scarcely an active organic chemist who has not turned his attention in one way or another to carbene chemistry. Synthetic chemists have been attracted not only by the simple route to cyclopropanes available through carbenes, but also by the ability of divalent carbon to form bridged or multicyclic compounds by intramolecular insertions or cycloadditions. It is no exaggeration to claim a major role for carbenes in the modern chemist's attitude that he can very probably make anything he wants.

A second group of chemists has been drawn to carbene chemistry by the well-deserved aura of mystery surrounding the mechanisms of even the simplest reactions of divalent carbon. Despite their efforts, there is today little understanding of the cycloaddition and, especially, the insertion reaction of carbenes. This attention of the chemical community has manifested itself in roughly two to three reviews a year since the books of Kirmse[1] and Hine[2] in 1964.[3–18a]

The breadth of our subject, the properties of carbenes formed from diazo compounds, requires or at least permits us to be selective both in broad outline and in detail. We shall be arbitrary in both respects, ignoring for the most part metal-catalyzed decompositions of diazo compounds but including under the umbrella of diazo compounds such progenitors as tosylhydrazones and diazirenes. An occasional ketene will creep in as well, especially in the discussion of methylene.

No attempt to be comprehensive has been made. Rather, we have tried to select typical examples to make our points. Throughout the review we have distinguished between percentage yields and percent compositions by using a percent sign (%) for the former and parentheses ( ) for the latter.

II. METHYLENE

We shall continually contrast the chemistry of methylene in solution with that in the gas phase. Solution chemistry is relatively simple because of the extreme reactivity of singlet methylene and the absence of rearrangements of the "hot" molecules produced by the extremely exothermic reactions of methylene. The singlet state is trapped by virtually anything before it is able to undergo intersystem crossing to the ground triplet state. Herzberg and his collaborators[19–21] provided the first experimental evidence that a metastable, bent singlet was initially produced and decayed to a ground state triplet. It was first thought that the triplet was linear [19–21,35] but both theory and experiment now agree that triplet methylene is bent at an angle of approximately 137°.[22–34] There has been substantial debate over the energy difference between the lowest singlet and the ground state triplet. Many values have been calculated, but the most recent and complete efforts agree with the experimental estimate of Herzberg and Johns[21] of ca. 0.5 eV.[36]

Much smaller values had been previously suggested by two groups on thermodynamic grounds[37] and on the basis of RRKM calculations.[38] The latter estimate may not be accurate as it depends both on the assumption that methylene can survive many collisions before reaction and the accuracy of the RRKM calculations of the rate of dissociation of activated ethane.[36] The former work has been reevaluated by Frey[38a] who came to the conclusion that the data are best reconciled with an energy gap of ca. 8 kcal/mole.

Interpretation of work done in the gas phase is most difficult because of the isomerization of products of the reactions of methylene, and because both spin states (produced, perhaps, in several different ways) are often present. By taking 92 kcal/mole as an average estimate of the heat of formation of methylene,[39–42] it can be seen how the initially formed molecules might well isomerize if not rapidly stabilized, as they are in solution. For example, consider cyclopropane formation:

$$\begin{array}{ccc} CH_2=CH_2 & + :CH_2 & \rightarrow \triangle \\ +12.5 & +92 & +12.7 \\ & \Delta H \approx 92 & \end{array}$$

Here ΔH is about 92 kcal/mole, and the activation energy for isomerization of cyclopropane to propylene is about 65.9 kcal/mole.[43] That for geometrical isomerization is even lower, 65.1.[44]

In addition, the gas phase chemistry of methylene is enlivened by several strange reactions which are most difficult to explain. We shall contrast the reactions of methylene in the liquid and gas phases as we discuss the individual reactions.

A. Addition to Olefins

Early work in this area which demonstrated the complications just noted included that of Frey,[45,46] who decomposed diazomethane in the 2-butenes and varying pressures of inert gas at 278–2100 torr. As the pressure increased, the yield of cyclopropanes increased, as did the stereospecificity of the addition. Presumably increased pressure produced more collisional deactivation of the initially formed "hot" cycolpropanes. A parallel investigation of additions to hexene-3 showed similar results.[47] When higher pressures were used, new results appeared, and in 1960 it was discovered that between 2100 and 3200 torr, where collisional deactivation should be very efficient, the reaction was quite nonstereospecific.[48,49] This was attributed to collisionally induced intersystem crossing from the initially formed singlet to the more stable triplet. Small amounts of added oxygen increased the stereospecificity remarkably as, presumably, the triplet was selectively scavenged by oxygen.

As these experiments suggest, it was later found that two sources of the "wrong" cyclopropane existed and that its production went through a minimum.[50] As the pressure of inert gas increases, initially "hot molecule" rearrangements are suppressed and the stereospecificity increases. At high pressures of inert gas, intersystem crossing appears and the stereospecificity declines again. Different inert gases are of varying efficiency in these processes.

Entry to the triplet can be gained directly by the mercury-photosensitized decomposition of ketene. Here, as expected, a species is produced which is nonstereospecific in the addition reaction,[51] although it is not certain that pure triplet methylene is produced.[52,53] The mercury-sensitized decomposition of dideuterioketene has been used to resolve a question regarding the source of the nonstereospecific addition of triplet methylene.[54] This can be explained in three different ways. The first involves what has come to be known as the Skell hypothesis. This explanation posits that a triplet carbene will be prevented from forming two new bonds simultaneously with the paired π-electrons of the double bond and will therefore lead to a diradical, which must change the spin of one electron before becoming a cyclopropane. The resulting time interval permits rotation to occur. An alternative proposes the formation of a triplet cyclopropane from triplet methylene and ethylene.[55]

A third formulation, by Hoffmann,[56] points out that singlet methylene plus ethylene correlates with a trimethylene in its lowest singlet state and thus with the ground state of cyclopropane. The net result is a rapid closing of an initially formed trimethylene. Triplet methylene, on the other hand, correlates with a triplet excited state of trimethylene in which rotation is most easy. Hoffmann goes on to comment that his calculation indicates that the lowest triplet state of cyclopropane should not retain the cyclopropane geometry but go to the triplet trimethylene.

A distinction has been made between the triplet cyclopropane mechanism and the others.[54] In the reaction of dideuteriomethylene with ethylene, several products beside cyclopropane are formed, among them *n*-butane, pentene-1, and hexadiene-1,5. These arise from the various combinations of ethyl and allyl radicals. The allyl radical itself is formed by isomerization of the product of addition of CD_2 and ethylene (either a triplet cyclopropane or trimethylene). As predicted by the trimethylene mechanism the *n*-butane

$CH_2{=}CH_2$ ← $:CD_2$ → $CH_2{=}CH_2$

H· + D, D allyl radical —($CH_2{=}CH_2$)→ $C_2H_5\cdot$ + $C_5H_7D_2\cdot$

n-butane-D_0 1-pentene-D_2

H· + D, D allyl radical; D· + D allyl radical —($CH_2{=}CH_2$)→ $C_2H_5\cdot$, $C_2H_4D\cdot$; $C_5H_7D_2$, $C_5H_8D\cdot$

n-butane-D 1-pentene (less than D_2)

produced is 99.3% D_0 and the pentene-1 95% D_2. A *symmetrical* triplet cyclopropane cannot be an intermediate.

These aforementioned experiments give the broad outlines of the reactions of methylene with double bonds in the gas phase. More recent work has concentrated on mechanistic details and little agreement is found here. For instance, a variety of experiments has shown that the mix of singlet and triplet methylene is dependent on the wavelength, and hence energy, of the incident light. As the energy increases, the proportion of singlet methylene also increases.[47,57–63] The estimates of the percentage of triplet methylene formed from ketene at 3130 Å range from 14[62] to 37[59] %, with a recent estimate at 29%.[60]

Much of the foregoing work has been with ketene. A recent study of the photolysis of diazomethane at 3660 and 4358 Å shows that the portion of ground state triplet molecules produced is independent of the wavelength.[64]

Generally, the ratio of *cis*- and *trans*-1,2-dimethylcyclopropane formed from methylene and 2-butene has been used as a measure of the singlet/triplet mix.[65,66] This has been criticized,[67] and it was pointed out that the distribution of paths followed by the diradical **1** produced from triplet methylene and a butene varied with pressure even under conditions thought to give 100% triplet methylene. However, the concept of conditions producing 100%

$:CH_2$ + → $\cdot CH_2$ **1** → (H~) olefins

triplet has been attacked by Eder and Carr,[60] who find that a residual component of singlet methylene exists, even at 1000 to 2000 times excess inert gas concentrations. These authors also reaffirm the utility of using the cis–trans cyclopropane ratio in estimating spin state mixtures.

The presence of triplet carbenes and their formation of relatively long-lived diradicals is indicated by the work of Elliot and Frey[47,68] on the reaction with cyclobutene. Singlet methylene gives, as expected, bicyclo[2.1.0]pentane, insertion products, and at low pressures further rearrangement products. The major product from triplet methylene is none of these but vinylcyclopropane. It was thought that diradical **2**, rather than close, severs a carbon–carbon bond to give vinylcyclopropane. However, the photosensitized

$^1:CH_2$ + → + +

$^3:CH_2$ + → **2** → →

decomposition of **3** gives only 1,4-pentadiene and bicyclo[2.1.0]pentane. No substantial amount of vinylcyclopropane appears.[69]

It was thought that the decomposition of **3** primarily involved formation of diradical **2**, and that a cleavage to **4** followed by formation of carbene **5** was not responsible for the formation of the bicyclic product. It is true (see Section III) that singlet **5** would not give much bicyclic product, but there is no evidence that triplet **5** would not do so.

N—CH_2 / N **3** —hν, $PhCOCH_3$→ **2** → (89.7) + (10.3)

N_2CH— **4** → H—C̈— **5** (?)

Another reaction used by the trimethylene produced from triplet methylene and olefins to achieve tetravalency is methyl shift.[70] Reaction of triplet CHT formed from tritioketene with *trans*-2-butene followed by analysis of the 3-methylbutene formed as one of the products showed that methyl shift was among the paths followed by the intermediate trimethylene. Tritium is found in both the vinyl and methyl positions, implicating a 1,2 methyl shift.

$$\text{trans-2-butene} + :CHT^3 \longrightarrow CH_3{-}\underset{\cdot CHT}{CH}{-}\dot{C}H{-}CH_3 \xrightarrow{H\sim} CH_3{-}\underset{CH_2T}{CH}{-}CH{=}CH_2$$

$$\downarrow CH_3\sim$$

$$CHT{=}CH{-}CH(CH_3)_2$$

The liquid phase chemistry of methylene is by comparison benignly simple. The addition reaction is stereospecific.[71,72] Dilution with a two-hundredfold excess of inert perfluoropropane induces some intersystem crossing in either diazomethane or, more likely, methylene itself, as the following product distributions reveal:[66,67]

$$:CH_2 + \text{cis-2-butene} \longrightarrow$$

	cis-1,2-dimethylcyclopropane	trans-1,2-dimethylcyclopropane	3-methyl-1-butene	2-methyl-1-butene	1-pentene (trans)	cis-2-pentene	2-methyl-2-butene
$C_3F_8 = 0$	(47.5)	(0.4)	(0.2)	(0.3)	(0.0)	(39.1)	(12.5)
$= 200\times$	(60.4)	(13.3)	(6.1)	(1.9)	(7.1)	(9.3)	(1.9)

Similarly, the benzophenone-photosensitized decomposition of diazomethane induces a limited nonstereospecificity in the addition reaction.[73] Here less certainty exists as to which species is undergoing intersystem crossing, and the change in stereospecificity is not impressive.

		cis-1,2-dimethylcyclopropane	trans-1,2-dimethylcyclopropane
cis-2-butene	$+ CH_2N_2 \xrightarrow[h\nu]{Ph_2C{=}O}$	(1.9	(1)
trans-2-butene	$+ CH_2N_2 \xrightarrow[h\nu]{Ph_2C{=}O}$	"trace"	"mainly"

Fluorine substitution reduces the activity of the carbon-carbon double bond toward methylene.[74] The decreased reactivity was attributed to the lowering of π-electron density by the electronegative fluorines. It seems to us that the hostility of fluorine toward the cyclopropyl position (see Section III) might well also be important.

Olefin	*Relative reactivity*
 $CH_2{=}CH_2$	1.0
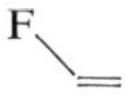 $CHF{=}CH_2$	0.6
 $CF_2{=}CH_2$	0.33
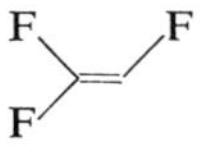 $CF_2{=}CHF$	0.16
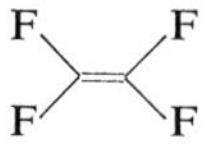 $CF_2{=}CF_2$	0.10

In synthesis, methylene has been used on countless occasions to make three-membered rings. Generally, however, diazomethane has been either decomposed catalytically by the Gaspar-Roth method[75] or bypassed entirely by means of the Simmons-Smith[76] procedure. We discuss here neither of these processes.

Uncatalyzed addition of methylene generally offers little or no advantage over these catalytic methods and so is usually used only in studies of mechanism. The exothermicity of addition frequently is used to generate molecules in vibrationally excited states from which their decomposition can be studied. For examples see references 77 to 81.

B. Carbon-Hydrogen Insertion Reactions

Let us look first at the liquid phase. The insertion reaction was discovered by Meerwein, Rathjen, and Werner,[82] who found ethyl-*n*-propyl ether and ethyl isopropyl ether on irradiating diazomethane in diethyl ether. The scope of the reaction was extended by Doering et al.,[83] who found that the reaction did not depend upon the ether oxygen and proceeded in a statistical fashion in hydrocarbons. The statistical nature was confirmed by Richardson, Simmons, and Dvoretzky[84] with highly branched hydrocarbons.

$$n\text{-}C_5H_{12} + CH_2N_2 \xrightarrow{h\nu} n\text{-}C_6H_{14} + \text{2-methylpentane} + \text{3-methylpentane}$$

The mechanism of this reaction has not been closely probed. We deal with the question of ylids when we speak of insertion into heteroatomic single bonds, but two experiments and two theoretical efforts should be mentioned here. First is the classic experiment of Doering and Prinzbach[85] which demonstrated that the carbon-hydrogen insertion reaction was direct and not a radical-abstraction-recombination process. These authors concentrated on the product of "insertion" of methylene into the allylic position of isobutylene. In the event of a radical process, two differently labeled products should be obtained; a direct insertion should yield one. These results show that in the liquid and gas phases the predominant reaction mechanism is

$:CH_2$

$:CH_2$

$+$
$\cdot CH_3$

Liquid phase	(>99)	(<1)
Gas phase	(92)	(8)

direct insertion. In the gas phase up to 16% of the reaction involves radicals of some sort. It is not known, however, whether "hot molecule" rearrangements or abstraction by triplet methylene is responsible, nor is it known to what extent cage effects reduce scrambling in the liquid phase.

Second is the astonishing report by Franzen[86] that methylene reacts with the *tertiary* carbon-hydrogen bond in *trans*-1,2-cyclopentanediacetate with inversion. This result is surprising for several reasons. The initial description of this reaction[87] reported retention. Moreover, the bridgehead carbon-hydrogen bond in bicyclo[2.2.1]heptane behaves normally toward methylene[88] and here inversion is impossible. We shall see in Section V that carboethoxycarbene inserts with retention.[89]

H OAc OAc H $\xrightarrow{:CH_2}$ OAc OAc CH_3 H

Finally, Kirmse and Buschoff[90] found that methylene inserts in the carbon-hydrogen bond of paraldehyde with retention. Conceivably the acetoxy groups confer special properties on the reaction, but more work clearly is necessary.

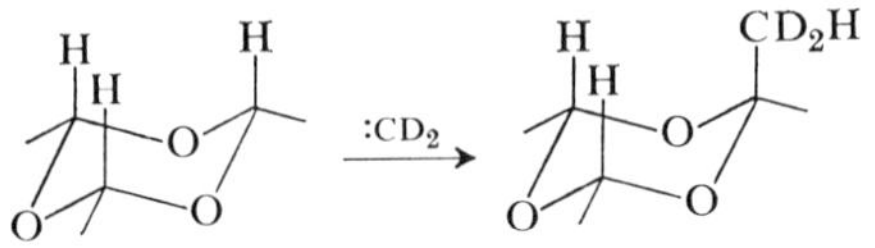

Dobson, Hayes, and Hoffmann[91] used the extended Hückel method to calculate the potential surface for the insertion reaction of singlet CH_2. They found an initial end on approach of CH_2 with the empty p orbital "impinging . . . on the H atom." In the next stage of the reaction, the hydrogen is transferred and finally the two halves collapse to ethane. The pathway is similar to that initially proposed by Benson[5,92] and no true abstraction reaction was found.

Bader and Gangi[93] used an SCF calculation of the potential surface for the reaction of singlet and triplet oxygen atoms to relate the difference in chemistry to a polarization of spins in the hydrogen-hydrogen bond by an approaching triplet which leads to abstraction.

As before, analysis of work on gas-phase systems is complicated. It is claimed that for triplet methylene two mechanisms are operative, an abstraction-recombination process and a direct insertion.[94,95] The abstraction mechanism was previously known. Ho and Noyes[61] had shown, for instance, that the reaction of triplet methylene with propane involved abstraction. The isolation of the dimeric products hexane and 2,3-dimethylbutane speaks strongly for an abstraction mechanism.

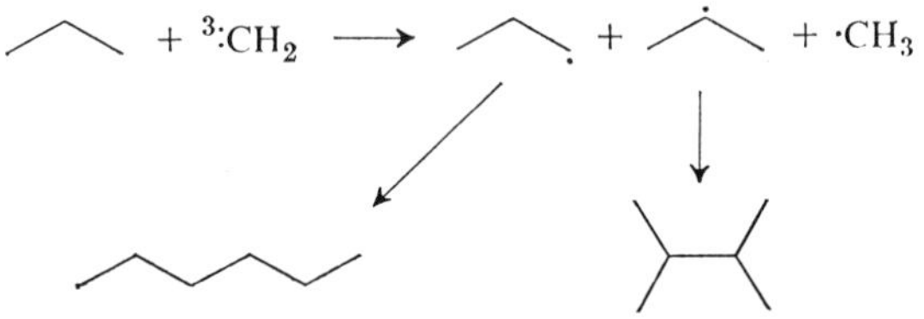

Ring and Rabinovitch found that the relative proportions of R· and R·′ can be deduced from reactions 2, 3, and 4 (or from reaction of R· and R·′ with ethyl radicals).

$$RH + {}^3CH_2 \rightarrow CH_3\cdot + R\cdot \quad \text{primary H} \tag{1a}$$

$$R'H + {}^3CH_2 \rightarrow CH_3\cdot + R\cdot' \quad \textit{sec}\text{ H or }\textit{tert}\text{ H} \tag{1b}$$

$$R\cdot + R\cdot \rightarrow R{-}R \tag{2}$$

$$R\cdot' + R\cdot' \rightarrow R'{-}R' \tag{3}$$

$$R\cdot + R\cdot' \rightarrow R{-}R' \tag{4}$$

$$2CH_3\cdot \rightarrow C_2H_6 \tag{5}$$

$$CH_3\cdot + R\cdot \rightarrow R{-}CH_3 \tag{6}$$

$$CH_3 + R\cdot' \rightarrow R'{-}CH_3 \tag{7}$$

One can thus deduce the relative rates of abstraction by triplet methylene. Average values are primary/secondary/tertiary = 1/14/150. From the relative amounts of R—CH_3 and R′—CH_3 one would expect to find the same values. Instead one gets primary/secondary (propane) = 1/3; primary/secondary (*n*-butane) = 1/4; and primary/tertiary (isobutane) = 1/15. Further, the total amounts of these compounds far exceeded what the radical recombination would dictate, assuming the formation of ethane (reaction 5) as a base. The authors explain the "extra" amounts of RCH_3 and $R'CH_3$ by invoking a triplet direct insertion reaction. The following relative rates are calculated for triplet insertion: primary/secondary/tertiary = 1/2/7.

At least two contradictory pieces of work are available. McKnight, Lee, and Rowland[96] found conditions where the same mixture of cyclopropanes is formed from either *cis*- or *trans*-2-butene and assumed that full conversion to triplet methylene was achieved. Methyl iodide was added to scavenge free radicals. Here very little insertion was noted.

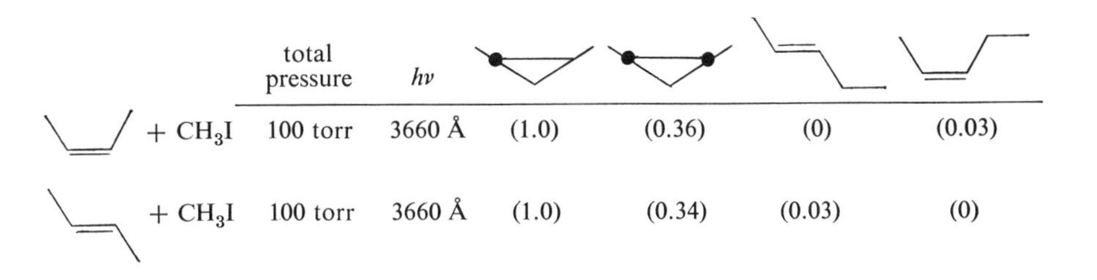

	total pressure	$h\nu$				
+ CH_3I	100 torr	3660 Å	(1.0)	(0.36)	(0)	(0.03)
+ CH_3I	100 torr	3660 Å	(1.0)	(0.34)	(0.03)	(0)

It might be argued that in such a system triplet insertion would be unable to compete with addition. However, the estimates of the relative rates of reaction with the carbon-hydrogen bond for addition make it almost certain that more products should have been seen.

Eder and Carr, it is recalled,[60] claimed that some singlet methylene persisted even at enormous pressures of added inert gas. One may legitimately ask if such a remnant might account for the "extra" insertion products. The answer is apparently no. Herzog and Carr[97,98] found the relative rates of insertion for singlet methylene to be primary/secondary/tertiary = 1/1.3/1.5 in a system using isopentane as substrate and oxygen as triplet scavenger. Halberstadt and McNesby[38] found nearly indiscriminant insertion using NO as a scavenger, and Simmons, Marzac, and Taylor[99] found a secondary/primary ratio of 1.35. These values do not compare favorably with the 1/2/7 estimated by Ring and Rabinovitch, as they should if a singlet is responsible for the "extra" insertion products.

This[95] and other[100–103] work allow a collection of isotope effects to be noted (see Table 1).

TABLE 1
Isotope Effects for CH_2 Insertions

k_H/k_D	Substrate	Type	Reference
~1.3	Propane	2° CH ins	101
1.96	*cis*-2-Butene	Allylic CH ins	102
1.55	*cis*-2-Butene	Vinyl CH ins	102
1.94	*cis*-2-Butene	Allylic CH ins	103
1.84	*cis*-2-Butene	Vinyl CH ins	103
1.29	Cyclopropane	2° CH ins	100
1.32	Cyclohexane	2° CH ins	100
~3.9	Propane ($^3CH_2$)	1° CH abs	95
~2	Propane ($^1CH_2$)	1° CH ins	95

In any event, the triplet species is far less reactive than the singlet. It was recently estimated, for instance, that triplet methylene survived up to 10^9 collisions with CD_4.[104]

Flash photolysis[105] has allowed the determination of a number of absolute rate constants for reactions of CH_2, and these are given in Table 2.

TABLE 2
Rate Constants (cm^3 molecule^{-1} sec^{-1})

$^3CH_2 + {}^3CH_2 \rightarrow CH_2{=}CH_2$	$5.3 \pm 1.5 \times 10^{-11}$
$^1CH_2 + H_2 \rightarrow CH_4 \rightarrow CH_3 + H$	$7.0 \pm 1.5 \times 10^{-12}$
$^1CH_2 + H_2 \rightarrow {}^3CH_2 + H_2$	$<1.5 \times 10^{-12}$
$^1CH_2 + CH_4 \rightarrow C_2H_6 \rightarrow 2CH_3\cdot$	$1.9 \pm 0.5 \times 10^{-12}$
$^1CH_2 + CH_4 \rightarrow {}^3CH_2 + CH_4$	$1.6 \pm 0.5 \times 10^{-12}$
$^1CH_2 + He \rightarrow {}^3CH_2 + He$	$3.0 \pm 0.7 \times 10^{-13}$
$^1CH_2 + Ar \rightarrow {}^3CH_2 + Ar$	$6.7 \pm 1.3 \times 10^{-13}$
$^1CH_2 + N_2 \rightarrow {}^3CH_2 + N_2$	$9.0 \pm 2.0 \times 10^{-13}$
$^3CH_2 + H_2 \rightarrow CH_3 + H$	$<5 \times 10^{-14}$
$^2CH_2 + CH_4 \rightarrow 2CH_3\cdot$	$<5 \times 10^{-14}$

C. Reactions with Other Single Bonds

Methylene reacts with hydrogen to give methane[106,107] and a rate constant has been measured[106] [$0.9 \times 10^{12} \exp(-32000/RT)$ sec^{-1}].

The carbon-carbon bonds of spiropentane are not attacked, at least in solution.[108] An even weaker carbon-carbon bond, however, the central one in bicyclo[1.1.0]butane, is attacked.[109] Irradiation of a solution of diazomethane in 1,3-dimethylbicyclo[1.1.0]butane gave, in addition to **6** and **7**,

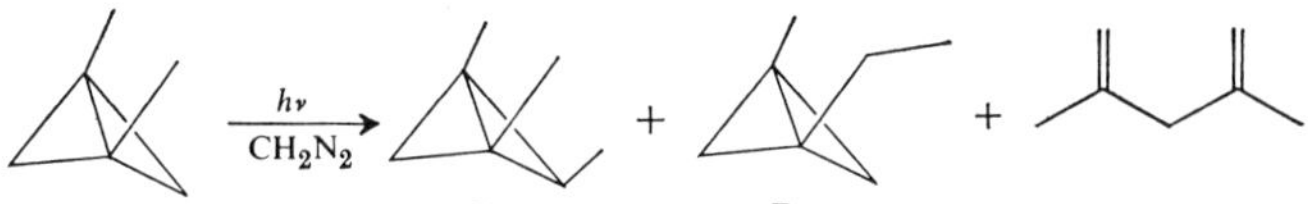

$$\xrightarrow[CH_2N_2]{h\nu}$$ **6** + **7** +

28% of 2,4-dimethylhexa-1,4-diene. This compound probably results from a stepwise addition of methylene (singlet?!) to the central bond, followed by further bond reorganization. A less likely path involves thermal rearrange-

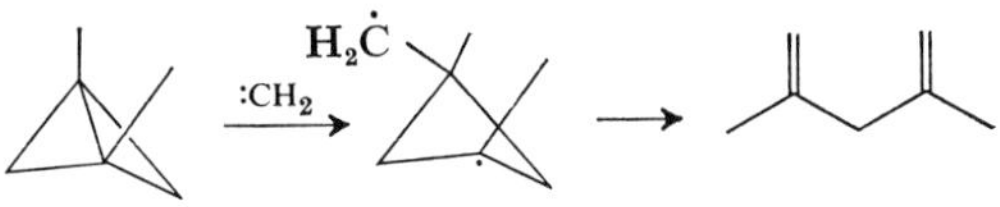

$$\xrightarrow{:CH_2} H_2\dot{C} \longrightarrow$$

ment of a bicyclo[1.1.1]pentane on work-up. The activation energy of 49.6 kcal/mole for isomerization of the parent compound to hexadiene-1,4[110] would not be lowered sufficiently by the addition of two methyl groups to make this a likely path. Moreover, a labeling experiment reveals that no symmetrical intermediate can be present.[109]

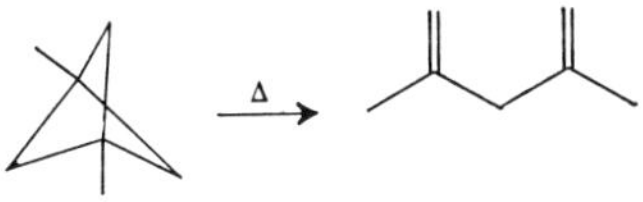

$$\xrightarrow{\Delta}$$

A somewhat analogous reaction of triplet methylene has been discovered by Frey and Walsh.[111] A methyl radical is abstracted from neopentane to give ethyl and t-butyl radicals. The latter product is detected by its disproportionation and dimerization products.

$$^{3}{:}CH_2 + H_3C{-}\overset{\displaystyle CH_3}{\underset{\displaystyle CH_3}{C}}{-}CH_3 \longrightarrow CH_3CH_2\cdot + CH_3{-}\overset{\displaystyle CH_3}{\underset{\displaystyle CH_3}{C}}\cdot$$

It was recognized early[112,113] that some of the products observed by Meerwein[82] could have been formed in more than one way. For example, ethyl-n-propyl ether could have as its source either insertion into a carbon-hydrogen bond or insertion into the carbon-oxygen bond. The second process could be direct or could occur via an ylid. In Section V we see several examples of similar ylid rearrangements. Two groups[114,115] performed tests. Franzen and Fikentscher[115] used methylene-C-14 to show that the ethyl propyl ether arose exclusively via carbon-hydrogen insertion, and Doering and co-workers found less than 0.5% of tetrahydropyran from tetrahydrofuran.[114]

$$CH_3CH_2-O-CH_2-CH_3 \xrightarrow{:CH_2^*} CH_3CH_2-O-CH_2-CH_2-{}^*CH_3$$

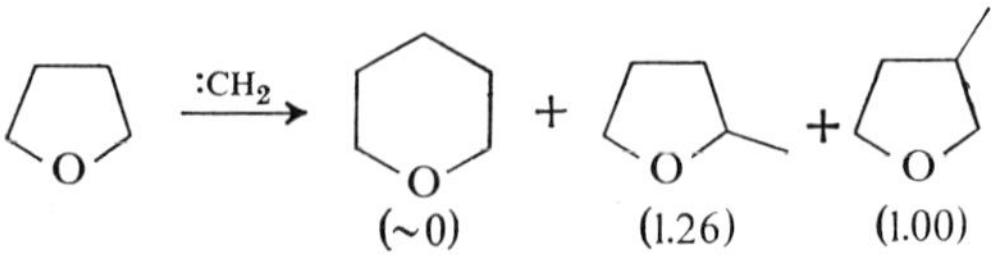

(~0) (1.26) (1.00)

On the other hand, some products probably do arise from ylids. Thus methyl ethers often are obtained in decompositions of ethereal solutions of diazomethane.[115]

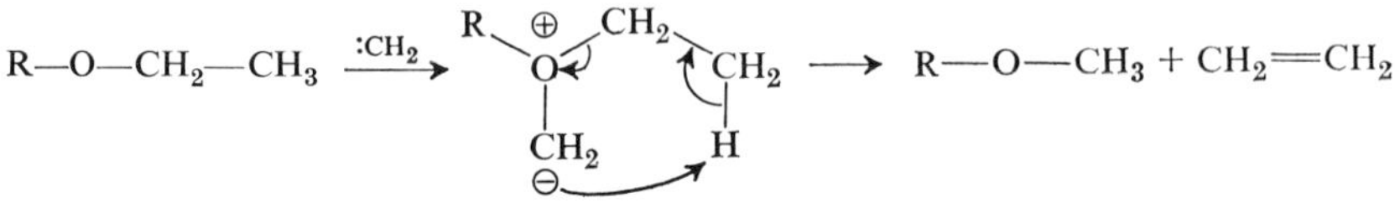

$$R-O-CH_2-CH_3 \xrightarrow{:CH_2} \text{ylid} \longrightarrow R-O-CH_3 + CH_2{=}CH_2$$

Work in the gas phase reveals some similarities and some remarkable differences. Photolysis of diazomethane in methyl-*n*-propyl ether gave the following products:[116]

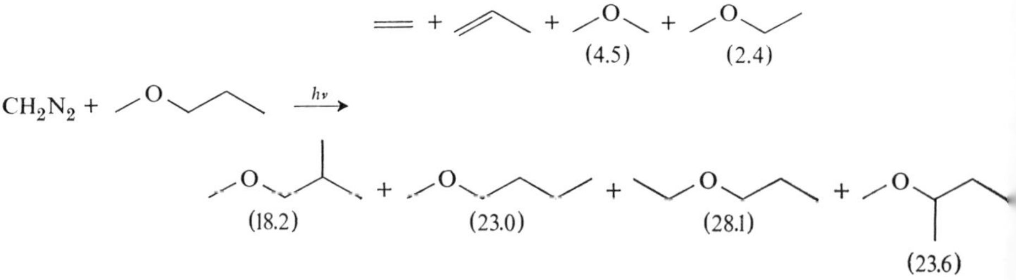

CH_2N_2 + methyl *n*-propyl ether $\xrightarrow{h\nu}$

(4.5) (2.4)

(18.2) (23.0) (28.1) (23.6)

The presence of dimethyl ether was explained in terms of the β-hydrogen transfer mechanism just mentioned and methyl ethyl ether by methyl transfer.

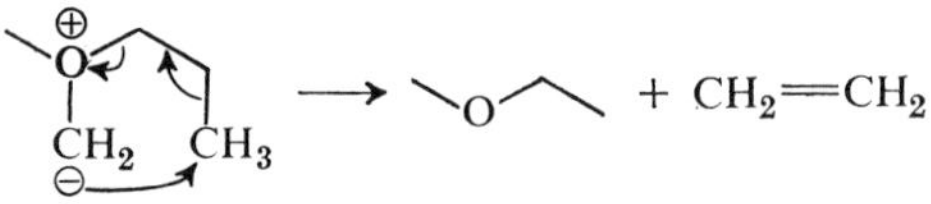

$$\longrightarrow \text{methyl ethyl ether} + CH_2{=}CH_2$$

More details became available in 1968.[117,118] It was found that the singlet state underwent near random insertion while the triplet showed a greater preference for the α-positions.

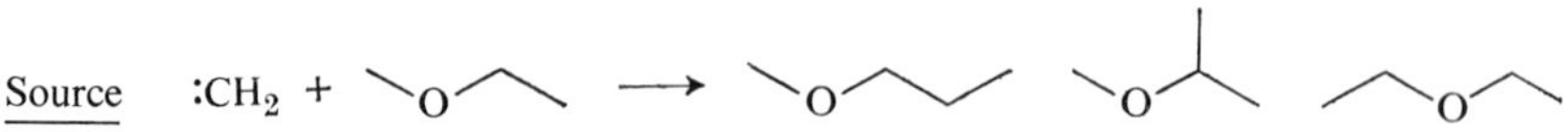

Source	:CH₂ + methyl ethyl ether →	methyl n-propyl ether	methyl isopropyl ether	diethyl ether
$CH_2C{=}O$/ Hg/*hν*	$^1CH_2$ (gas)	(30.2)	(31.8)	(38)
	$^3CH_2$ (gas)	(0)	(85)	(15)
	$^1CH_2$ (liquid)	(34.5)	(30)	(35.5)
	statistical	(3)	(2)	(3)

TABLE 3
Reaction of Methylene with Acetals and Ortho Esters

Compound	Phase	α-Hydrogen relative reactivities compared to: CH_3—C	CH_3—O	O—CH_2—CH_2—O
$CH_3OCH_2OCH_3$	Gas		0.73	
	Liquid		0.72	
$CH_3CH(OCH_3)_2$	Gas	1.56	1.20	
	Liquid	1.23	1.13	
$HC(OCH_3)_3$	Gas		0.93	
	Liquid		1.16	
1,3-dioxolane	Gas			0.98
	Liquid			1.30
2-methyl-1,3-dioxolane	Gas	3.56		1.40
	Gas + O_2	3.02		1.21
	Liquid	2.16		1.60
2-methoxy-1,3-dioxolane	Gas		0.67	0.54
	Liquid		0.54	0.71

In 1969 Kirmse and Buschhoff[119] examined the reaction of methylene with a number of acetals and ortho esters. Notice from Table 3 that only the α-positions bearing methyl groups are especially activated. The increased reactivity of the methyl-substituted cyclic compound over the acyclic counterpart is attributed to better overlap of the nonbonding orbitals on oxygen with the breaking carbon-hydrogen bond.

H, CH_3 —:CH_2→ CH_3, CH_3

CH_3O, C—O—CH_3, CH_3, H —:CH_2→ CH_3O, C—O—CH_3, CH_3, CH_3

The most striking reaction was uncovered by Frey and Voisey in 1966.[120] In contrast to the liquid-phase photolysis of diazomethane in tetrahydrofuran,[114] the gas-phase reaction produces large amounts of tetrahydropyran.

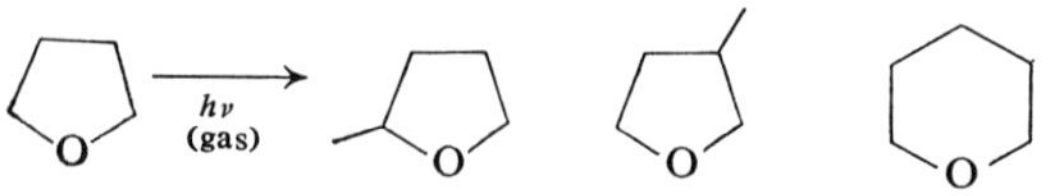

$h\nu$ (gas)

CH_2N_2 (55) (34) (11)
CH_2CO (50) (29) (21)

Triplet methylene is not responsible, since sources of triplet methylene fail to give any tetrahydropyran. The authors postulate a singlet methylene which has lost some of its excess energy. We are baffled by this reaction because it seems that a methylene lower in energy should be more rather than less selective.

Irradiation of diazomethane in the gas phase in the presence of various alcohols showed that insertion into the oxygen-hydrogen bond was very rapid compared to insertion into the primary carbon-hydrogen bond[121] (Table 4).

TABLE 4
Reaction of Methylene with Alcohols

Relative reactivities	OH/CH, 1°	Alcohol
1.0	10.9	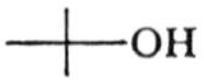 —OH
1.37	14.9	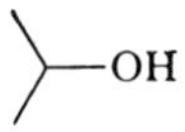 —OH
1.95	21.2	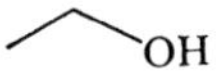 OH
2.01	21.8	CH_3OH

Insertion into the carbon-sulfur bond does take place but is not a major contributor to the products.[122]

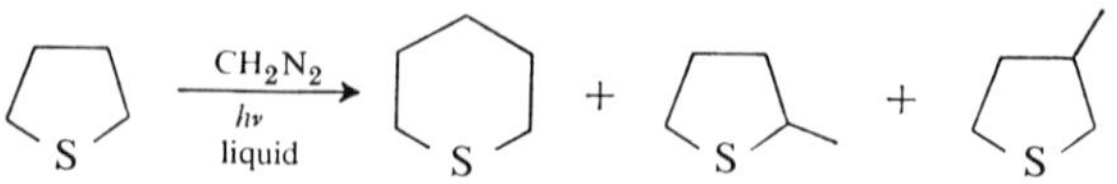

CH_2N_2, $h\nu$, liquid

2% 49% 49%

The carbon-halogen bonds are quite labile toward methylene, and much evidence that radicals are involved exists. With carbon tetrachloride, carbon tetrabromide, and other bromochloromethanes free radical chain reactions abound.[123–126] Quantum yields are high and inhibitors are effective. We deal here mainly with the mono-halo compounds, and radical reactions dominate.

With both isopropyl chloride and isopropyl bromide insertion into the carbon-halogen bond is the main process.[127] It is apparent, however, that the mechanism of this reaction is not the direct process found for the carbon-hydrogen bond. Gaspar[128] established that the hydrogens on methylene maintain their identity in the reaction with isopropyl chloride and Doering and Sampson[129] showed that reaction with optically active *sec*-butyl chloride gave "insertion" product which was 90% racemic! A radical reaction is thus

$$(CH_3)_2CH{-}Cl \xrightarrow{:CD_2} (CH_3)_2CH{-}CD_2Cl$$

$$CH_3CH_2\overset{*}{C}H(Cl)CH_3 \xrightarrow{:CH_2} CH_3CH_2CH(CH_2Cl)CH_3 \quad 90\%\ \text{racemic}$$

implicated and further confirmed by Doering and Wiegandt's[130] observation of a large amount of rearrangement in the reaction of methylene with labeled methallyl chloride. It seems clear that even in simple systems carbon-chlorine "insertion" is a two-step process.

$$\overset{*}{C}H_2{=}C(CH_3)CH_2Cl \xrightarrow[h\nu]{CH_2N_2} \left[\overset{*}{C}H_2{=}C(CH_3)\dot{C}H_2 \leftrightarrow \cdot\overset{*}{C}H_2C(CH_3){=}CH_2\right] + \dot{C}H_2Cl \longrightarrow \overset{*}{C}H_2{=}C(CH_3)CH_2CH_2Cl + ClH_2C\overset{*}{C}H_2C(CH_3){=}CH_2$$

More recently, the use of both CIDNP[131,132] and CO as a triplet scavenger[133] have implicated the singlet state in chlorine abstraction and the triplet state in hydrogen abstraction.

Reaction of methylene with methyl chloride in the gas phase is also a radical reaction which goes primarily (~70%) by chlorine abstraction.[134–140] The abstraction process is sensitive to both the energy of methylene and the spin state. The singlet prefers chlorine abstraction by a factor of 22, but the triplet is much less discriminating.[135,137,139,140] This was determined by utilizing the finding of DeGraff and Kistiakowsky that triplet methylene reacts with CO roughly 20 times as fast as the singlet.[141] Triplet methylene even reacts with CO twice as fast as it does with *cis*-2-butene.[142] This allowed isolation of the singlet and large amounts of nitrogen were used to produce the triplet. It must be noted, however, that CO is not an effective triplet scavenger below

50 torr,[143] and that CH_2 will react with CO to form an oxirene.[144,145] These results correlate well with solution work in which the singlet state was found to show a preference for chlorine abstraction and perhaps reflect the ability of the singlet (but not the triplet) to form an ylid with chlorine. Section V gives examples of ylid formation by substituted carbenes.

Similar results are found with ethyl chloride[146] and 1-chloropropane.[147] The carbon-fluorine bond is apparently inert toward methylene.[74,148,149]

Reaction with the carbon-iodine bond seems to have received very little attention, although the room temperature "double insertion" shown below has been reported.[150]

The carbon-nitrogen bond has also received little attention. Insertion is not observed in *N*-methylpyrolidine.[151]

$$ClCF_2{-}CClF{-}CF_2CF_2{-}I \xrightarrow{CH_2N_2} Cl{-}CF_2{-}CClF{-}CF_2{-}CF_2{-}CH_2CH_2I$$

$$\text{N-methylpyrrolidine} \xrightarrow[h\nu]{CH_2N_2} \text{no N-methylpiperidine}$$

liquid-phase

The carbon-silicon bond is not attacked,[108] but the silicon-hydrogen bond is. Kramer and Wright[152] found that silicon-hydrogen insertion occurred in solution and was at least 100 times as fast as carbon-hydrogen insertion. In the gas phase Mazac and Simons found a Si—H/C—H insertion ratio of 8.9 for singlet methylene by using oxygen as a radical scavenger.[153] The isotope effect for silicon-hydrogen insertion was found to be 1.15.[153] Abstraction to give radicals accompanies insertion and accounts for at least 27% of the primary reaction.[153,154]

Methylene reacts both in a solid matrix[155,156] and in the gas phase[156,157] with molecular nitrogen:

$$CH_2N_2 \xrightarrow[^{15}N_2]{h\nu} CH_2N_2^{15}$$

D. Additions to Acetylenes

Both liquid- and gas-phase decomposition of diazomethane in acetylenes give cyclopropenes,[157a,158–160] but only allene and methyl acetylene were

$$\text{4-octyne} \xrightarrow[h\nu]{CH_2N_2} \text{1,2-dipropylcyclopropene} \quad 23\%$$

$$HC{\equiv}CH \xrightarrow[h\nu]{CH_2CO} \underset{(1.5)}{CH_3{-}C{\equiv}CH} + \underset{(1)}{CH_2{=}C{=}CH_2}$$

isolated on irradiation of diazomethane in acetylene-argon matrix.[161] A mechanism involving addition of triplet methylene followed by hydrogen shift was postulated and seems reasonable.

E. Additions to Aromatic Compounds

Decomposition of diazomethane in benzene leads to tropilidene and toluene in the ratio of ~3.5.[162,163] Lemmon and Strohmeier determined that labeled methylene did not appear in the ring of toluene,[164] and Russell and Hendry[165] found that the gas-phase reaction proceeded in largely the same fashion as in solution. The age of these references points up the need for additional work in this area.

A large number of copper-catalyzed additions of methylene to aromatic compounds has been studied by Müller[166] and his collaborators. A case of

CH_2N_2, $h\nu$

(1) (3.5)

special interest is the reaction with anthracene in which a product of formal 1,4 addition is isolated. As pointed out by the authors, however, it need not be a primary product of the reaction.[166]

CuBr, CH_2N_2

(9)

+

(11)

+

(80)

III. REACTIONS OF ALKYL-, ALKENYL-, AND ALKYNYLCARBENES AND THEIR CYCLIC COUNTERPARTS

A. Acyclic Alkyl- and Alkenylcarbenes

The chemistry of these carbenes is dominated by intramolecular capture of the divalent carbon. This is natural, because intramolecular pairs of electrons have a great advantage in proximity over those external to the reacting molecule. With few exceptions it can be stated that if a singlet carbene has an internal insertion or cycloaddition available, it will not be possible to trap it efficiently with an external reagent. One method used to stabilize alkyl carbenes is halogen substitution. Perfluoroalkyl groups are often used, and we discuss such compounds in a separate section. Ethylbromodiazomethane yields a carbene stable enough to give, along with hydrogen-shifted products, some dimers.[167,167a] We first consider intramolecular insertion or hydrogen shifts and then go on to cyclopropane formation.

Br N_2 −30° or $h\nu$ Br Br + Br Br

The simplest alkylcarbenes, methyl- and dimethylcarbene, might be expected to be the most easy to capture intermolecularly since their primary carbon-hydrogen bonds should be the most resistant to insertion. Small amounts of cyclopropanes are formed from propene[168] and cyclohexene.[169] Phenylsilane is attacked by ethylidene to give phenylethylsilane[170] and F_2HC—CF inserts in both the Si—H and Si—Cl bonds.[171] Branched carbenes should be even more difficult to trap and their reactions are described almost entirely in terms of intramolecular reactions.

CH_3—C̈H → $PhSiH_3$ → Ph—SiH_2—CH_2CH_3

Typical of the behavior of branched alkylcarbenes are the reactions of isopropylcarbene. In this case[172] hydrogen shift occurs roughly 12 times as fast (per bond basis) as insertion to give cyclopropane. Alkyl shifts (or insertion into carbon-carbon bonds) are not nearly so easy and *t*-butylcarbene prefers cyclopropane formation to methyl shift by a factor of about 4.[172]

C̈H → + (2) (1)

C̈H → + + (92) (7) (1)

The deuterium isotope effect is very similar for both 1,2 and 1,3 carbon-hydrogen insertion,[173] and is estimated at 1.1–1.4. This isotope effect is higher in a carbenoid reaction; the value for cyclopropane formation from **8** was reported as 1.71 ± 0.06.[174]

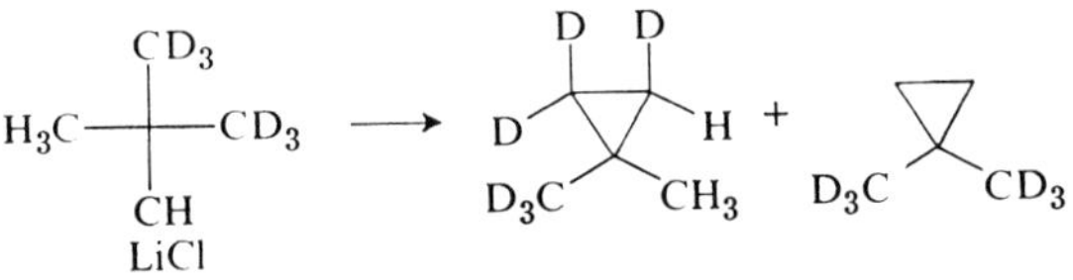

H_3C—C(CD_3)(CD_3)—CH LiCl → D, D, D, H, D_3C, CH_3 + D_3C, CD_3

8

Optical activity is preserved in the cyclopropane-forming insertion reaction, although the data quoted are not from the decomposition of a diazo compound.[175] Alkyllithium-catalyzed decomposition of R(+)-iso-amylchloride gave optically active (1R,2R)-*trans*-1,2-dimethylcyclopropane.

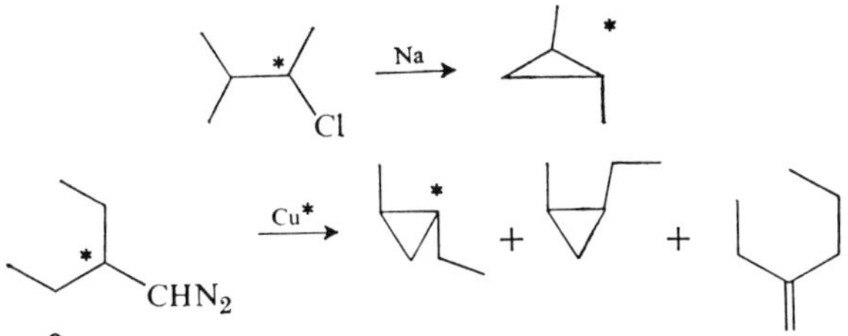

Cl —Na→

CHN_2 —Cu*→ + +

9

Similarly, decomposition of diazo compound **9** with an optically active catalyst[176] gave optically active cyclopropane. Thermal decomposition of racemic **9** in an optically active solvent gave racemic product unless a weak proton donor was used as solvent, in which case a small amount of optical activity was induced.

In cyclopropane formation there is a slight (1.3–1.9) preference for secondary over primary hydrogen.[177,178] Thus *sec*-butylcarbene gives the following products:

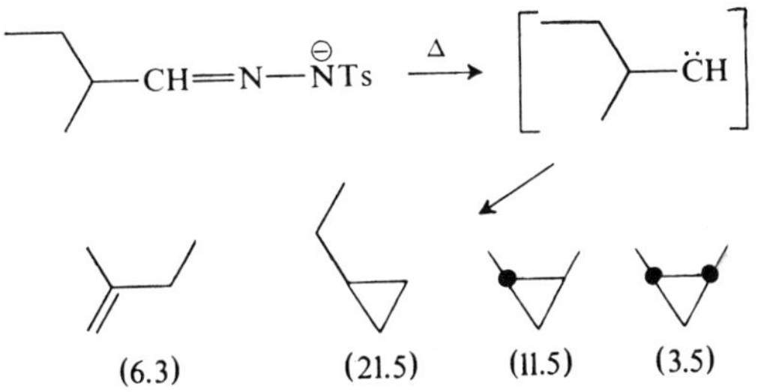

—CH=N—$\overset{\ominus}{N}$Ts —Δ→ [—$\ddot{C}$H]

(6.3) (21.5) (11.5) (3.5)

As has been pointed out,[178] one must be careful in interpreting such data, since steric effects must be very influential. Newman projections for the conformations leading to the three cyclopropanes point this out and explain why the *trans*-cyclopropane is favored over the *cis*.

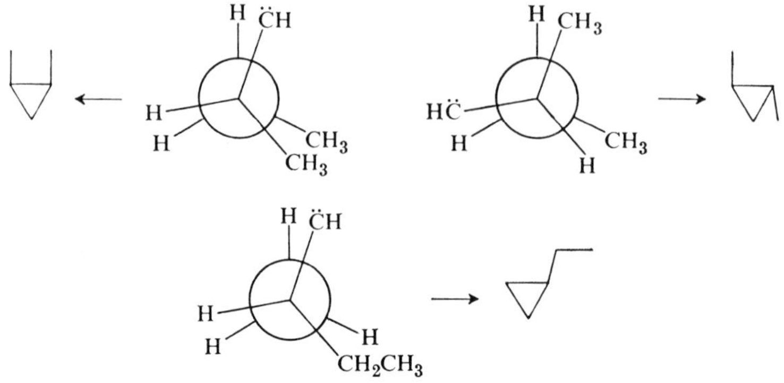

Insertion into adjacent carbon-hydrogen bonds appears to be much more selective. Methylethylcarbene gives the following products:[172]

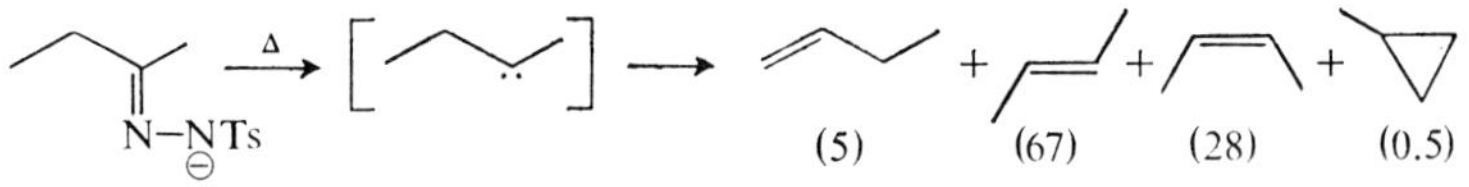

(5) (67) (28) (0.5)

Secondary hydrogen is favored over primary by a factor of roughly 30, although the stability of the products may be an important factor here.

Comparisons with data such as the foregoing have led Frey, Stevens, and co-workers to postulate that photolysis of diazirenes leads to an excited and therefore less selective carbene. Thus for the immediately preceding case the following product ratios were obtained:[179]

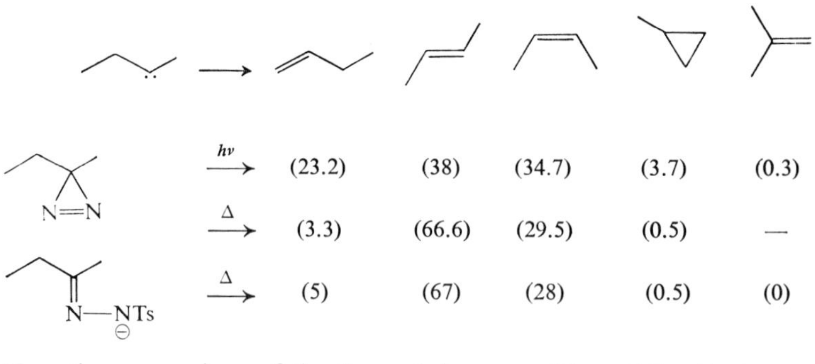

Diazirene	$h\nu$	(23.2)	(38)	(34.7)	(3.7)	(0.3)
Diazirene	Δ	(3.3)	(66.6)	(29.5)	(0.5)	—
Tosylhydrazone salt	Δ	(5)	(67)	(28)	(0.5)	(0)

Here the comparison of the thermal decomposition of the diazirene with the thermal decomposition of the tosylhydrazone salt reveals very similar product ratios, but the photochemical decomposition of the diazirene is quite different and appears to yield a less selective intermediate. Similar work reveals much the same trend.[180–182] It would be most interesting to see the results of a photochemical decomposition of the tosylhydrazone salt. A similar finding for diazo compounds has been made by Kirmse and Horn[182] in a study of the catalytic, thermal, and photolytic decompositions of diazoalkanes in various solvents. Here, too, the photochemical decomposition gave a less selective intermediate than did the thermal reaction.

The prolific Kirmse has provided much information on the reactions of alkylcarbenes substituted mainly but not exclusively with methoxy groups. A β-methoxy group, although not prone to migration itself, accelerates the shift of its adjacent hydrogens.[183,184] Methoxy is slow to move with respect not only to hydrogen but to alkyl as well. Note that again the photochemically generated intermediate is less selective.

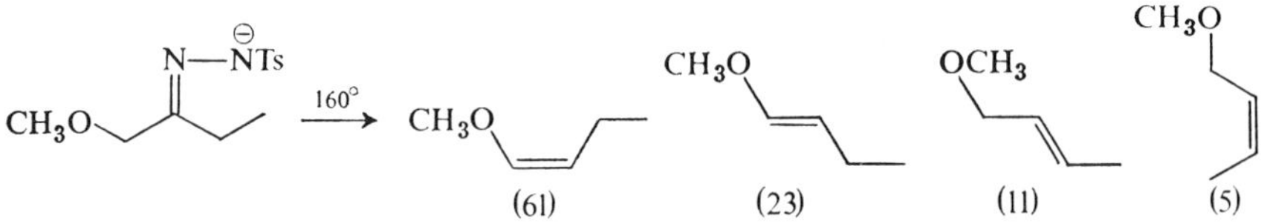

CH_3O ... $N—\overset{\ominus}{N}Ts$ $\xrightarrow{160°}$ CH_3O (61) CH_3O (23) OCH_3 (11) CH_3O (5)

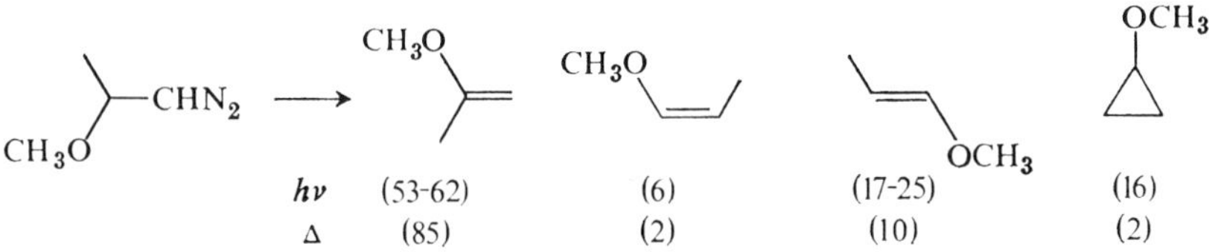

CH_3O ... $—CHN_2$ ⟶ CH_3O ; CH_3O ; OCH_3 ; OCH_3

$h\nu$	(53-62)	(6)	(17-25)	(16)
Δ	(85)	(2)	(10)	(2)

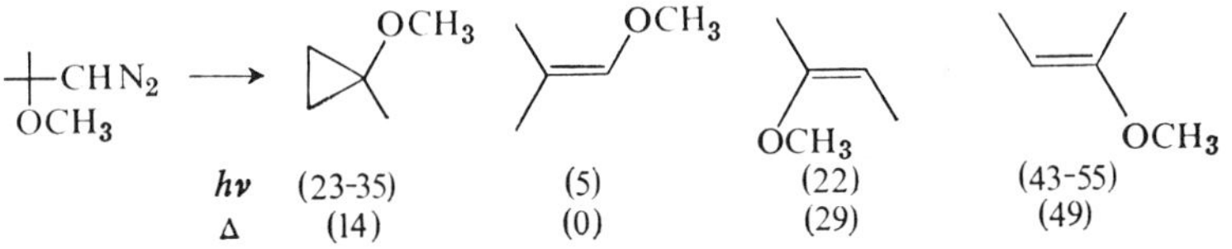

$+CHN_2$ / OCH_3 ⟶ OCH_3 ; OCH_3 ; OCH_3 ; OCH_3

$h\nu$	(23-35)	(5)	(22)	(43-55)
Δ	(14)	(0)	(29)	(49)

In contrast, substitution at the γ-position by a variety of atoms makes insertion more difficult.[185] Two methoxy groups reduced insertion into the proximate carbon-hydrogen bond to zero.[186] Perhaps this is a steric effect.

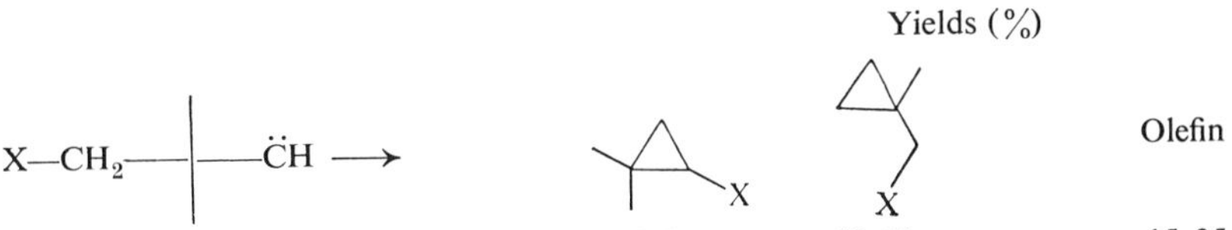

$X—CH_2—\overset{|}{\underset{|}{C}}—\ddot{C}H$ ⟶

	Yields (%)		
	(cyclopropane, X)	(cyclopropane, X)	Olefin
$X = OCH_3$	1.5	60–80	15–35
Cl	1.1	3.7	
$N(CH_3)_2$	6–8	92–94	
Ph	9–13	60–70	

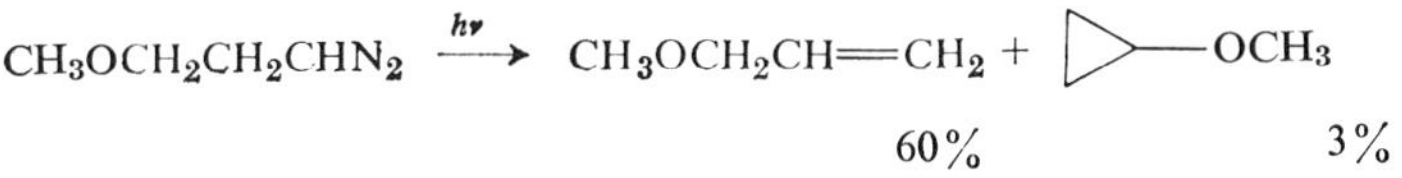

$$CH_3OCH_2CH_2CHN_2 \xrightarrow{h\nu} CH_3OCH_2CH{=}CH_2 + \triangleright—OCH_3$$

60% 3%

The conformation necessary for insertion into the carbon-hydrogen bond α to the X group seems significantly more crowded than that required for insertion into a methyl group.

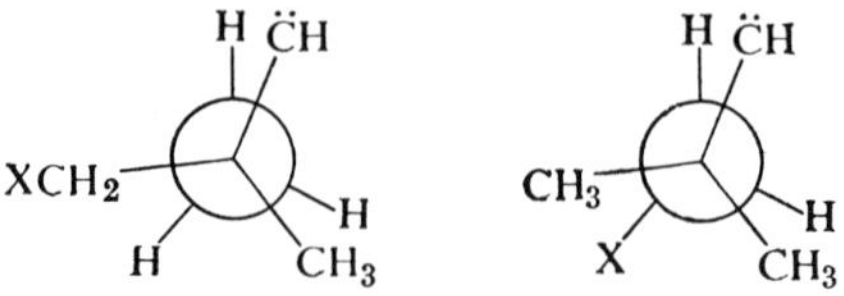

A similar effect on 1,2 insertion appears in diazirene thermolysis. The sole product of pyrolysis of **10** was propanal.[187]

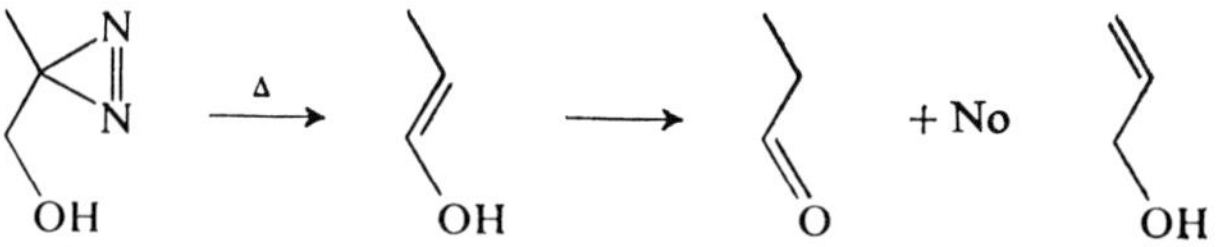

Intramolecular cycloaddition is not as frequently encountered in uncatalyzed decompositions for the simple reason that easy insertion reactions usually are effective competitors. Thus butadiene is favored by a factor of 5 over bicyclobutane in the photochemical decomposition of **11**.[188] Kirmse and Grassmann[189] find that 1,2 hydrogen shift dominates the reactions of ω-alkenylcarbenes. The small amount of intramolecular cycloaddition (**14**) may come from decomposition of **12**, but it was shown that photolytic decomposition of **12** was slow compared to that of **13**. At any rate very little **14** can be formed from the carbene.

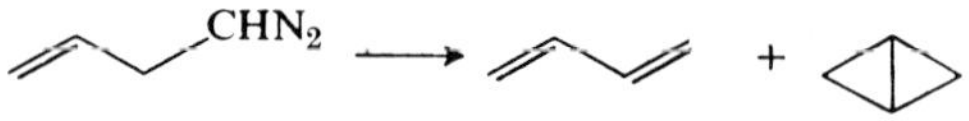

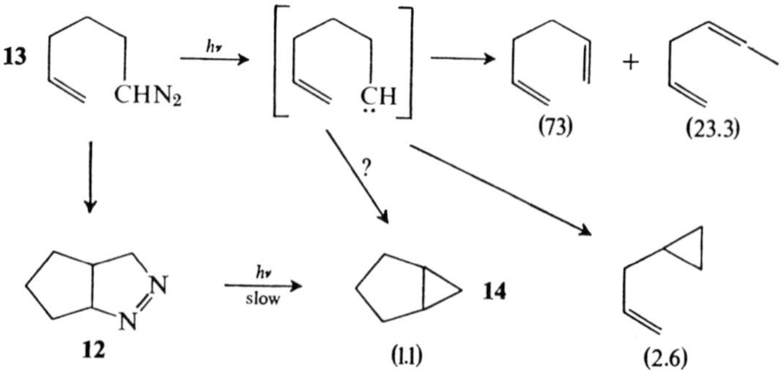

Similarly, bicyclo[2.1.0]pentane is not a product of the following tosylhydrazone salt decomposition:[69]

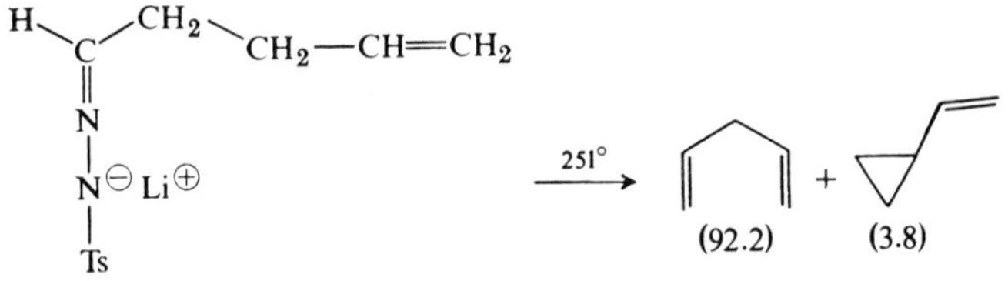

$$\text{H}(\text{C}{=}\text{N}{-}\overset{\ominus}{\text{N}}(\text{Ts})\,\text{Li}^{\oplus})\text{CH}_2\text{CH}_2{-}\text{CH}{=}\text{CH}_2 \xrightarrow{251^\circ} (92.2) + (3.8)$$

The situation changes drastically when the intramolecular double bond is α,β to the carbene or otherwise held in proximity by the structure of the molecule. Now internal cycloaddition is easy and cyclopropenes or cyclopropanes are formed in good to excellent yield. The latter case is exemplified by the decomposition of tropone acetoaldehyde tosylhydrazone.[190] Formation of **15** gives rise to the product of addition to the remote double bond, barbaralane (**16**), and bicyclo[4.2.1]nonatriene ("Grimme's hydrocarbon," **17**),[191] which may arise via cycloaddition to the proximate double bond to give **18** followed by rearrangement. Most surprisingly, no vinyltropilidene was reported. It seems likely that the double bonds are directly attacking the diazo compound, and, indeed, in the phenyl series **19** and **20** were isolated.

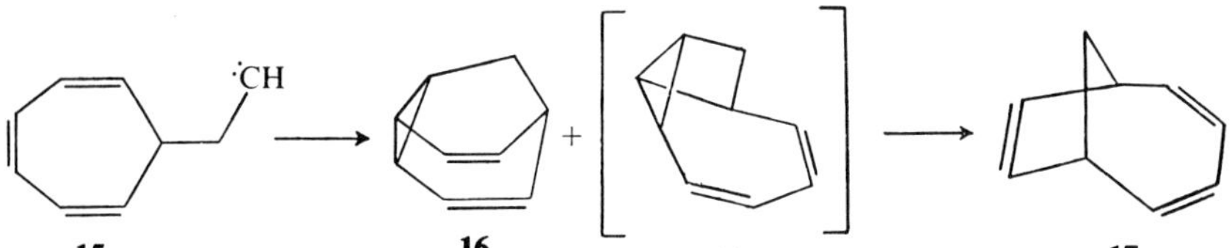

15 → **16** + [**18**] → **17**

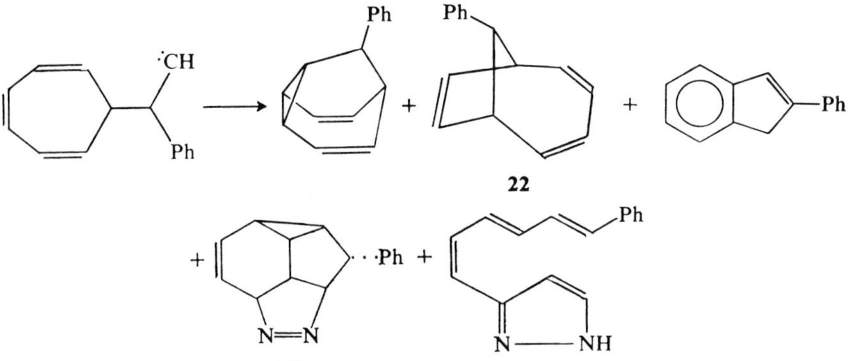

Ph → Ph + Ph + Ph

22

+ ··Ph + Ph

N=N N—NH

19 **20**

Compound **20** probably is the result of decomposition of **21**. Compound **19** does not give **22**, however, and it is thought that the norcaradiene **23** is the source of **22** as shown.[192]

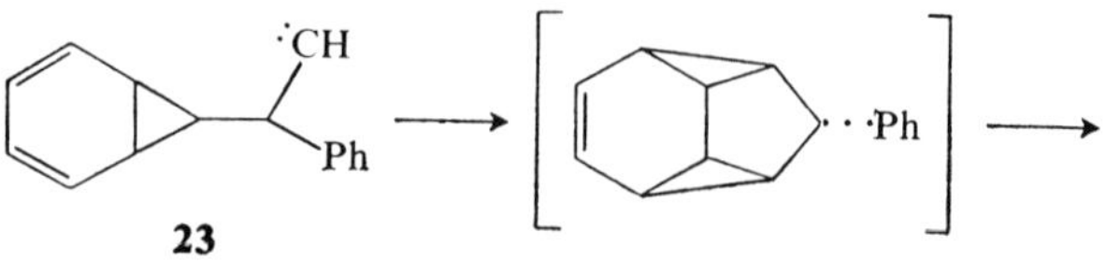

23

22

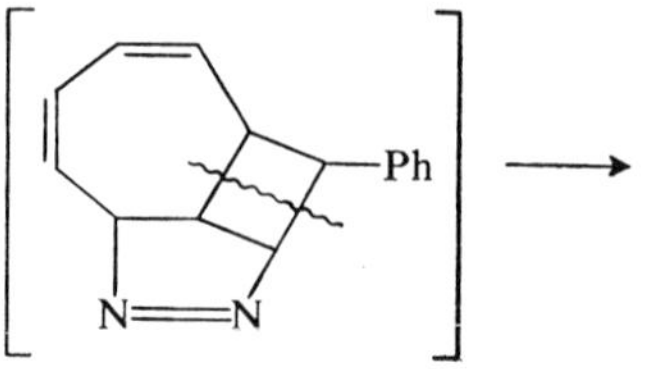

21

20

An earlier total synthesis of the racemic counterpart of the natural product thujopsene **24** also took advantage of the addition of a carbene to a remote double bond.[193] Although cyclopropene formation (**25**) is the major path followed, some addition to the remote double bond occurs and may be favored by the photolytic rather than thermal decomposition.

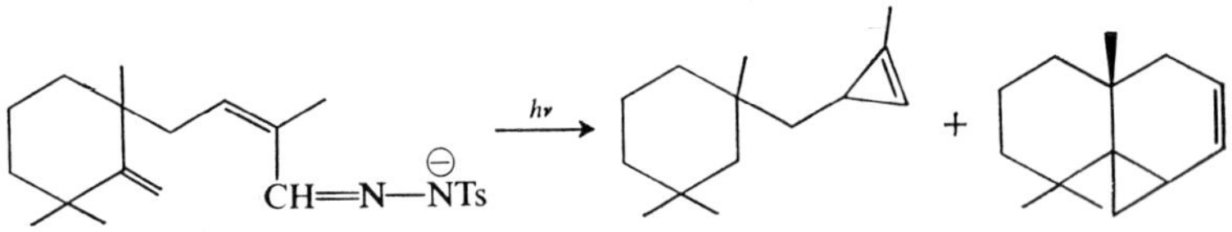

25 ; 10%   **24** ; 4%

The synthetic utility of α-β unsaturated carbenes as cyclopropene sources was demonstrated by Closs, Closs, and Böll,[194] who made many cyclopropenes in relatively good yield by this method. Only one example (**26**) is shown. Since this work, this method has been used repeatedly and extended to many photolytic decompositions.[195,196]

Vinylcarbene itself is a slightly studied species, no doubt because of the tendency of vinyldiazomethane to form pyrazole[197] and, with external olefins present, pyrazolines.[198] Irradiation of vinyldiazomethane in olefins does give very small yields of vinylcyclopropanes, however.[199] Their origin may lie in the addition of vinylcarbene to the olefin, although the photochemical decomposition of a pyrazoline is certainly a strong possibility.

Geibel and Mäder have studied the decompositions of a variety of bistosylhydrazones.[200] A typical example is shown.

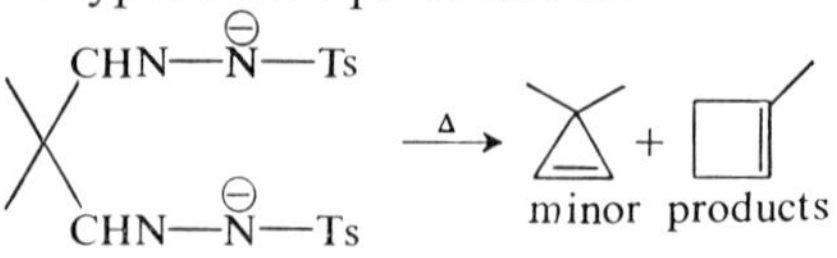

minor products

A general route to large ring acetylenes has also been developed using the decomposition of bistosylhydrazones.[201]

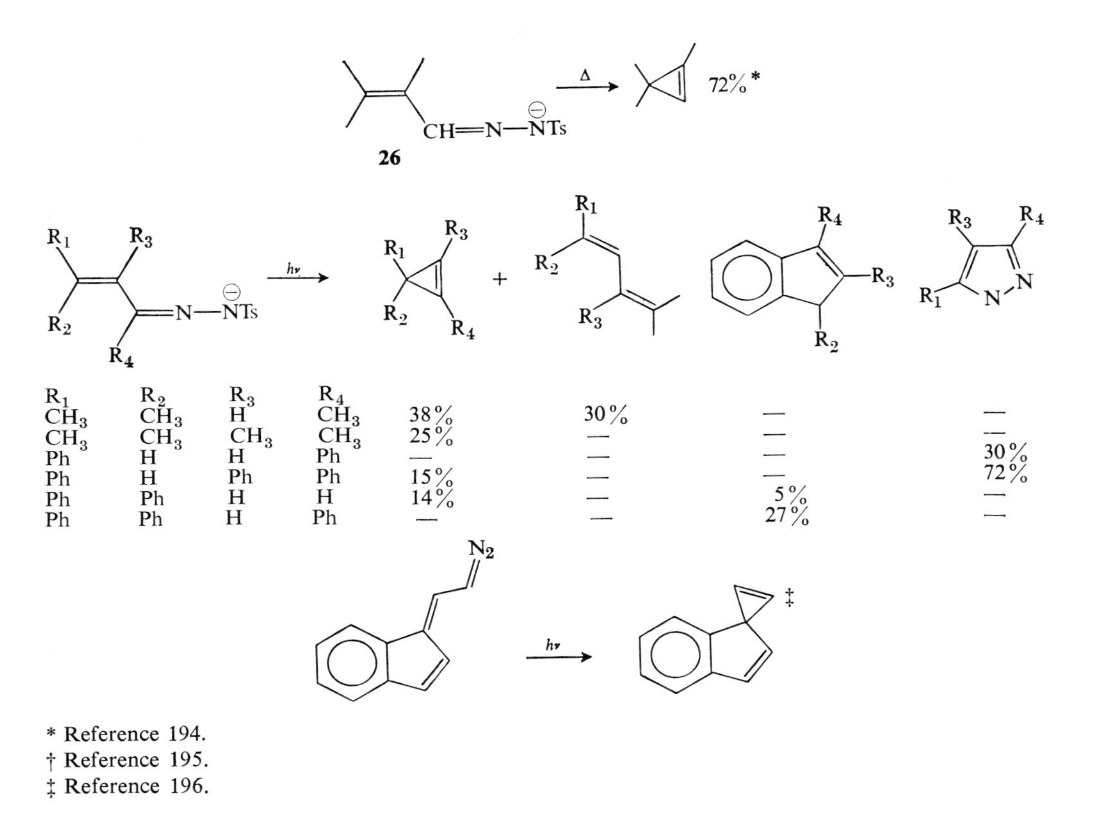

R_1	R_2	R_3	R_4	Cyclopropene	Diene	Indene	Pyrazole
CH_3	CH_3	H	CH_3	38%	30%	—	—
CH_3	CH_3	CH_3	CH_3	25%	—	—	—
Ph	H	H	Ph	—	—	—	30%
Ph	H	Ph	Ph	15%	—	—	72%
Ph	Ph	H	H	14%	—	5%	—
Ph	Ph	H	Ph	—	—	27%	—

* Reference 194.
† Reference 195.
‡ Reference 196.

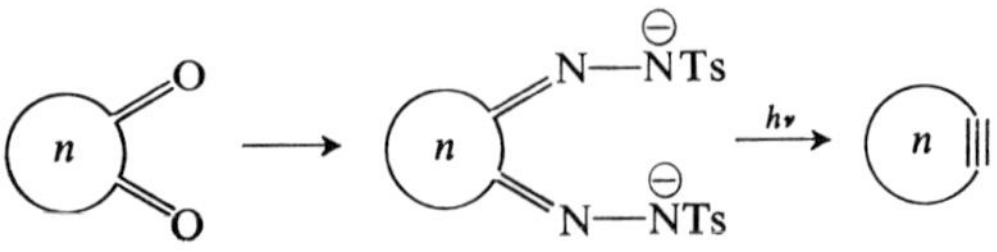

$n \geqslant 8$

Also studied was the bis-tosylhydrazone of fumaraldehyde.[199] Only a 3% yield of cumulene could be obtained. This obvious route to tetrahedrane had been tried earlier utilizing the bishydrazone, which gave only butadiene(!) on oxidation.[202]

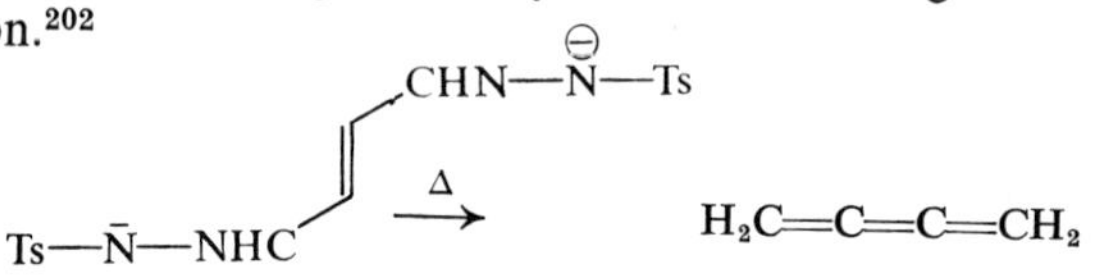

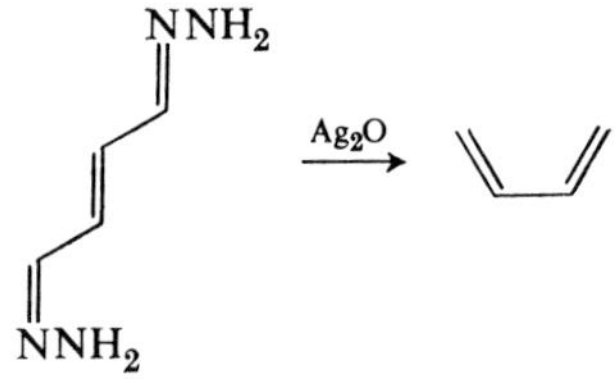

$$H_2C{=}C{=}C{=}CH_2$$

B. Alkynylcarbenes

Relatively little has been done with these species. An early report by Skell and Klebe[203] has been amplified in the thesis of Gramas.[204] Photolysis of diazopropyne in *cis*- and *trans*-2-butene gives a carbene that is slightly nonstereospecific in its addition. Scavenging of the triplet state with isoprene

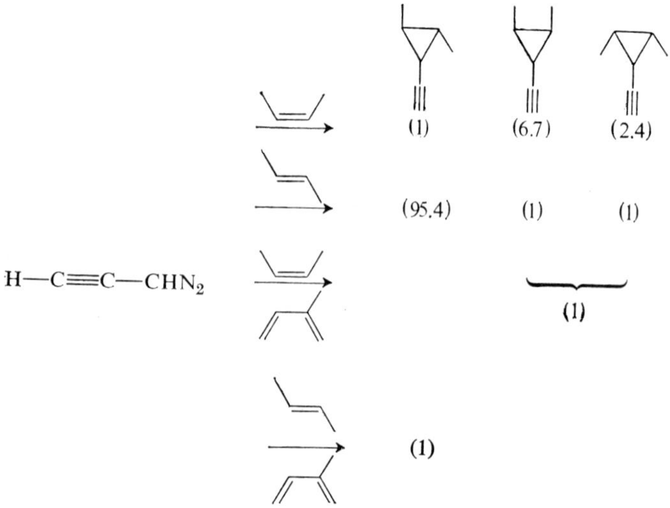

$$H{-}C{\equiv}C{-}CHN_2$$

(1) (6.7) (2.4)

(95.4) (1) (1)

(1)

(1)

led to stereospecific addition. The linear, *sp*-hybridized triplet state originally proposed by Skell and Klebe[203] predicts equal reactivity at either end of the molecule. This is achieved in the gas-phase reaction of labeled propargylene with styrene, as shown.[204] In the liquid phase only 14% of C_3 addition occurred, which indicates that only 28% of a linear intermediate was formed.[205] The alternate explanation—in the gas phase the singlet state has time enough to achieve equilibration but reacts too quickly in solution—cannot be ruled out, however. That the triplet is linear was confirmed by ESR studies, which showed the zero-field splitting parameter E to be near-zero for a variety of propargylenes.[206] Similar results have been found for cyanocarbenes.[207,208]

$$H—C≡C—\overset{*}{C}HN_2 \xrightarrow[h\nu,\ \text{gas phase}]{\text{PhCH=CH}_2} \text{(46)} + \text{(54)}$$

Phenylpropargylene is implicated in the formation of benzylacetylene, phenylpropyne, and phenylallene on irradiation of the diazo ketone shown below:[209]

$$Ph—C≡C—C(=O)—CHN_2 \xrightarrow{h\nu} Ph—C≡C—C(=O)—\ddot{C}H \longrightarrow Ph—C≡C—C=C=O \xrightarrow{-CO} Ph—C≡C—\ddot{C}H$$

$$Ph—C≡C—\ddot{C}H \xrightarrow{\text{cyclohexane}} Ph—C≡C—CH_3 + Ph—CH=C=CH_2 + Ph—CH_2—C≡CH$$

C. Fluorinated Alkylcarbenes

One might expect that perfluoroalkylcarbenes would be somewhat protected from the intramolecular escape that has been noted so often here. This protection is not complete; Fields and Haszeldine have observed a fluorine shift in trifluoromethylcarbenes,[210] but in competition with fluorine shift the rearrangement of hydrogen or a fluoroalkyl group wins out wherever possible.[210,211]

$$\underset{\mathbf{27}}{CF_3—CHN_2} \xrightarrow{h\nu} [CF_3—\ddot{C}H] \longrightarrow \underset{30\%}{CF_2=CHF} + \underset{50\%}{CF_3—CH=CH—CF_3}$$

$$CF_3CF_2CF_2—\ddot{C}H \rightarrow CF_2=CHCF_2CF_3$$
$$CF_3CF_2CF_2—\ddot{C}—CH_3 \rightarrow CF_3CF_2CF_2CH=CH_2$$
$$CHF_2CF_2—\ddot{C}—H \rightarrow CF_2=CH—CHF_2$$

In addition, products of bimolecular reactions can be found. Trifluoromethylcarbene reacts with its precursor to form butenes[210] and has been trapped in fair yield by external olefins.[212] Reaction of the electrophilic carbene with trifluoroethylene is slowed by the poor electron-donating properties of the olefin.[212]

$$CF_3{-}CHN_2 \ (\mathbf{27}) \xrightarrow[h\nu]{CH_2{=}CH_2} CF_3\text{-cyclopropane} \quad 26\%$$

$$CF_3{-}CHN_2 \ (\mathbf{27}) \xrightarrow[h\nu]{CF_2{=}CHF} CF_3\text{-trifluorocyclopropane} \quad 9\%$$

$$CF_3{-}CHN_2 \ (\mathbf{27}) \xrightarrow[h\nu]{\text{cyclohexene}} \text{7-}CF_3\text{-norcarane} \quad 46\%$$

Trifluoromethylcarbene also undergoes external carbon-hydrogen insertion in preference to intramolecular fluorine shift.[213] Irradiation of **27** in cyclohexane gives the insertion product **28** in 76% yield. Irradiation in isobutane gives a relative rate ratio of tertiary/primary = 1.3, and thus insertion is nearly random.

$$CF_3CHN_2 \ (\mathbf{27}) \xrightarrow[\text{cyclohexane}]{h\nu} C_6H_{11}CH_2CF_3 \ (\mathbf{28})$$

The stereochemistry of addition has received attention and has been virtually the only tool available to study the spin states of alkylcarbenes. Spectroscopic measurements have shown a variety of fluorinated alkylcarbenes to have bent (approximately 160°) triplet ground states.[214] Addition of trifluoromethylcarbene to *trans*-2-butene is largely stereospecific, but *cis*-2-butene gives substantial amounts of the "wrong" isomer.[215] Similarly, bistrifluoromethylcarbene gives only *trans*-cyclopropane from *trans*-2-butene, but small amounts of *trans*-cyclopropane are formed from *cis*-2-butene as well.[216] The cyclopropanes are accompanied by products of evident carbon-hydrogen insertion (not, apparently, abstraction), and so a mixture of spin states seems likely. A dilution experiment showed increasing nonstereospecificity with increasing concentration of inert solvent, which bears out the previous conclusion.

Addition products are also formed with acetylenes,[215] benzene,[216,217] and hexafluorobenzene.[218]

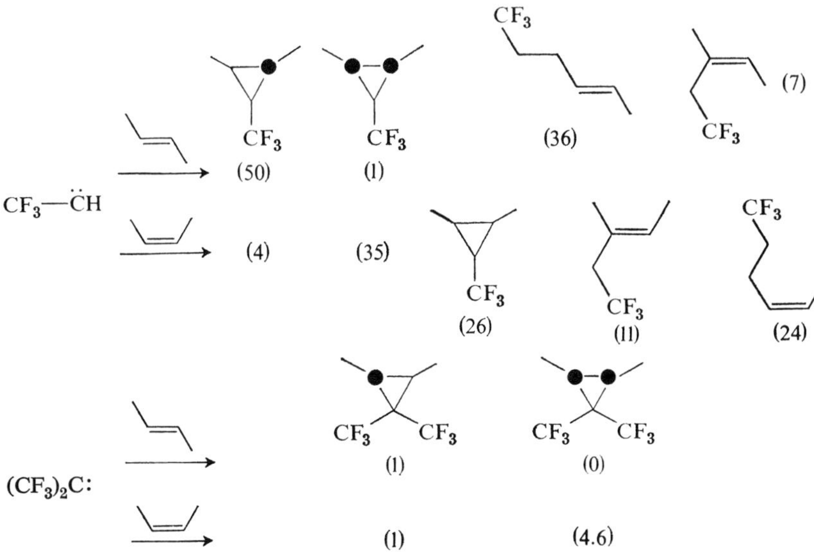

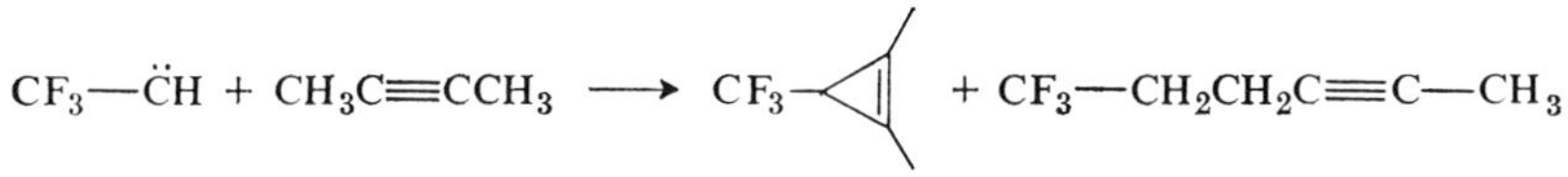

CF_3 CF_3 CF_3 CF_3 (7)

(50) (1) (36)

$CF_3—\ddot{C}H$

(4) (35) CF_3 CF_3 CF_3

(26) (11) (24)

$(CF_3)_2C$:

CF_3 CF_3 CF_3 CF_3

(1) (0)

(1) (4.6)

$$CF_3—\ddot{C}H + CH_3C{\equiv}CCH_3 \longrightarrow CF_3\text{-cyclopropene} + CF_3—CH_2CH_2C{\equiv}C—CH_3$$

49% 14%

$(CF_3)C$: CF_3 CF_3 + $CH(CF_3)_2$

(88) (12)

CF_3 CF_3 20%

F_6 F_6

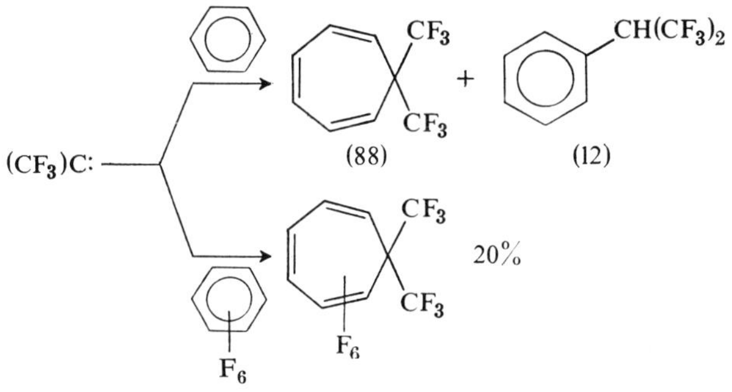

Trifluoromethylcarbenes have been found to deoxygenate carbonyl groups of various kinds.[219]

$$(CF_3)_2C: \xrightarrow[160^\circ \text{or } h\nu]{F_3P{=}O} (CF_3)_2CO$$

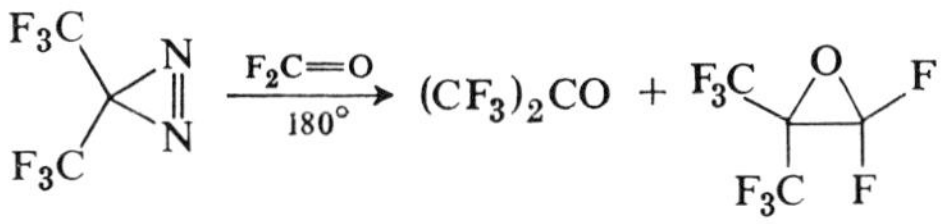

$$F_3C,\ F_3C\text{-diazirine} \xrightarrow[180^\circ]{F_2C{=}O} (CF_3)_2CO + F_3C,\ F_3C,\ F,\ F\text{-oxirane}$$

D. Cycloalkyl- and Cycloalkenylcarbenes

With such compounds the shift of an alkyl group becomes a ring expansion and is quite facile. In solution cyclopropylcarbene itself gives about 65% cyclobutene and about 22% fragmentation products (ethylene and acetylene).[220] The gas-phase reaction[221] gives much more fragmentation, presumably by reaction of "hot" cyclobutene or direct fragmentation of the carbene. The influence of ring size can be seen from the following data in which ring-expanded products drop from ~ 100% in the cyclopropyl compound to 59% in the cyclohexyl (as measured against phenyl migration).[222,223]

Ph CH → Ph (90) + Ph (10)

Ph CH → Ph + Ph (8) Ph + Ph (92)

Ph CH → Ph + Ph (20) + Ph (80)

Ph CH → Ph + Ph (41) + Ph (59)

Ph CH → Ph + Ph (43) + Ph (57)

The stereochemistry of the fragmentation has been briefly explored by Guarino and Wolf.[224] These authors find a slight nonstereospecificity at low pressures, which becomes greater when nitrogen is added. The authors interpret these data to mean that the singlet carbene reacts at low total pressure to give (nearly) stereospecific fragmentation and (less nearly) stereospecific ring expansion to *trans*-3,4-dimethylcyclobutene, which opens in a conrotatory sense to give the *trans,trans*-diene. At high pressures, the

Pressure (torr)	Yields (%)				
11	41	43	2	20	4
23	40	41	2	25	6
21 + 739N_2	21	10	9	8–15	7–13

authors suggest, a triplet carbene is formed which loses stereochemistry in both the fragmentation and ring expansion reactions. This explanation seems reasonable, at least in broad outline, since we would expect a triplet carbene to give a lower yield of fragmentation products and to do so in a non-stereospecific fashion. Similarly, loss of stereochemistry in the ring-expansion reaction is not surprising.

In solution, ring expansion of *trans*-2,3-dimethylcyclopropylcarbene is stereospecific,[225,226] as is the fragmentation to *trans*-2-butane.[225]

When a choice exists in the ring expansion one might expect the weaker bond to break. This turns out not to be so; it is always the stronger that migrates.[227,228] It seems strange to us that in these and simpler cases no vinylcyclopropanes are formed. Why does hydrogen migration not take place? Decomposition of the tosylhydrazone salt of methyl cyclopropyl ketone gives 92% ring expansion and only 1–2% vinylcyclopropane.[220] This reaction is very sensitive to conditions, however, and vinylcyclopropane can be made to be the major product. Thermal decomposition of the dry salt of

R_1	R_2	(bond 1)	(bond 2)
CH_3	CH_3	(2.5)	(97.5)
CH_3	H	(4.4)	(95.6)
H	CH_3	(27.9)	(72.1)

R_1	R_2	R_3	R_4	(bond 3)	(bond 4)
CH_3	CH_3	H	CH_3	(97.5)	(2.5)
H	CH_3	H	CH_3	(95.6)	(4.4)
CH_3	H	H	CH_3	(72.1)	(27.9)
CH_3	H	H	H	(49.3)	(50.7)

the tosylhydrazone still gives ring-expanded compound as the major product, but only when the salt is made with one equivalent of base.[229] Decomposition of the salt formed by using three equivalents of base gives predominantly vinylcyclopropane.[229] Tosylhydrazone decompositions always must be viewed cautiously, and the formation of vinylcyclopropane may well involve an elimination reaction similar to those already known.[230–232]

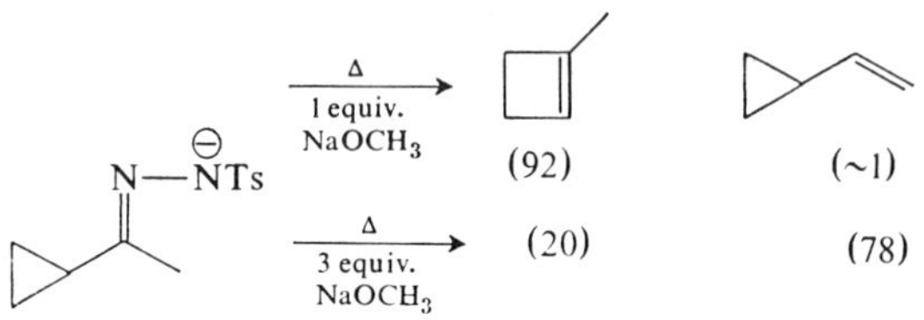

Δ, 1 equiv. $NaOCH_3$: (92) (~1)

Δ, 3 equiv. $NaOCH_3$: (20) (78)

Dicyclopropylcarbene,[233] presumably formed via the tosylhydrazone salt **29**, also gives largely the ring expanded product **30**.[234]

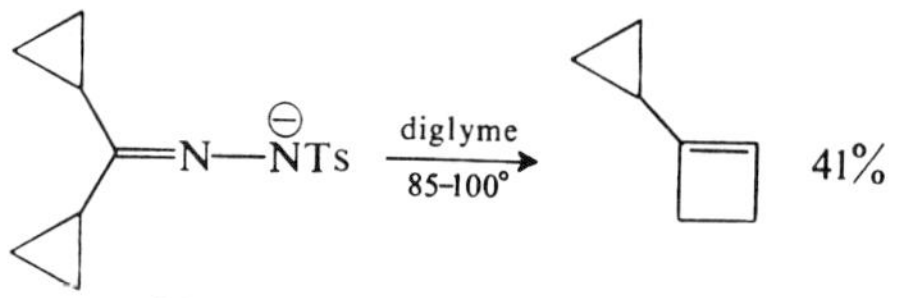

29 → (diglyme, 85–100°) **30** 41%

Saturated bicyclic systems also undergo ring expansion, although the fragmentation reaction is important.[235] Ring expansion occurs in spiro systems as well, and some unusual olefins have been so synthesized by Kirmse and Pook[236] and later by Wiberg, Burgmaier, and Warner.[237] Here, as before, it seems remarkable that no hydrogen shift was seen and again it is possible to vary conditions to get a shift. Decomposition of the dry tosylhydrazone salt formed with three equivalents of base gives the product of hydrogen migration in comparable yield to that of ring expansion.[229]

When the bicyclo[n.1.0] system contains unsaturation in the n ring, not only ring expansion but deep-seated rearrangements occur. In the thermal decomposition of the tosylhydrazone salt of bicyclo[6.1.0]nona-2,4,6-triene-9-carboxaldehyde (**31**), the presence of the ring-expanded product **32** was inferred[238] and the major rearranged product of the reaction, bicyclo[4.2.2]-deca-2,4,7,9-tetraene (**33**), was isolated. Masamune et al.[239] isolated the ring-expanded product **32** and confirmed its rearrangement to *trans*-9,10-dihydronaphthalene.

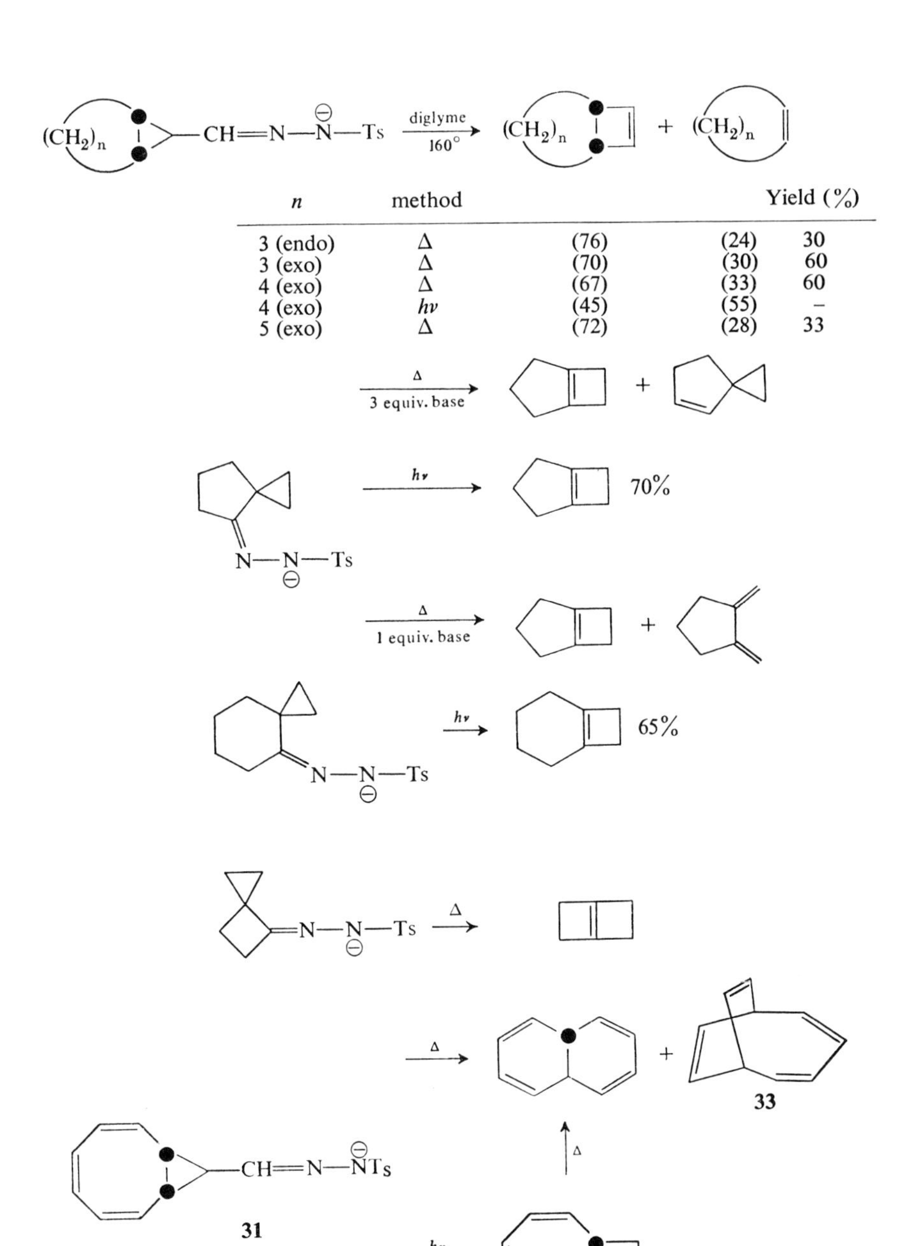

n	method	Yield (%)		
3 (endo)	Δ	(76)	(24)	30
3 (exo)	Δ	(70)	(30)	60
4 (exo)	Δ	(67)	(33)	60
4 (exo)	$h\nu$	(45)	(55)	–
5 (exo)	Δ	(72)	(28)	33

The decomposition of the tosylhydrazone salt of bicyclo[5.1.0]octa-2,4-dien-8-carboxaldehyde is similar: the ring-expanded product **34** and two rearranged products (**35** and **36**) were isolated.[238] The mechanism proposed is illustrated for this case:

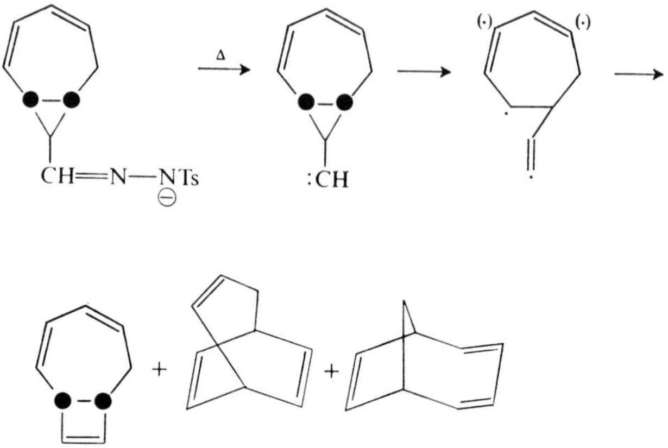

CH=N—NTs ⊖ —Δ→ :CH → →

34 + **35** + **36**

Such rearrangements do not require two double bonds in the *n* ring, since bicyclo[3.1.0]hept-2-en-6-ylidene undergoes a similar rearrangement.[240]

Cycloheptatrienylcarbene yields only benzene, acetylene, heptafulvene, and cycloöctatetraene.[241] It might have been hoped that the related norcaradienylcarbene would give barrelene, but none has been reported.[241]

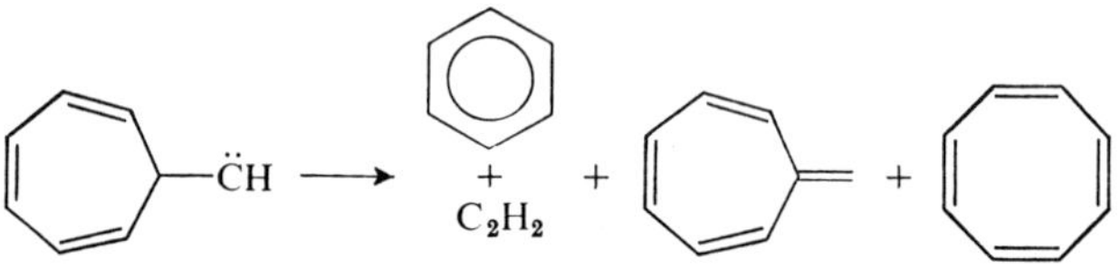

$\ddot{C}H$ → + C_2H_2 + +

Cyclopropenylcarbenes are a member of that class of carbene that has nothing to do, or perhaps only very interesting things to do. Normal ring expansion would lead to cyclobutadiene, whereas addition to the remote double bond would give tetrahedrane. This latter objective has not been

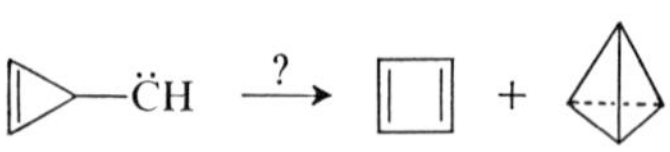

$\ddot{C}H$ —?→ +

37 Ph, Ph, CHN_2 —Δ→ Ph—C≡C—Ph 10–14% + **38** 0.1%

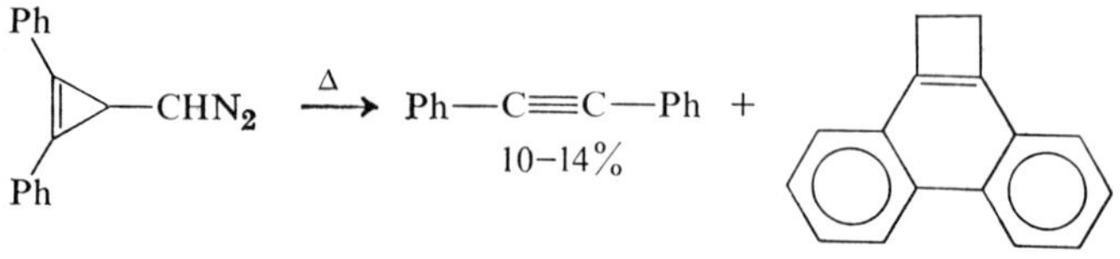

achieved from a diazo compound, since tolan is the major product of the decomposition of **37**.[242] A minor (0.1%) product, originally thought to be diphenyltetrahedrane, turned out to be **38**.[242,243]

Apparently the major path followed by the cyclopropenylcarbene is fragmentation. Reaction of recoil C^{11} with cyclopropene, which should generate cyclopropenylcarbene, does lead to a mixture of products for which tetrahedrane is a plausible intermediate.[244] Irradiation of carbon suboxide in cyclopropene-3,3-d_2 gives a mixture of acetylenes consistent with the formation and subsequent fragmentation of tetrahedrane.[245]

The ring expansion to cyclobutadiene has not been achieved from a diazo compound, but the carbenoid **39** gives rise to a number of dimers of trimethylchlorocyclobutadiene.[246]

CCl_2Li → [Cl] → Dimers

39

It seems that the ease of the fragmentation reaction to acetylenes should make isolation of a tetrahedrane a difficult task. The fragmentation reaction should be a singlet state process, however, and might well be severely retarded (*vide infra*) by formation of the triplet state.[224] It seems to the authors once the triplet is formed it has little else to do but to start the crucial cycloaddition.

Considerably less work has been done on cyclobutylcarbenes, but existing evidence[247] seems to indicate that their behavior will parallel that of the cyclopropyl compounds,[222] with ring expansion the main process. The formation of methylcyclobutane is a bit puzzling; perhaps it results from an overall Wolff-Kishner reduction of cyclobutanecarboxaldehyde.[248]

=N—NTs ⊖ —diglyme, Δ→ + + +

(13.6) (17.7) (56.5) (12.2)

Although ring expansion still dominates the reactions of α-1-phenylcyclopentylcarbene,[222] in simpler systems internal additions and insertions are more important. Cyclopentylcarbene itself gives 72.5% insertion to yield bicyclo[3.1.0]pentane.[249] But Lemal and Shim[250] have found the following products from cyclopentenylcarbene:
Similarly, Closs and Larrabee[251] found mainly internal cycloaddition from the pyrolysis of the diazo compound **40**. The cyclobutenylcarbene **41** also gave some polycyclic hydrocarbon, which seems to result from an intramolecular cycloaddition.[251]

72.5%

:CH

(52) (9) (29) (10)

CHN_2

Δ

40

:CH

H⊕

41

Exo- and *endo*-bicyclo[3.1.0]hexan-3-ylcarbenes gave a variety of products of carbon-hydrogen insertion but little ring expansion.[249] As the Newman projections below show, cyclopentylcarbene can easily achieve the conforma-

$—CHN_2$ hν (38) (6.5)

(47) (8.5) (0)

(52) (3.3) (0.7)

$—CHN_2$ hν (44)

tion required for insertion. Ring expansion, on the other hand, appears much more difficult, especially for small rings. In the case of cyclopropylcarbene, where ring expansion occurs to a large extent, overlap of the empty *p* orbital with the bent bonds of the ring may favor the carbon migration.

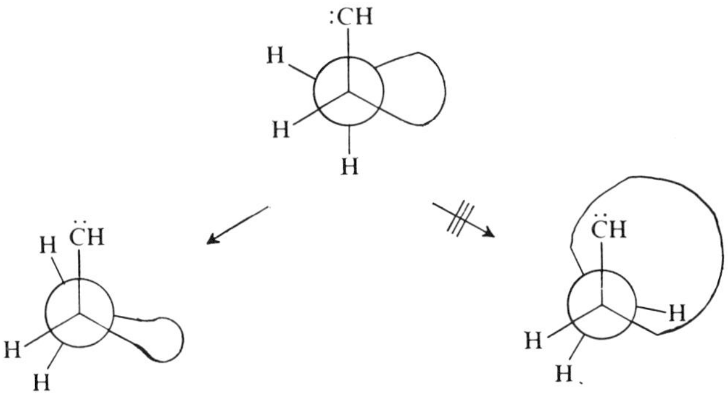

Cyclohex-3-enylcarbene (**42**) also gives a mixture of products of 1,2 and 1,3 carbon-hydrogen insertion.[252,253] Decomposition of the diazo compound was not sensitive to changes in solvent, causing one group of workers[252] to postulate intermediate carbenes in all solvents, regardless of polarity.

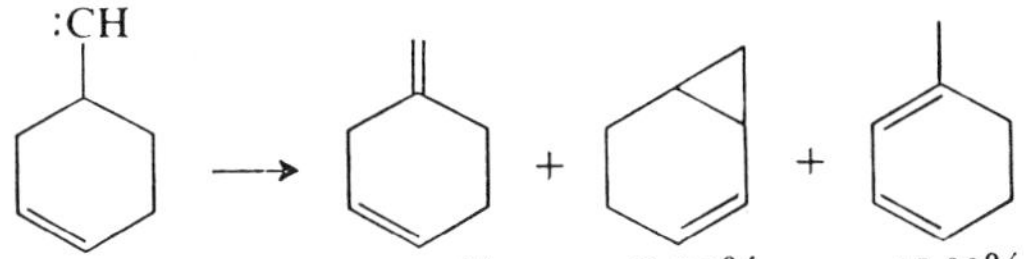

42 35-70% 18-22% 15-30%

Cycloheptyl- and cyclooctylcarbenes gave mainly 1,2 and 1,3 insertion, although here[254] and in larger rings[255] the ring is large enough so that *trans*-cyclopropanes can be isolated and even predominate in the eight-membered ring.[254]

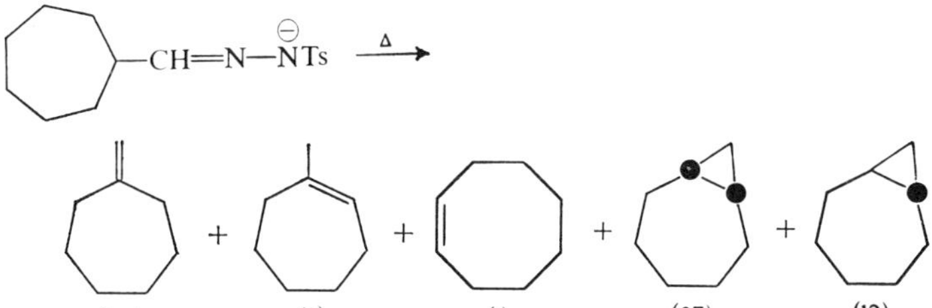

(48) (2) (1) (37) (12)

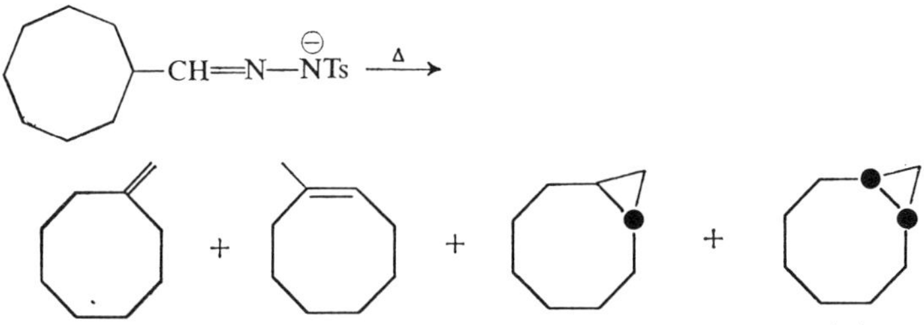

(44.5) (1.5) (33) (18)

When a double bond is well disposed, both cyclopropane and pyrazoline formation can occur.[256]

CH=N—$\overset{\ominus}{N}$Ts $\xrightarrow{\Delta}$ +

An interesting question arises: What will carbenes attached to the bridgehead positions of bicyclic hydrocarbons do? The first investigators of this problem, Wilt et al.[257] found no products attributable to intramolecular reactions of carbenes. Alcohols were formed, it was thought, via protonation of a long-lived carbene or diazo compound followed by tosylate addition (see the example that follows). It is scarcely likely, however, that ring-expanded bridgehead olefins could have survived either the decomposition or isolation conditions.[258,259] When the divalent carbon is not attached to the

CH=N—$\overset{\ominus}{N}$Ts $\xrightarrow{180°}$ OH + OH

(6) (1)

bridgehead position, "normal" reactions occur. The compounds shown below give products of 1,2 and 1,3 carbon-hydrogen insertion as well as of ring expansion.[260] There are preliminary indications that 1-adamantylcarbene and 1-adamantylphenylcarbene formed by photolysis of the appropriate tosylhydrazone salts ring expand to a thus far nonisolable bridgehead olefin, which subsequently dimerizes.[261]

E. Cycloalkanylidenes

Doering and LaFlamme[262] originally noticed the conversion of 1,1-dihalocyclopropanes to allenes by treatment with metals or bases and speculated that a carbene might be involved. The likelihood that this is a carbenoid process and the necessary absence of a diazo compound allows us to bypass this reaction. Similar problems (i.e., what is or are the active intermediates?) have attended further work on cyclopropylidenes. The diazo compounds themselves seem capable of undergoing the reaction, as do the carbenes. Originally the dependence on solvent of the ratio of allene formation to external addition was thought to favor two intermediates (diazo compound plus carbene) or possibly implicate a solvent effect.[263] More recently, it has

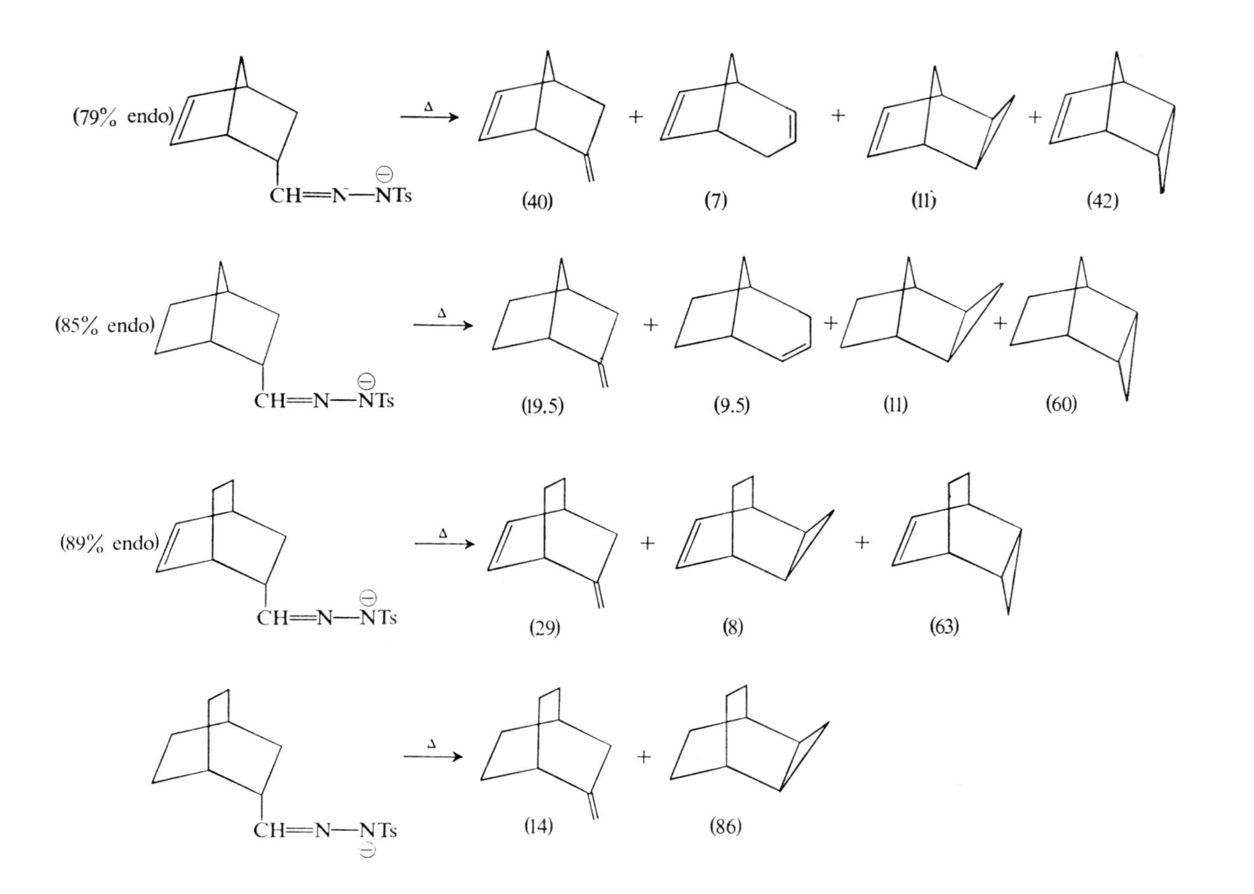
(79% endo)
CH=N–NTs
Δ
(40)
(7)
(11)
(42)
(85% endo)
CH=N–NTs
Δ
(19.5)
(9.5)
(11)
(60)
(89% endo)
CH=N–NTs
Δ
(29)
(8)
(63)
CH=N–NTs
Δ
(14)
(86)

been shown that the allene formed from optically active **43** was equally optically active when generated in either cyclohexane or cyclohexene.[264] It is likely that a solvent effect was operating in the early work[263] and that the carbene is the source of both allene and cyclopropane.

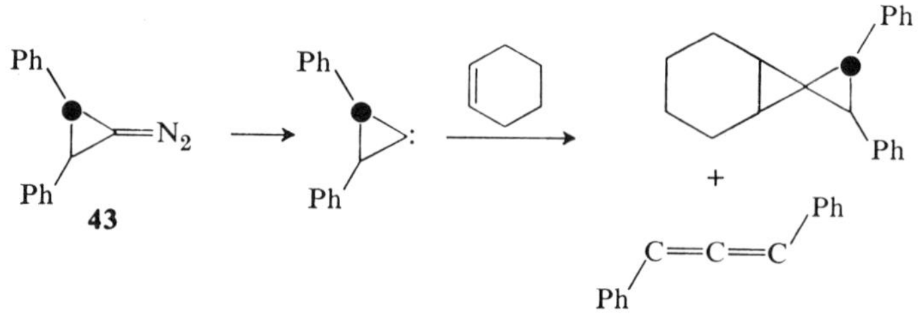

Optically active cyclopropylidenes do give optically active allenes, however.[265,266] A general method for the synthesis of optically active 1,3-disubstituted allenes has been described.[267]

Cyclobutylidene primarily undergoes ring contraction to give methylenecyclopropane.[220] The ring-contraction reaction has been used in a clever

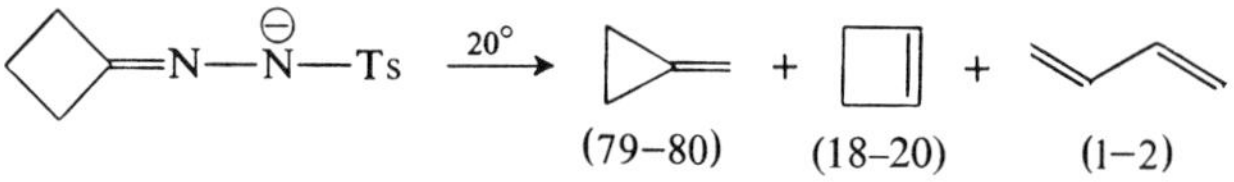

(79–80) (18–20) (1–2)

synthesis of homofulvenes.[268] The initially formed methylenecyclopropane participates in a vinylcyclopropane rearrangement which yields the product. The intermediate methylenecyclopropane can be isolated if a molecule such as **44** is constructed. Substitution on the ring by methyl groups leads to little

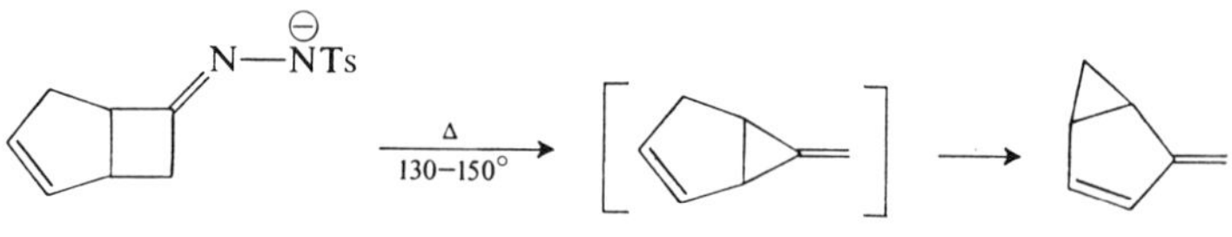

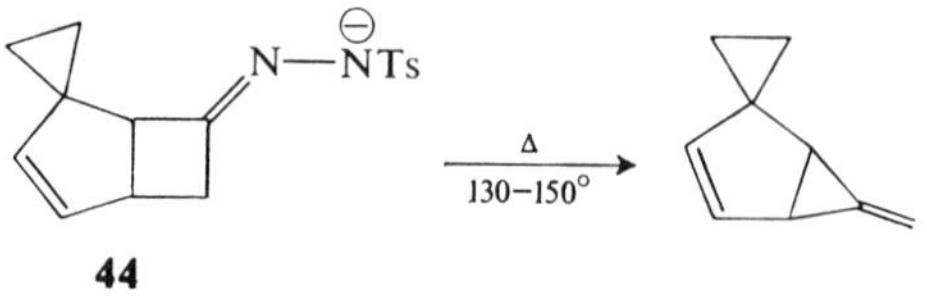

44

change as **46** is the major product from **45**,[269] but the cyclopropyl compound **47** gives a carbene that ring expands.[237,270]

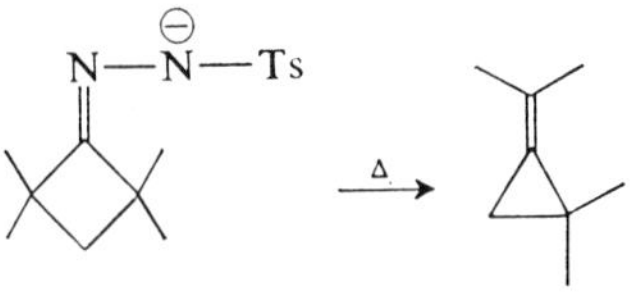

N—N⊖—Ts

45 → (Δ) 46

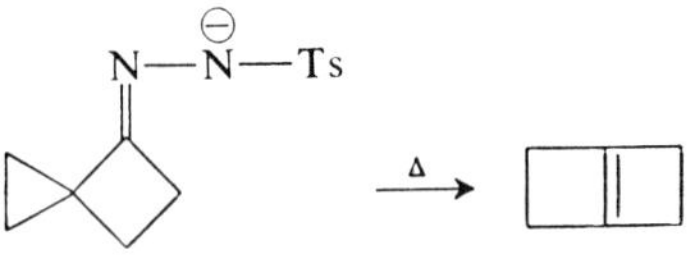

N—N⊖—Ts

47 → (Δ)

Some more extensively substituted cyclobutylidenes resist the ring contraction found by Meinwald[269] or undergo further reaction. Thus **48** gives carbon-hydrogen insertion and no ring contraction.[271] Compound **49** gives the product of a methyl shift, **50**, and the hexatriene **51**, which is presumably formed by further reaction of an initially formed methylenecyclopropane.[272]

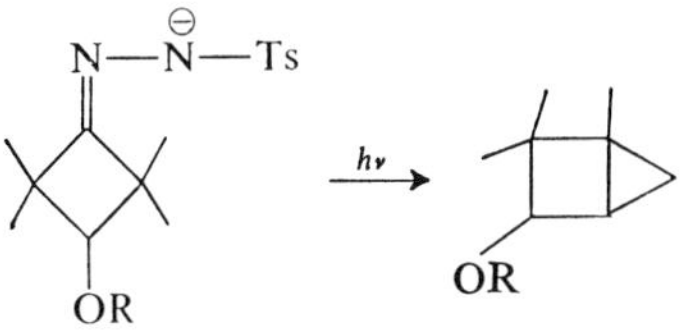

N—N⊖—Ts

OR

48 → (hν) OR

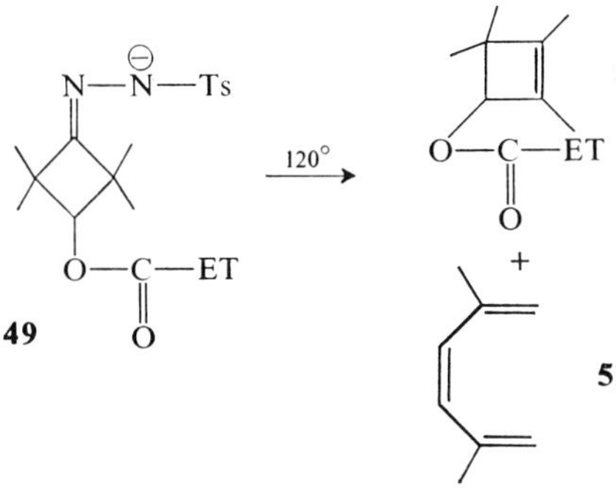

N—N⊖—Ts

O—C(=O)—ET

49 → (120°) 50 O—C(=O)—ET + 51

Tetramethylallene is formed from **52**, although the mechanistic details of the reaction are not known.[273] Similarly, the di-sodium salt of the

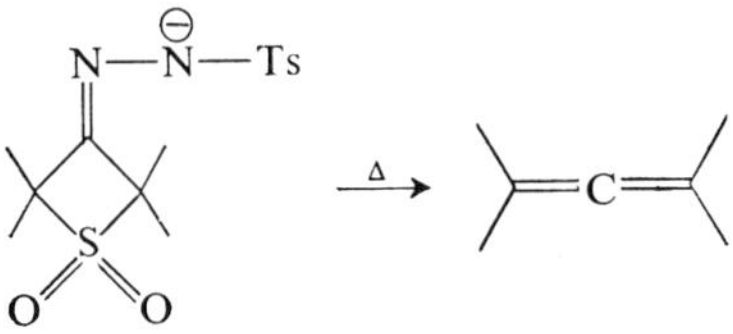

N—N⊖—Ts

S, O, O

52 → (Δ) C 55%

bishydrazone **52a** has been found to give tetramethylbutatriene on pyrolysis with[274] and without[275] copper powder. The compound without the four methyl groups is reported to give noncarbenic products.[276]

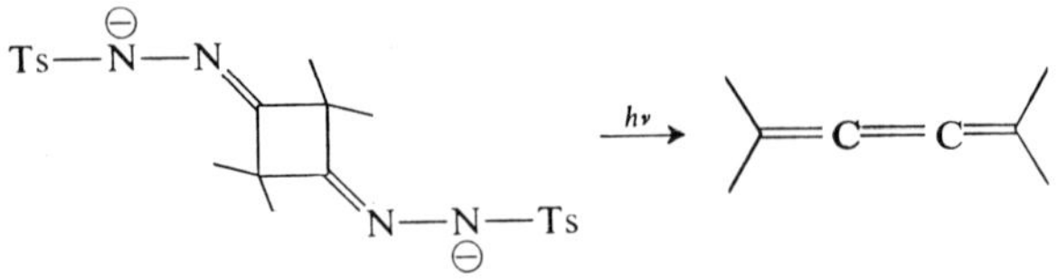

Ts—N⊖—N= ... =N—N⊖—Ts $\xrightarrow{h\nu}$ >=C=C=<

52a

Cyclopentylidene shows little tendency toward intramolecular reaction beyond hydrogen shift. Tosylhydrazone salt **53** gives a carbene which undergoes almost exclusive 1,2 hydrogen shift.[277]

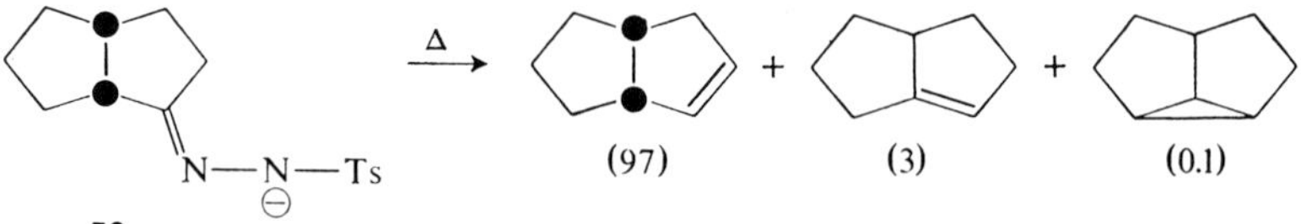

N—N⊖—Ts $\xrightarrow{\Delta}$ (97) + (3) + (0.1)

53

Diazocyclopentane and diazocyclohexane give only cyclopentene and cyclohexene, respectively, but cyclopentyl- and cyclohexyldiazirene give different products, depending upon the method of decomposition. Photolysis gives more products than thermal decomposition and thus is thought to yield a more energetic intermediate.[181] It is worthy of note that carbenes **54** and **55** give products resulting from exclusive shift of secondary hydrogen.[279,280]

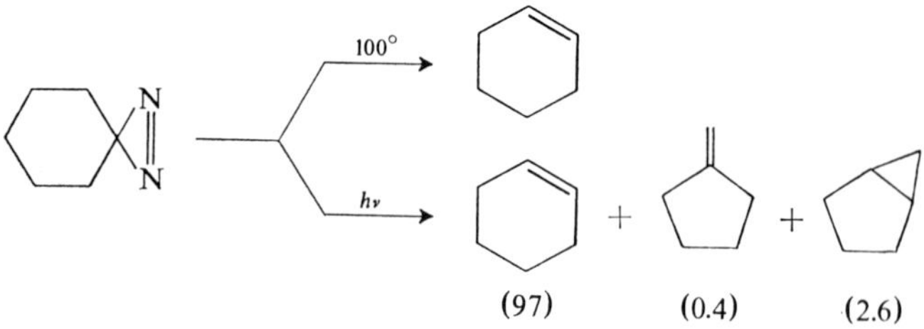

N=N $\xrightarrow{100^\circ}$

$\xrightarrow{h\nu}$ (97) + (0.4) + (2.6)

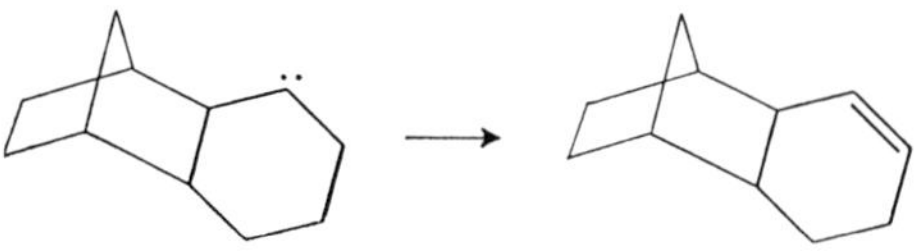

54 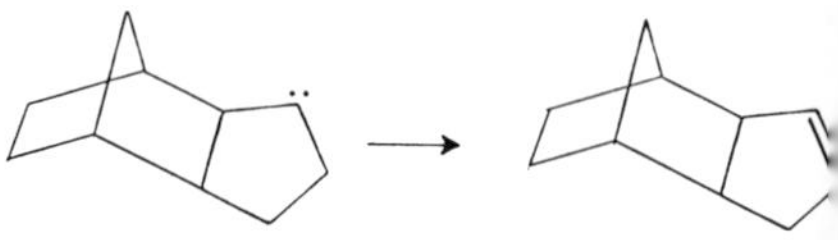 **55**

Various spiroketones have been used as sources of spirenes. Of the several combinations of the five- and six-membered ring compounds only the spiro[4.4] compound showed a tendency to rearrange[281] or undergo 1,3 insertion.

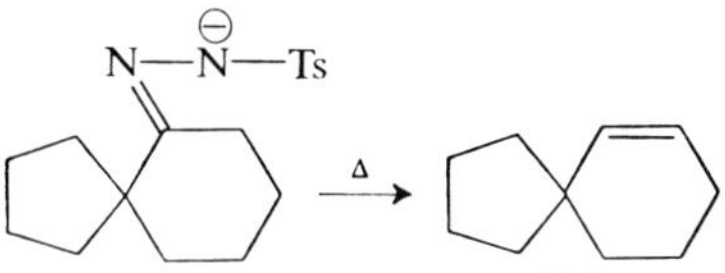

(99)

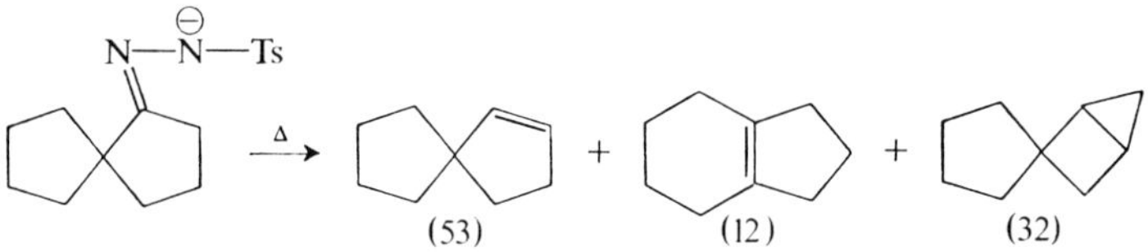

(53) (12) (32)

Products formed from bicyclo[3.1.0] carbenes depend upon the relative orientation of the three-membered ring. When the carbene is remote to the cyclopropyl ring then the reactions are "ordinary" and hydrogen shift predominates.[282–284] When the carbene is α to the cyclopropyl ring various fragmentation reactions occur.

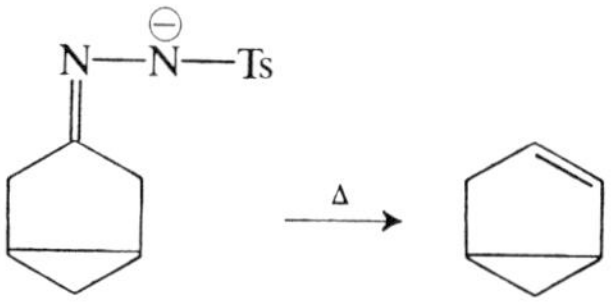

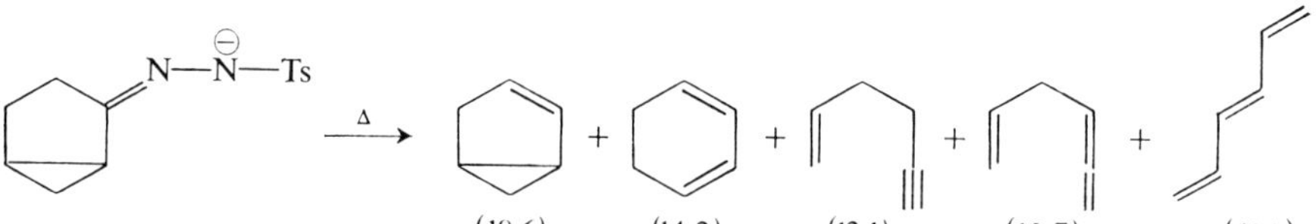

(18.6) (14.2) (13.1) (12.7) (41.8)

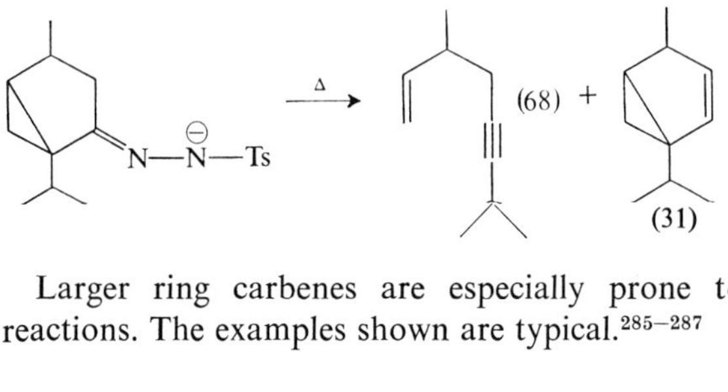

(68) (31)

Larger ring carbenes are especially prone to transannular insertion reactions. The examples shown are typical.[285–287]

(46) (45) (9)

(60) (40) (<1)

(51-55) (35-40) (3-6) (4-6)

25%

Bicyclic carbenes generally give multicyclic products. Much attention has been devoted to the synthetic utility of such species. The bicyclo[2.1.1]hexene system can be generated in 25% yield,[288] but benzvalene apparently rearranges further under the conditions of its generation.[289]

A variety of norbornylidenes has been made, and both in the parent[290,291] and substituted[292,293] cases nortricyclanes are the major products.

N—N⊖—Ts Δ 67%

N—N⊖—Ts Δ + (97) (3)

Norbornan-7-ylidene favors carbon shift over hydrogen shift on its way to tetravalency.[294] The divalent carbon atom is far from the available hydrogen and the migrating carbon-carbon bond seems well aligned for movement.

N—N⊖—Ts 180° 20 torr. + (1) (6.2)

Bridged norbornanes have also been studied and unsaturation can be increased either by introduction of a double bond or formation of a new ring. The tosylhydrazone **56** yields **57**, and **58** forms a new cyclopropane ring to give **59**. In addition, since the carbene is adjacent to an already existing three-membered ring, fragmentation occurs.[295]

Tosylhydrazone **60** yields a carbene which inserts in two but not three places.[296]

Tricyclylcarbene largely undergoes fragmentation and is apparently so unreactive that tosylate ion attacks to give tricyclyltosylate.[297]

Adamantylidene gives 2,4-dehydroadamantane.[298] Similarly, homoadamantylidene gives the dehydrocompound **62** and the olefin **61**.[299] A bistosylhydrazone salt gives the bisdehydroadamantane **62a**.[299a]

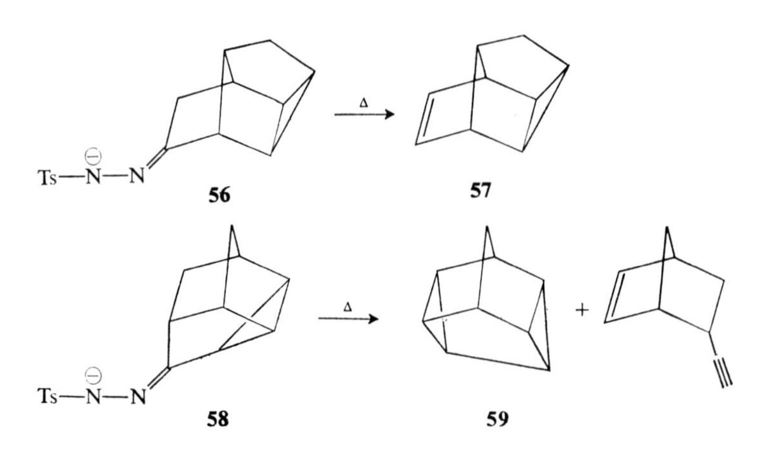

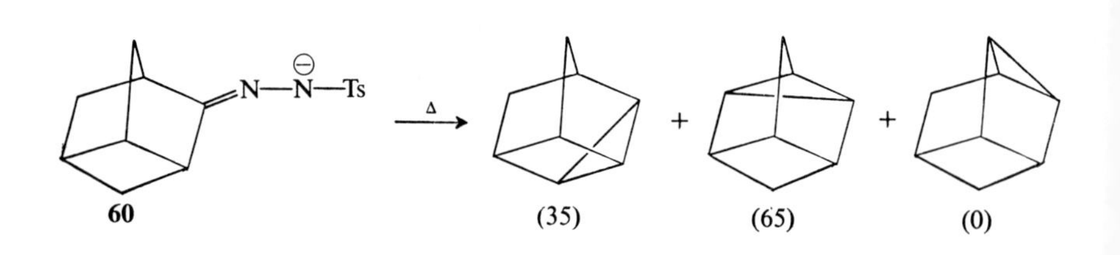

Δ

N—N—Ts

hν

Ts

+

Ts

N—N

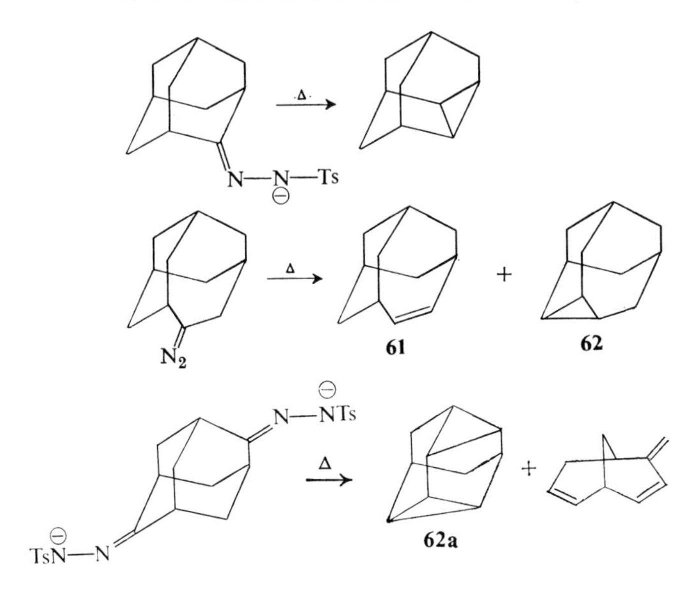

The azabicyclo[3.3.1] system yields tetracycles and 1,2 hydrogen-shifted products, depending on the position of the carbene[300,301] and the mode of decomposition. In particular, the remarkable compound **63** is claimed on photolysis.

N—N—Ts

Δ

N

N

Δ

diglyme

N

N

Δ

ethylene glycol

N

+

N

N

N

$\ominus$N

Ts

hν

ether

N

+

N

63

Noradamantane (**64**) itself can be formed by a similar reaction.[302,303]

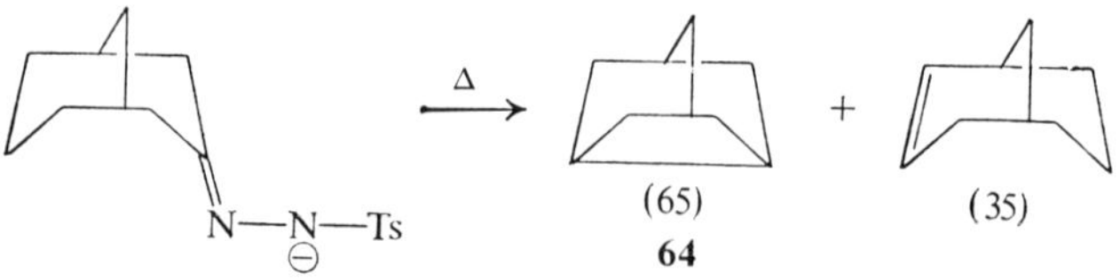

N—N—Ts
⊖
Δ
(65)
64
(35)

Although bicyclo[3.3.1]nonan-9-ylidene gives both carbon-hydrogen and carbon-carbon (bond migration) insertion,[304,305] **65** does not give the "expected" products of insertion into and migration of the allylic position **66** and **67**; instead it gives **68**, **69**, and **70**.[304] The authors interpret this in terms of an initial "interaction" of the divalent carbon with the double bond, which distorts the molecule and permits the observed products. Carbene **65** may be a "foiled methylene."[306] "Interaction" and "foiled methylene" are broad terms indeed and encompass effects ranging from a leaning of the divalent carbon toward the double bond to formation of one bond (or even two bonds) as in **71**. Intermediate **71** could easily give **70** through simple bond breaking, but **68** and **69** must be products of insertion into the "wrong" carbon-hydrogen bonds. The authors thus may be correct in their interpretation.

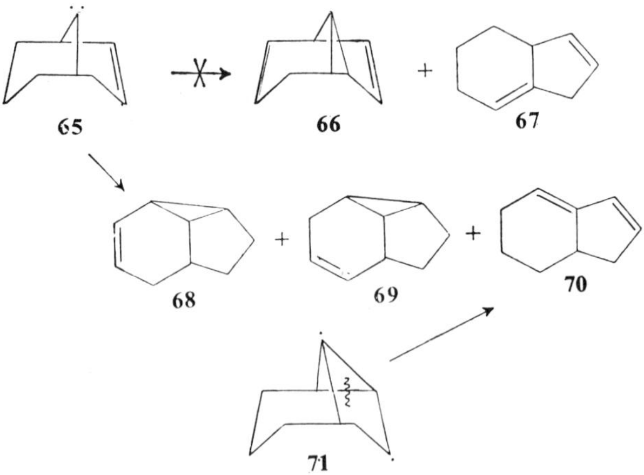

65 **66** **67**
68 **69** **70**
71

Norbornenylidene **71a** shows similar effects.[307] Here the vinyl "wing" migrates at least about 10 times as fast as the saturated wing. A similar

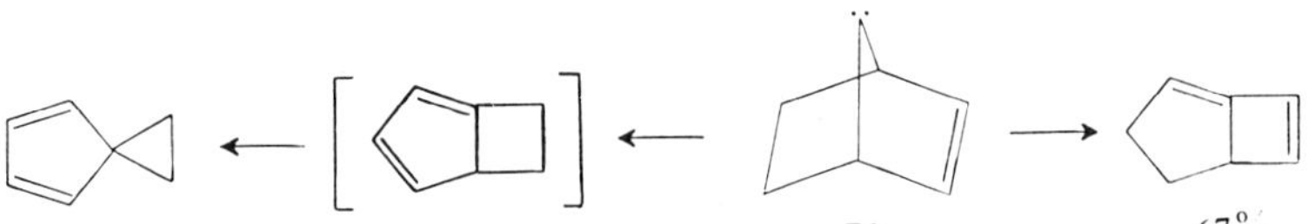

6.9% 71a 67%

"participation" has been invoked to explain the different products formed from **71b** and **71c**.[308]

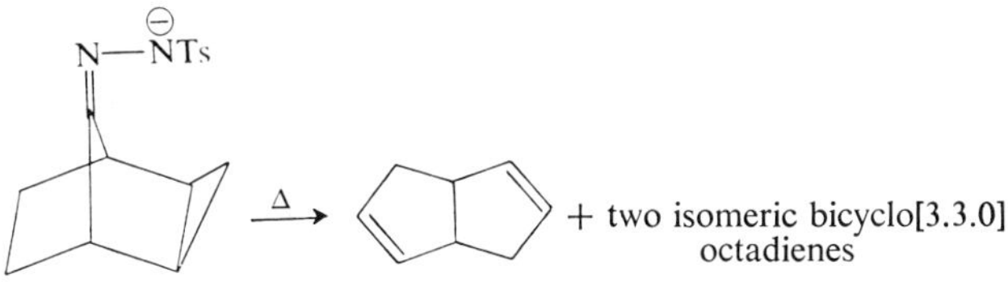

N—NTs

+ two isomeric bicyclo[3.3.0] octadienes

71b

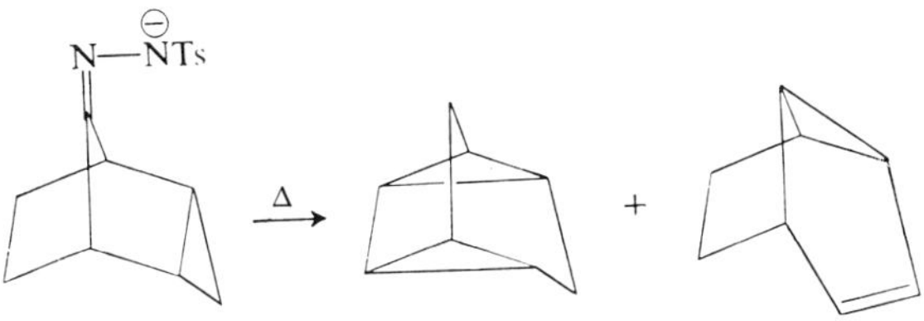

71c (81) (10)

Norbornadienylidene is unknown, but quadricyclanylidene **71d** apparently extrudes a carbon atom to give benzene.[309] Toluene is also observed and is thought to be a product of the reaction of benzene and carbon atoms. Labeling experiments rule out an intramolecular pathway for the formation of toluene.

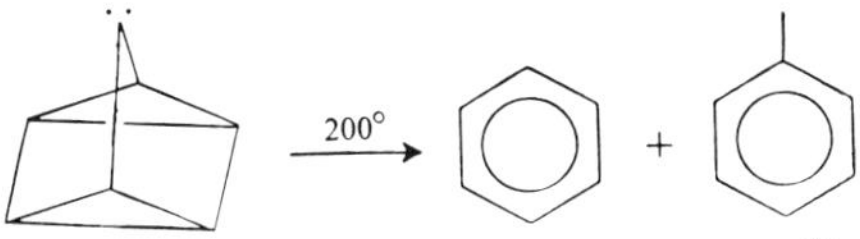

200°

71d 12% 26%

F. Cycloalkenylidenes

Cyclopropenylidene has eluded systematic study, although the persistence of W. M. Jones and co-workers has led to interesting information. Decomposition of the nitrosourethane **72** does not yield the diazo compound **73** but presumably proceeds via **73** directly to the carbene **74**.[310] It is scarcely surprising that **73** should be particularly elusive, since the major contributing resonance structure has the form of the antiaromatic cyclopropenyl anion, a species known to be exceptionally unstable.[311] The carbene **74** reveals itself via cycloaddition to dimethylfumarate to give the spiropentene **75**.[310] More conventional olefins such as the butenes do not yield adducts, presumably

ROOC
N—NO
Ph Ph
72
N_2
Ph Ph
73
Ph Ph
74
ROOC
COOR
ROOC Ph
75
ROOC Ph

due to the special nature of **74**. The divalent carbon would be expected to be especially electron rich and therefore more nucleophilic than usual. Fumaric ester is exactly the kind of olefin one would expect to react with a nucleophilic carbene. Presumably a dialkyl cyclopropenylidene would be even more nucleophilic,[312] since the stability of the system would be further enhanced by the change from aryl to alkyl groups. The yield of adduct is low (~10%) in any case and can be further diminished by the acid-catalyzed rearrangement to **76**.[313–314]

Ph COOR
Ph COOR
H^+
COOR
ROOC
Ph Ph
76

Abandoning logical progression and skipping directly to cycloheptatrienylidene **77**, we find another species that might be nucleophilic, and here properties similar to those of **74** are encountered. Adducts are formed only from dimethylfumarate[315,316] and the dicyanoethylenes.[315] The butenes,

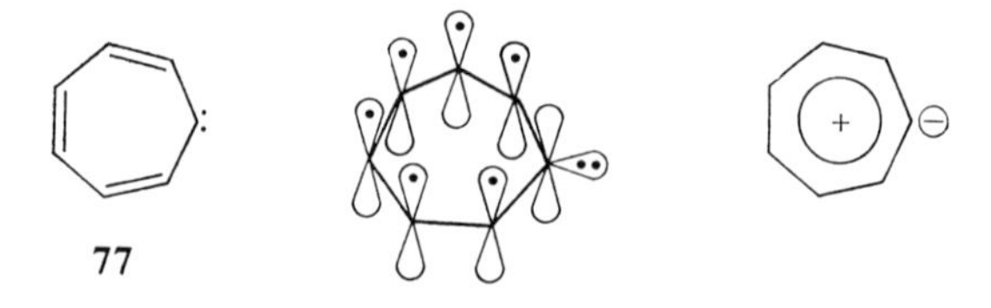

cyclohexene, and dibromoethylene are not attacked, and only heptafulvalene can be isolated.[315,316] Addition to the cyanoethylenes is completely stereospecific[317] and loss of stereochemical integrity during addition to dimethylmaleate[316] is probably due to olefin isomerization.[317] Assignment of a nucleophilic character to **77** seems correct, but a singlet ground state[317] is not demanded, although perhaps reasonable.

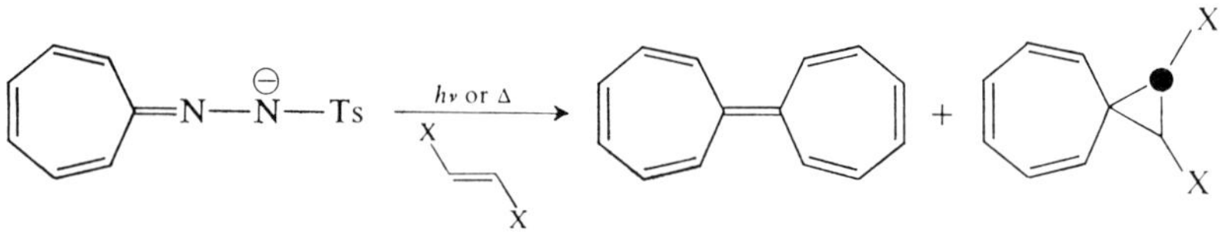

$h\nu$ or Δ

X = CN, $COOCH_3$

Conspicuous by its absence from the literature is reference to cyclobutenylidene. This potential precursor of cyclobutadiene seems worthy of study. A benzo analog gives no evidence of hydrogen shift, however[318] (see Section IV).

If **74** and **77** are especially nucleophilic because of a quasiaromatic structure, cyclopentadienylidene **78** might be either especially electrophilic or odd in some other way because of a similar aromatic structure. Here the question revolves around the contribution of structures like **79** in which two electrons have been promoted from the lower energy sp^2 orbital to the higher energy p orbital. By spending this energy, the carbene stands to gain whatever energy an aromatic structure is worth. Extended Hückel calculations[306] indicate that a structure such as **79** is not out of the question. Accordingly, a great deal of effort has been extended toward establishing the detailed properties of **78**.

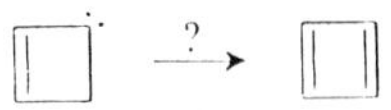

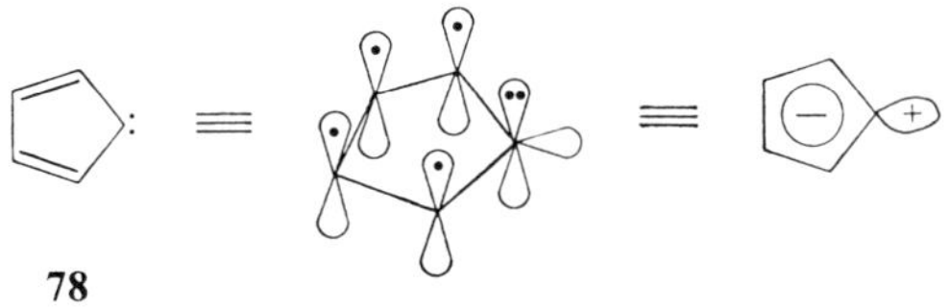

78

79

Very early work on **78** included the demonstration that the carbon-hydrogen insertion reaction occurred[319] and that **78** was very selective in this reaction.[320] More recent work has involved an extension of the work of Basinski,[320] a determination of the selectivity in the carbon-hydrogen insertion reaction,[321] and a more extensive investigation of the addition reaction.[321,322] Adducts are formed in good yield and addition is nearly but

57%

66%

78

(1) (23)

(1) (7.3)

not completely stereospecific. Addition of either hexafluorobenzene[323] or perfluorocyclobutane[322] has little effect on the stereochemistry. The reacting state seems to be the singlet, although the ground state, along with that of the related indenylidene, is known to be the triplet.[324] The relative rates of

78

(98) (2)

(0) (100)

addition of **78** to olefins reveal little difference between olefins of different substitution. This is a bit surprising, for an electrophilic singlet and was attributed, correctly we think, to a steric effect.

The reactions of **78** with butadiene and cyclooctatetraene are normal[325,326] and no 1,4 addition products are found. The further rearrangements of the initially formed adducts **80** and **81** have been studied.[325,326]

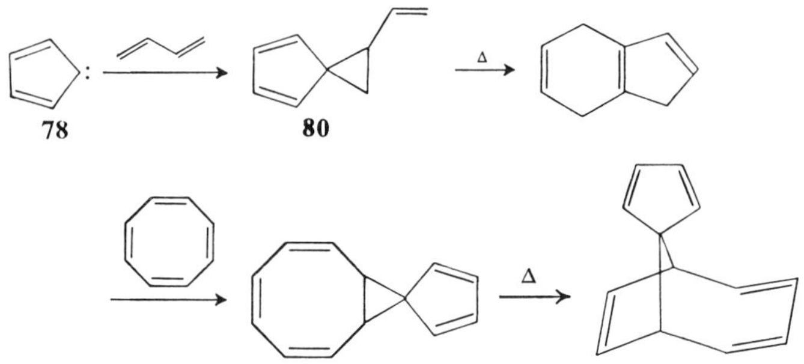

78 80 81

It is the reaction of **78** with benzene that is most interesting. Reaction with benzene itself yields the norcaradiene **82**,[325,327] whereas hexafluorobenzene gives the adduct **83** in the cycloheptatriene form.[328] Clearly the difference is due to the two fluorines that occupy either a vinyl or cyclopropyl position and reflects the growing evidence that fluorine abhors a cyclopropane ring even more than it does a double bond.[329]

78 82 83 F_6 F_6

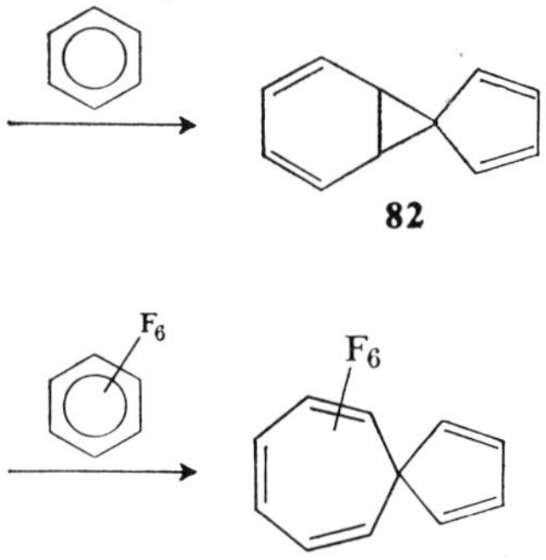

A variety of phenyl-substituted cyclopentadienylidenes has been examined and properties of **78** do not seem much changed by the phenyl groups. Addition and insertion[330,331] reactions are known for the triphenyl and tetraphenyl intermediates, and addition of the former is stereospecific.[331] Norbornadiene is attacked to give an initial adduct which rearranges further on photolysis.[332] Similarly, generation of tetraphenyl-**78** in alkynes gives cyclopropenes, which, though isolable, do rearrange further.[333,334]

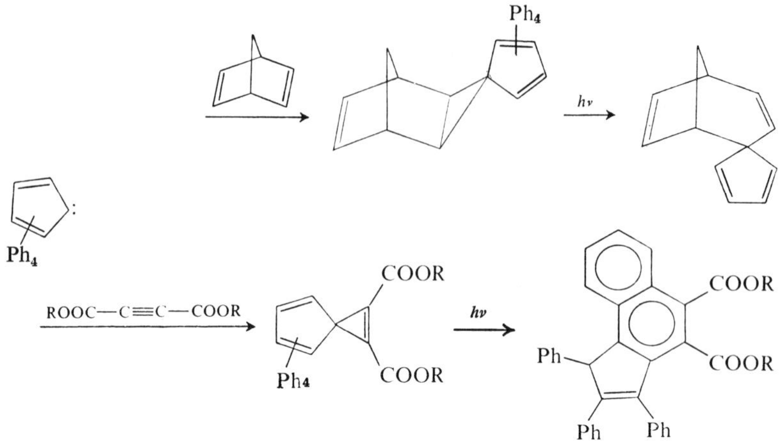

Benzene is attacked to give a product, presumably the norcaradiene, which rearranges further to **84** and **85**.[335,336]

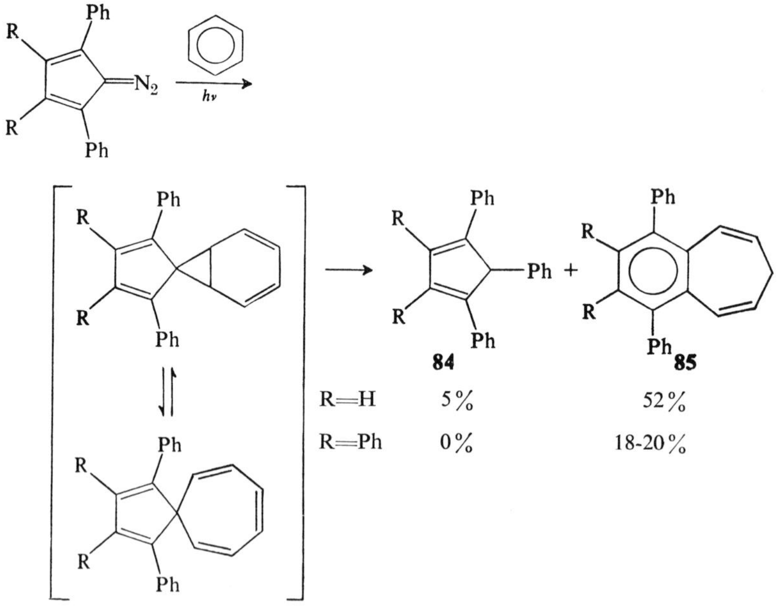

	84	**85**
R=H	5%	52%
R=Ph	0%	18-20%

The final product, **85**, which arises via ring expansion of the five-membered ring, is somewhat unusual, since ring expansion of the seven-membered ring usually occurs.[328] For example, the triphenyl compound **86** undergoes the following reaction with benzene:[337]

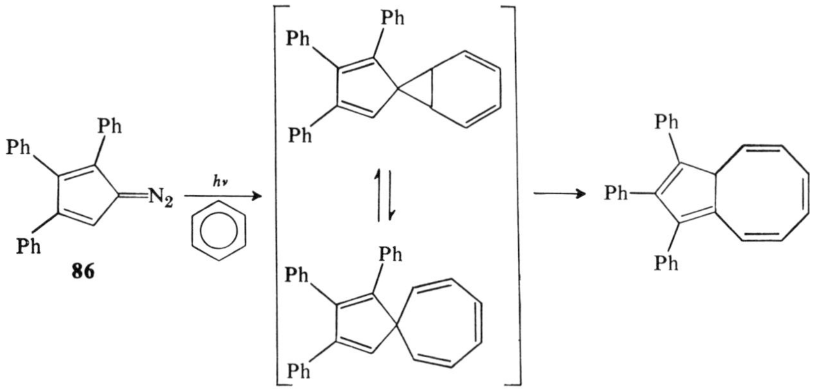

Tetrachlorocyclopentadienylidene inserts into the carbon-hydrogen bond and adds to olefins and is largely stereospecific in the latter reaction.[338] Cyclopropenes apparently are formed from acetylenes on copper-catalyzed decomposition of **87**, but these compounds rearrange further[339] to spiro[4.4]-nonatetraenes.

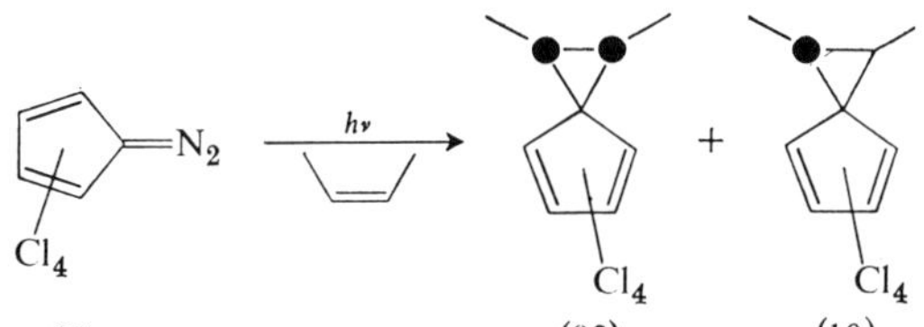

Tetraphenylcyclopentadienylidene forms stable arsenic,[340] sulfur,[341] antimony,[342] phosphorus,[343] nitrogen,[344,345] selenium,[346] and tellurium[347] ylids.

It is unfortunate that the report by Lloyd of the isolation of diazocyclononatetraene[348] has not yet been followed up by the description of the chemistry of cyclononatetraenylidene.

To return now to the question that led chemists to investigate the properties of cyclopentadienylidene: Are there data that suggest an aromatic structure? Virtually none of the aforementioned work gives any indication of odd behavior. The single exception, noted by Moss and Pryzbyla,[322] was the relative rate data which showed **78** to be sensitive to steric factors. The lack of a suitable model for the properties of **78** led one group of workers to investigate the properties of cyclohexadienylidene, which should approximate **78** in all respects save aromatic character.[349] Earlier Fry[350] had decomposed tosylhydrazone salt **88**, but no data on the conventional properties of the carbene were available; the observed products were attributed to decomposition of the azine and attack by tosylate ion on the carbene.

N—N—Ts

CCl$_3$

88

$\xrightarrow{\Delta}$ Azine ⟶

N=N + SO$_2$

48% 38%

4,4-Dimethylcyclohexadienylidene (**89**) undergoes typical carbene additions to unconjugated and conjugated olefins and to acetylenes.[349] The stereochemistry of addition approximates that of the five-membered ring compound.

(92) (8)

89

90

Benzene is attacked to give the norcaradiene **90**.[349,351] The major difference between **78** and **89** comes in the relative rate data given in Table 5. The six-membered ring carbene **89** gives a series typical of internally stabilized carbenes. A general electrophilic order is apparent, with the more substituted olefins the more rapidly attacked. Cyclopentadienylidene is rather different,

TABLE 5

Relative Rates of addition of **78** and **89**

Olefin	**78** (reference 321)	**89**
Tetramethylethylene	0.99	1.23
Trimethylethylene	1.00	1.00
Cyclohexene	1.33	
cis-4-Methyl-2-pentene		0.19
trans-4-Methyl-2-pentene		0.21
1-Pentene		0.24
1-Hexene	1.25	
t-Butylethylene	0.93	0.21
2,3-Dimethyl-1,3-butadiene		3.0

as noted previously, and **78** seems to be more sensitive to steric factors than **89**. However, it appears that the addition of the *geminate* dimethyl group should increase steric problems, not decrease them. One other difference appears: In its reaction with tetramethylethylene, **78** gives both insertion and addition in the ratio 37/63.[321] The cyclopropane is the only product of the reaction of **89** and tetramethylethylene. Carbene **78** is reacting at the periphery of the olefin while **89** penetrates to the center. Again steric problems seem to plague cyclopentadienylidene. This is understandable only if the quasi-aromatic structure **79** is important and cyclopentadienylidene is an electrophile via the empty sp^2 orbital.

78 (1.7) (1)

89

A search for the triplet state of **89** has proved fruitless. Neither deactivation with inert solvent nor photosensitized decomposition affected the cis/trans ratio.[349,352] This curious behavior has yet to be rationalized. A long-lived triplet ESR signal can be obtained from **89** and the *D* and *E* values are unexceptional.[349] Perhaps singlet and triplet are in thermal equilibrium and unsensitized and sensitized photolyses thus necessarily yield the same species.

An interesting question arises as to the fate of cycloalkadienylidenes **78** and **89** when attempts are made to force them to achieve tetravalency through intramolecular reactions. The inevitable sink for **89** seems to be the xylenes, although which and how many is not easy to predict. Perhaps surprisingly, when diazo-4,4-dimethylcyclohexadiene is evacuated into a 400° oven, only *para*-xylene and toluene are formed.[353] Stepwise mechanisms

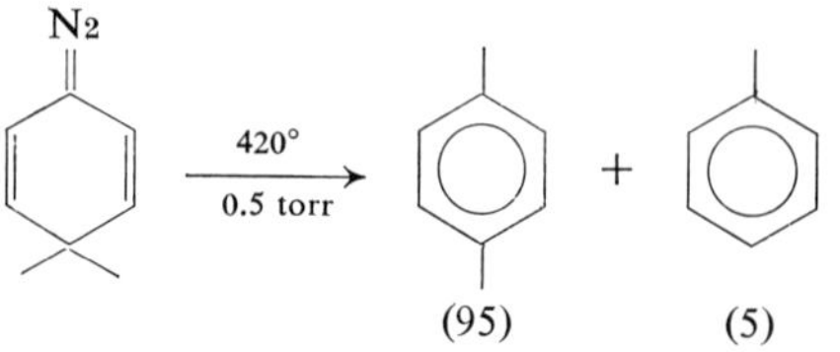

N_2

420°

0.5 torr

+

(95) (5)

inevitably, it seems to us, require that *ortho* and/or *meta*-xylene be formed. Accordingly, we favor a direct 1,4-methyl shift or insertion. Related work in steroidal systems was done some years ago by Dannenberg and Gross.[354] Here, 1,4-shift does not occur and presumably stepwise processes are

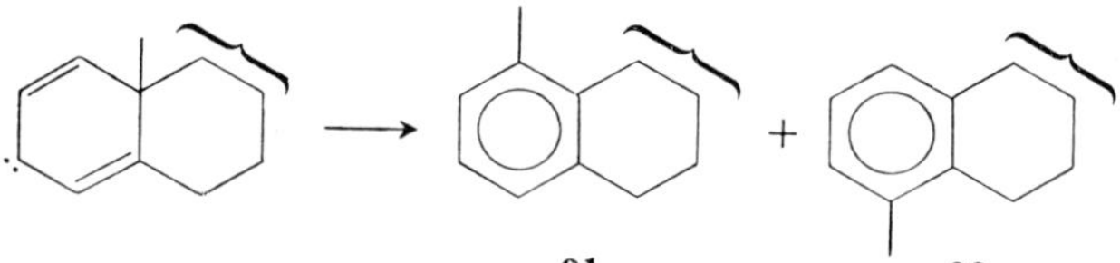

+

91 **92**

required. The authors favor a mechanism featuring dual paths to **91** and **92**, but the single path shown suffices.

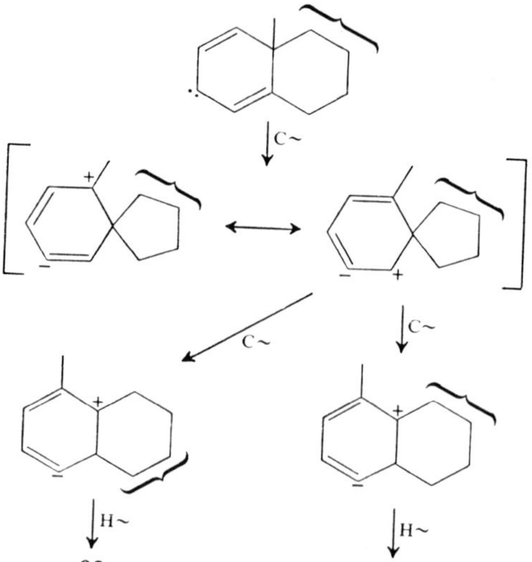

C~

+

↔

C~

C~

H~

92

H~

91

Cycloalkenylidenes have also received some study and here intramolecular reaction competes with the intermolecular processes. Indeed, published reports describe only diene syntheses,[355,356] but it is possible to achieve intermolecular cyclopropane formation as well.[357]

Ketocyclohexadienylidenes are also known, and Koser and Pirkle have shown that such carbenes do add to olefins (although they note that the intervention of a pyrazoline remains a possibility) in a 95–97% stereospecific manner.[358] The biphenyls reported by Dewar and Narayanaswami probably find their source in the noracaradiene **93**.[359]

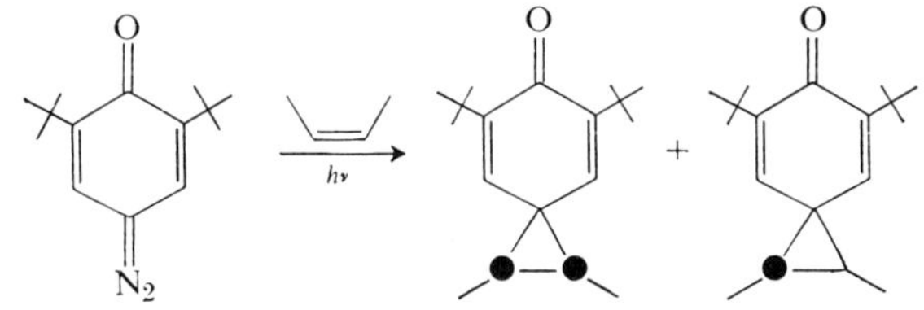

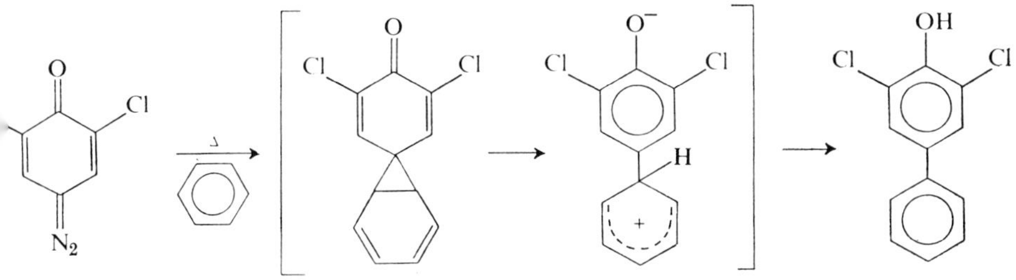

93

Bromonium salts, probably arising via the analogous ylid, have been isolated.[360]

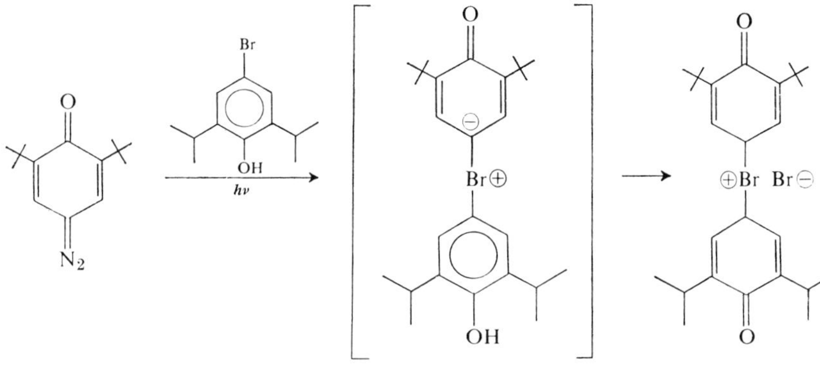

Finally, a methylenecyclohexadienylidene is a possible intermediate in the formation of arylmalononitriles from **94**, although a huge variety of mechanistic details is possible and cyclopropanes cannot be isolated from irradiation in olefins.[361]

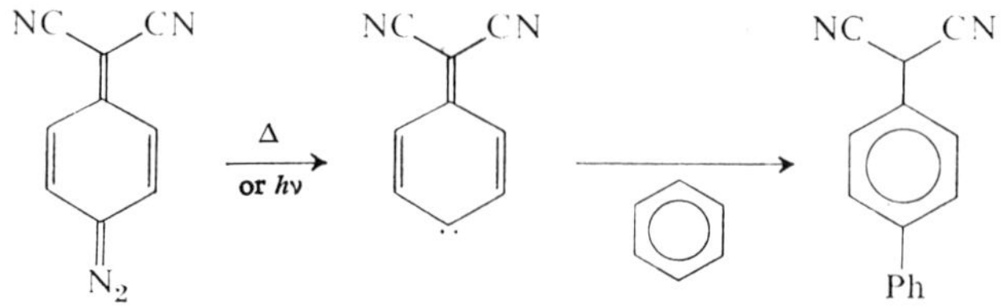

NC CN NC CN NC CN

Δ or $h\nu$

N_2 Ph

94

G. Alkylidenes

We have only a few recent references to add to the thoughtful discussion in Kirmse's book,[1] which concerned base-catalyzed reactions exclusively. We, too, have no diazocompounds to provide but the following reactions may involve carbenes.[362]

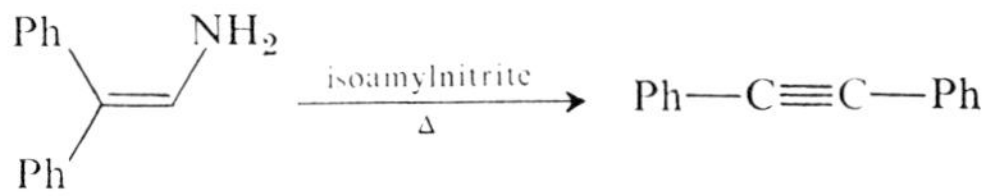

Ph NH_2 isoamylnitrite Δ Ph—C≡C—Ph

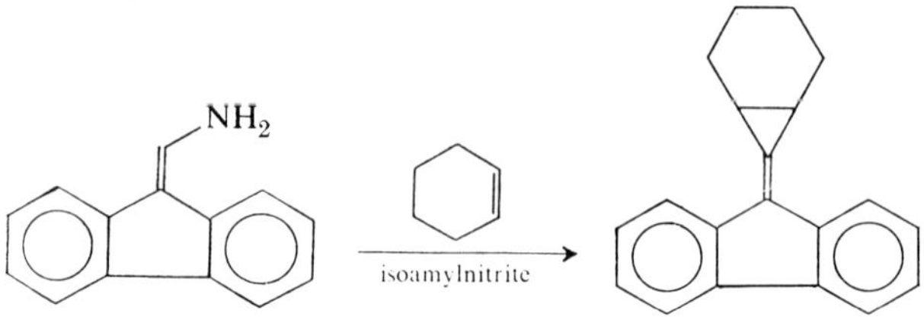

Ph

NH_2

isoamylnitrite

55–80%

Nitrosooxazolidinones have been shown to give species, probably carbenes, which add to olefins.[363] Although addition to *cis*- and *trans*-4-methyl-2-pentene is stereospecific, and reactions with substituted styrene reveal the

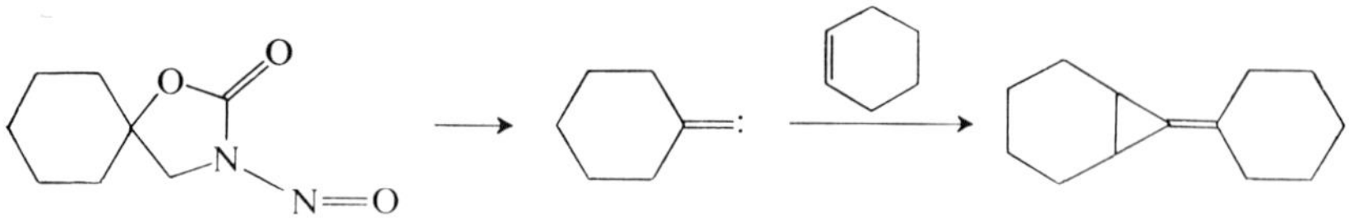

O O N N=O

carbene to be electrophilic ($\rho = -3.4$), relative rate studies show some very odd properties.[364] Isotetralin is attacked at the end rather than the middle double bond and cyclohexene is attacked more easily than tetramethylethylene.[365] The authors rationalize these data by postulating steric problems and a two-step singlet addition. It is true that should dimethylethylidene have structure **95**, the methyl groups would be exactly in the plane of the *p* orbital and steric problems would be likely. However, we see no reason to postulate a two-step singlet addition. It should be noted that the arguments in reference 365 given in support of a two-step addition of **95** were based on incorrect[366] structural assignments and have been retracted.

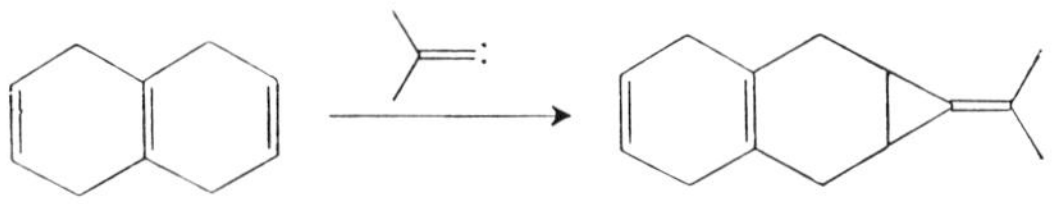

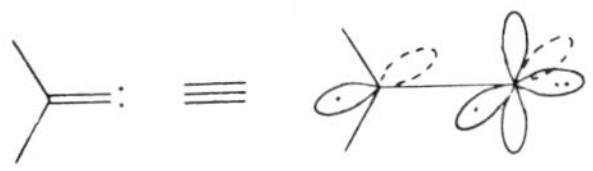

95

Reaction of **95** with alkoxyacetylenes gives allenic acetals as the final products.[367] Whether a rearrangement of an intermediate methylenecyclopropene or a two-step addition is involved has not been decided.

Diphenylallenylidene (**95a**) has been trapped in low yield by tetramethylethylene.[368] The diazo compound **95b** was not isolated, but relative rate

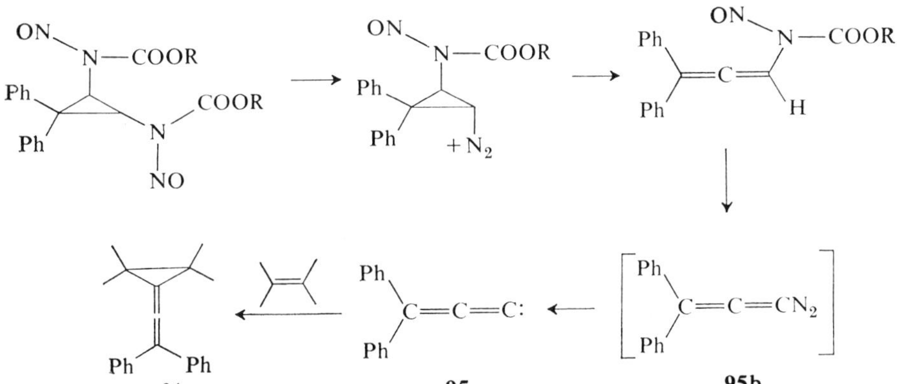

ON N—COOR Ph Ph COOR N NO → ON N—COOR Ph Ph $+N_2$ → Ph Ph C=C= ON N—COOR H

Ph Ph 1% ← Ph Ph C=C=C: **95a** ← [Ph Ph C=C=CN_2] **95b**

studies indicate that the reacting carbene must be **95a** and thus **95b** is strongly implicated.

IV. ARYLCARBENES

The chemistry of aromatically substituted carbenes is dominated by the question of spin state and the different properties of the singlet and triplet states. When one examines the available data several serious questions arise, and at best only tentative answers can be advanced.

A. Phenylcarbene

We first consider phenylcarbene, whose properties seem unexceptional. Decomposition of phenyldiazomethane in olefins leads to cyclopropanes,[369–371] and the species formed is capable of insertion into the carbon-hydrogen bond in both intramolecular and intermolecular systems.[369,372–374] The reaction with olefins is largely stereospecific (greater than 96% in all cases), and there is a slight preference for the phenyl group to become syn to the groups attached to the olefin. The picture that emerges is of a rather selective, electrophilic carbene. A number of questions now arise. What is the source of the 2–3% of the "wrong" stereoisomer in the addition to *cis*- and *trans*-2-butene, and why should the phenyl prefer the more sterically hindered position?

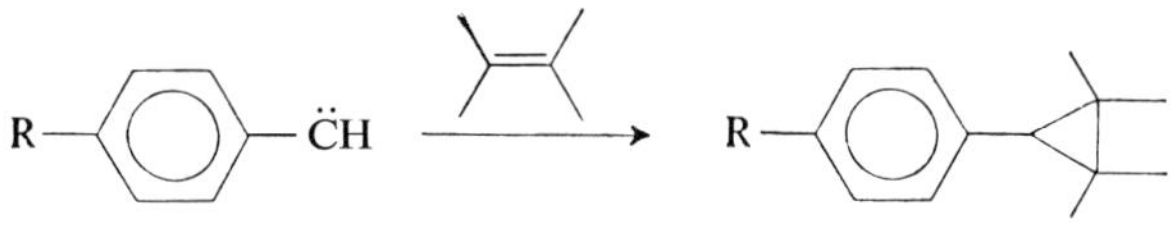

TABLE 6

Relative Rates of Addition of Phenylcarbenes to Olefins

Olefin		R: H	*m*-Cl	*p*-Cl	*p*-CH_3	*p*-OCH_3
1-Butene	*syn*	0.51	0.68–0.71	0.59–0.60	0.42–0.48	0.40–0.43
	anti	0.51	0.58–0.59	0.54–0.55	0.36–0.41	0.28–0.30
cis-2-Butene	*syn*	0.92–0.96	0.95–1.00	1.0–1.1	1.1–1.2	1.2
	anti	0.85–0.88	0.80–0.82	0.89–1.0	0.66–0.69	0.43–0.44
trans-2-Butene		1.00	1.00	1.00	1.00	1.00
Isobutene		0.91	1.2	1.1	0.83	0.77
2-Methyl-2-butene	*syn*	1.7			1.3	
	anti	1.6			1.1	

An obvious source of the "wrong" stereoisomer is triplet phenylcarbene. The phenyl group might make intersystem crossing more likely, relative to methylene, either by facilitating spin-orbit coupling or by extending the lifetime of the carbene. Another possibility is that excited phenyldiazomethane reacts with olefins to give a mixture of cyclopropanes. Assuming that the

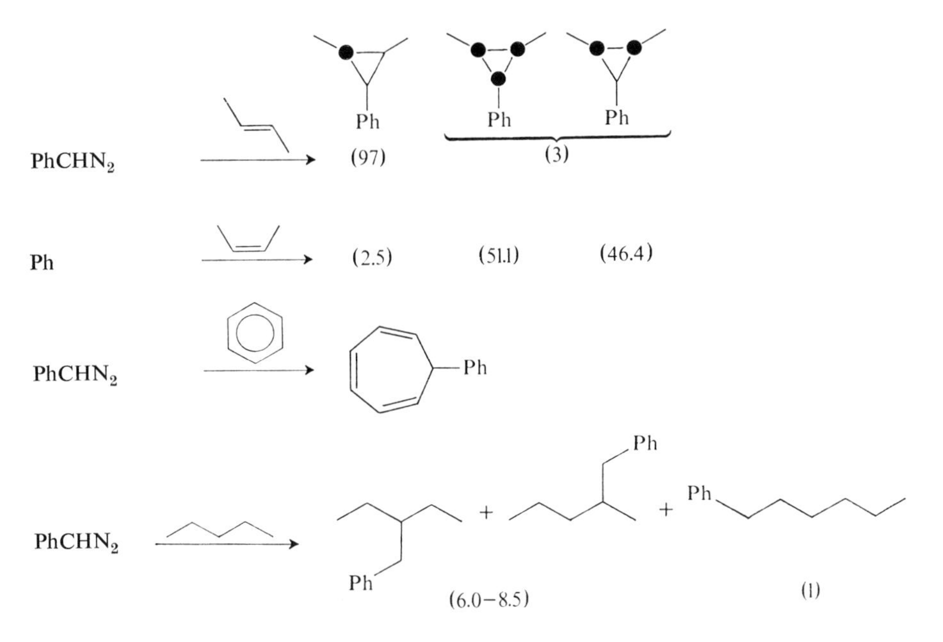

cis/trans ratio from such a reaction would be approximately 1, there would have to be about 6% of the reaction going through excited diazo compound, which does not seem an excessive amount.

$$
\begin{array}{ccccccc}
 & & & & \mathrm{PhCHN_2} & & \\
 & & & \swarrow h\nu & & & \\
\text{Products} & \longleftarrow & \mathrm{PhCH\overset{*}{N}_2}^{1} & \longrightarrow & \mathrm{Ph\ddot{C}H^{1}} & \longrightarrow & \text{Products} \\
 & & \updownarrow & & \updownarrow & & \\
\text{Products} & \longleftarrow & \mathrm{PhCH\overset{*}{N}_2}^{3} & \longrightarrow & \mathrm{Ph\ddot{C}H^{3}} & \longrightarrow & \text{Products}
\end{array}
$$

However, Kristinsson and Griffin[375] have found that phenylcarbene generated from stilbene oxide is identical to that formed by decomposition of phenyldiazomethane. It must be explicitly noted that the original report of the stereospecificity of the reaction with *cis*-2-butene is in error.[376] Thus an excited diazo compound is not a likely explanation for the ca. 3% of the "wrong" stereoisomer. Similarly, Griffin and co-workers[372] found that a variety of phenylcarbene precursors gave the same secondary/primary insertion ratios on photolysis in *n*-pentane.

The syn/anti ratio from *cis*-2-butene was initially reported[375] to be very different in the photolysis of stilbene oxide (0.6) from that achieved on the irradiation of the diazo compound (1.1). Note that this number has been

Source of phenylcarbene =

	trans-2,3-diphenyloxirane	2,2,3-triphenyloxirane	*trans*-1,2-diphenylcyclopropane	1,2,3-triphenylcyclopropane	$PhCHN_2$
2°/1°	8.3	8.45	7.7–8.6	8.7	8.3–8.6
C_2/C_3 insertion	1.4	1.5	1.4	1.4	1.3–1.4

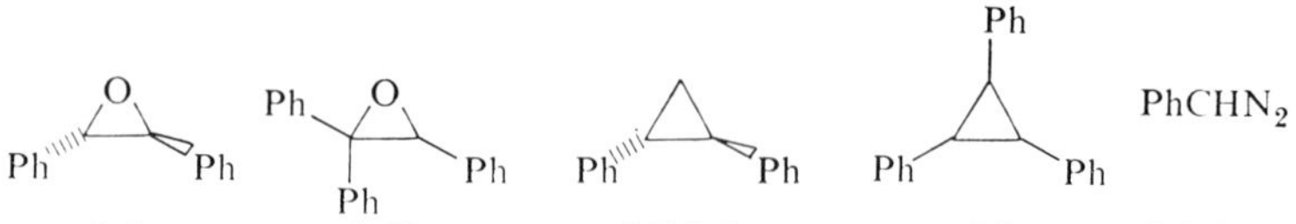

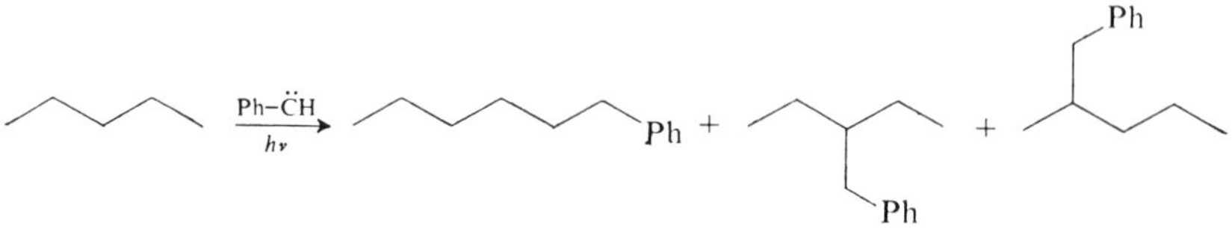

Ph–C̈H, $h\nu$

corrected in footnote 4 of reference 372. Still another source of phenylcarbene, photolysis of 5-phenyltetrazolide, gives a species with a syn/anti ratio of 1.0 for *cis*-2-butene and 0.8 for trimethylethylene.[377] These are *roughly* the values obtained by Closs and Moss,[371] although they are a bit low. Scheiner[377] favors a mechanism that involves phenyldiazomethane and thus the numbers should be similar to those found on decomposition of the diazo compound. The stereochemistry of addition is somewhat less specific, and Scheiner proposes that the phenyltetrazole anion is photosensitizing the decomposition of phenyldiazomethane, thus producing a slightly different stereochemical result. Identification of the "wrong" cyclopropane is by retention time only, however.

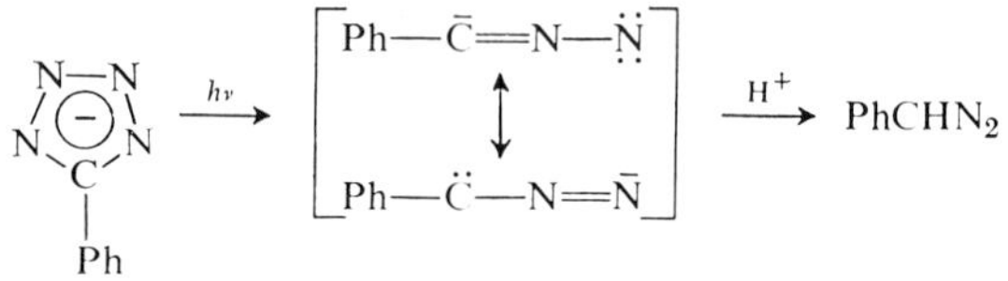

$h\nu$ → [Ph—C̄=N—N̈ ↔ Ph—C̈—N=N̄] $\xrightarrow{H^+}$ $PhCHN_2$

Baer and Gutsche[374] have studied the direct and photosensitized decomposition of the substituted phenylcarbene **96**. Little or no difference can be found! In light of the knowledge that the triplet is the ground state of phenylcarbene,[378,379] this is a striking observation indeed. The diazo compound is not being directly decomposed, because light is being delivered only to the sensitizer. Yet decomposition occurs, which seems to demand energy transfer from the sensitizing agent. Why, then, are the properties of phenylcarbene not changed? The following *a priori* explanations are possibilities:

1. Perhaps the properties of singlet and triplet phenylcarbene are identical, or nearly so.

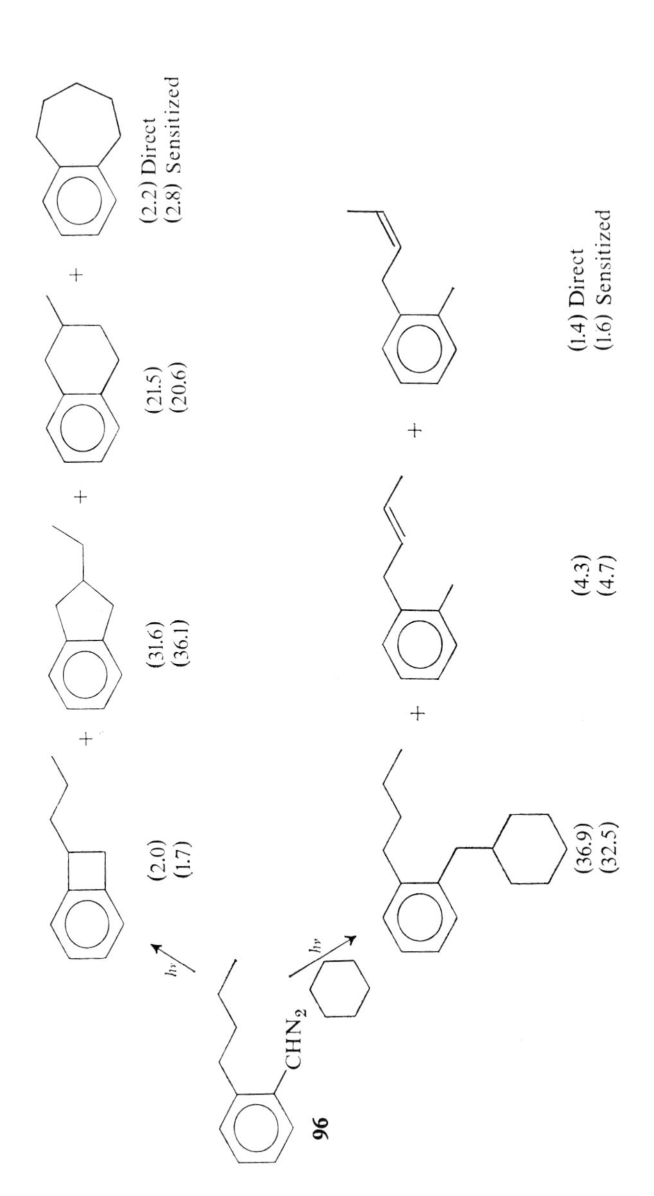

96
CHN$_2$
hν
hν
(2.0)
(1.7)
(31.6)
(36.1)
(21.5)
(20.6)
(2.2) Direct
(2.8) Sensitized
(36.9)
(32.5)
(4.3)
(4.7)
(1.4) Direct
(1.6) Sensitized

2. Perhaps the singlet and triplet state are very close in energy and are in rapid thermal equilibrium, and one observes the same mix of singlet and triplet carbene from any decomposition of phenyldiazomethane.

3. Singlet energy transfer is occurring and therefore only the singlet carbene is observed.

Explanation 1 seems only a very remote possibility, in view of the demonstrated great differences in singlet and triplet behavior (*vide supra et infra*). Explanation 2 has merit, but would seem to predict that the mix of spin states be heavy in singlet, and this conflicts with the supposition that the ground state is the triplet. Perhaps the singlet is far more reactive than the triplet.

Moss and Dolling have presented evidence that the properties of triplet phenylcarbene are quite different from those of the singlet. These authors find that irradiation of a frozen solution of phenyldiazomethane in *cis*- or *trans*-2-butene yields not only cyclopropanes but products of abstraction-recombination reaction as well. They suggest that triplet phenylcarbene reacts largely by hydrogen abstraction.[380] The products formed in the sensitized decompositions of Baer and Gutsche[374] could arise by hydrogen abstraction. However, abstraction is not the *usual* reaction of triplets with olefins (*vide supra et infra*), and the findings of Moss and Dolling seem most remarkable.

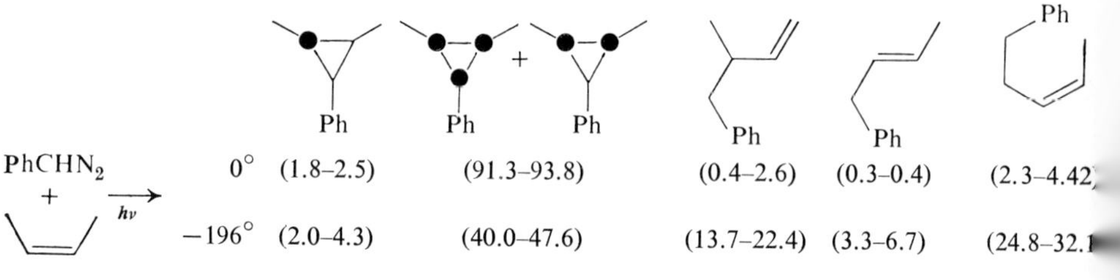

$PhCHN_2$ + (cis-2-butene) $\xrightarrow{h\nu}$

	Ph	Ph + Ph	Ph	Ph	Ph
0°	(1.8–2.5)	(91.3–93.8)	(0.4–2.6)	(0.3–0.4)	(2.3–4.42)
−196°	(2.0–4.3)	(40.0–47.6)	(13.7–22.4)	(3.3–6.7)	(24.8–32.1)

A possible alternative source of the olefinic products involves the intervention of the phenyldiazomethyl radical **96a**:

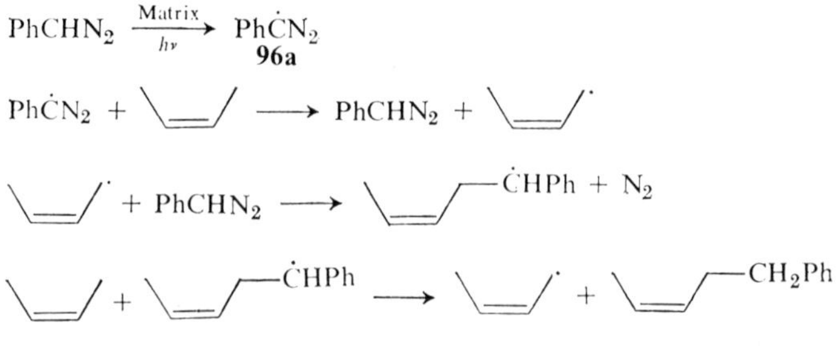

$$PhCHN_2 \xrightarrow[h\nu]{\text{Matrix}} Ph\dot{C}N_2 \quad (\textbf{96a})$$

$$Ph\dot{C}N_2 + \text{(2-butene)} \longrightarrow PhCHN_2 + \text{(butenyl radical)}^{\cdot}$$

$$\text{(butenyl radical)}^{\cdot} + PhCHN_2 \longrightarrow \text{(butenyl)}\text{—}\dot{C}HPh + N_2$$

$$\text{(2-butene)} + \text{(butenyl)}\text{—}\dot{C}HPh \longrightarrow \text{(butenyl radical)}^{\cdot} + \text{(butenyl)}\text{—}CH_2Ph$$

Diazomethane yields substantial amounts of the corresponding diazo radical on irradiation in an Argon matrix.[380a]

Photosensitization of some mono-phenyldiazomethanes has indicated differences in properties between the singlet and triplet states. In 1964 Overberger and Anselme[381] showed that methylphenylcarbene underwent hydrogen shift to give styrene, which in turn was attacked by carbene to give 1-methyl-1,2-diphenylcyclopropane. Moritani et al.[382] extended this early work to a study of the direct and photosensitized decomposition in olefins. In the unsensitized runs roughly equal amounts of hydrogen shift and external cycloaddition occur, and the formation of cyclopropane is very largely stereospecific. On photosensitization with benzophenone the amount

Ph, =N$_2$ $\xrightarrow{h\nu}$ Azine + Ph + Ph, Ph + Ph, Ph

(95)

of styrene (hydrogen shift) decreases, as would be expected for a triplet

N$_2$, Ph $\underset{-=-}{\xrightarrow{h\nu}}$ Ph (*a*) O, Ph (*b*) H, Ph (*c*) H, Ph (*d*) Ph (*e*)

Olefin	conc. (m/1) $Ph_2C{=}O$	*a*	*b*	*c* + *d* + *e*	*e*/*c* + *d*	*d*/*c*
trans-2-butene	0	7.0	2.0	7.0	91/9	0.5
cis-2-butene	0	6.0	2.3	6.9	4/96	1.2
cis-2-butene	0.36	3.7	20.2	7.1	8/92	0.8
cis-2-butene	1.33	1.7	45.9	6.0	15/86	0.8

species. However, the change in stereochemistry is relatively slight. If photosensitization is forming a triplet *carbene*, the stereochemistry should reflect this and change greatly, as it does, for instance, with biscarbomethoxycarbene (see Section V).

If an equilibrium between singlet and triplet carbene exists, and if the rates of reaction and equilibration are comparable, then the foregoing data seem reasonable. The cyclopropanes and styrene arise largely from the singlet and acetophenone from the triplet. Again we must allude to the curious result of Moss and Dolling,[380] who claim abstraction to be the major path of triplet phenylcarbene and note the absence of products of abstraction in the sensitized decomposition of phenylmethyl diazomethane.

The compound with the diazo group moved to the end of the side chain, benzyldiazomethane, has also been studied by Moritani.[383] Here the authors first confirmed Sargeant and Shechter's earlier observation[384] that hydrogen migrates faster than phenyl, and then studied the competition between hydrogen shift and external cycloaddition. The tiny yield of cyclopropane is

$$\text{Ph—CD}_2\text{—CHN}_2 \xrightarrow{\Delta} \text{Ph—CD=CHD} \quad (\text{not Ph—CH=CD}_2)$$

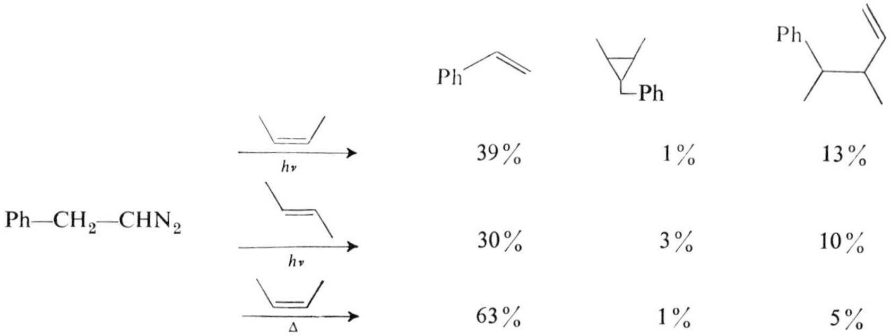

Ph—CH₂—CHN₂	Ph(vinyl)	cyclopropane–Ph	Ph, (97)
cis-2-butene, hν	39%	1%	13%
trans-2-butene, hν	30%	3%	10%
cis-2-butene, Δ	63%	1%	5%

consistent with a rapid intramolecular reaction overwhelming intermolecular cyclopropane formation. It is claimed that addition is stereospecific, although the detection of 5% of a 1% product seems a formidable task. No comment is made about the amazing compound **97**. It can scarcely be a carbene product, and has its apparent origin in an ene[385] reaction between styrene and 2-

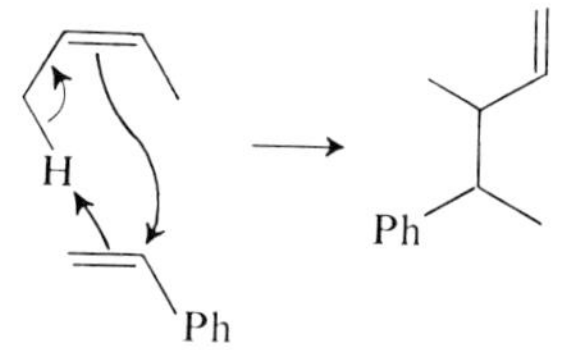

H → Ph Ph

97

butene. Alternatively, **97** may be formed via an abstraction-recombination reaction as shown, but then where is compound **98**?

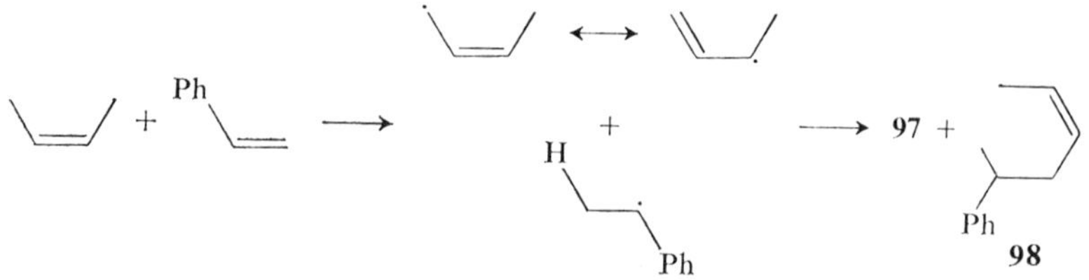

Ph + → ↔ + H Ph → **97** + Ph

98

It has also been noticed[386] that phenyl-*t*-butylcarbene prefers self-insertion to give **99** rather than addition to allene to give **100**. No product of methyl shift could be found.

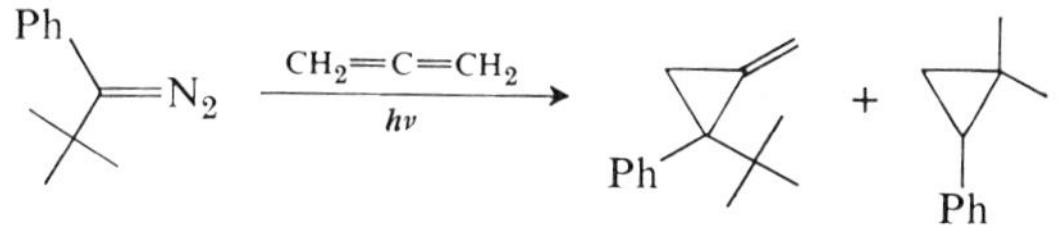

$CH_2{=}C{=}CH_2$, $h\nu$

100 (1) **99** (99)

The preference for intramolecular reactions noted above has been used by several groups as a synthetic tool, and a few examples follow. Aromatic tetrahydrofurans and pyrans have been made through the decomposition of **101**.[387] Less than 1.5% of **103** is formed leading the authors to suppose that direct insertion is responsible for **102** and not a radical abstraction-recombination reaction. Similarly indoles have been made from the thermal

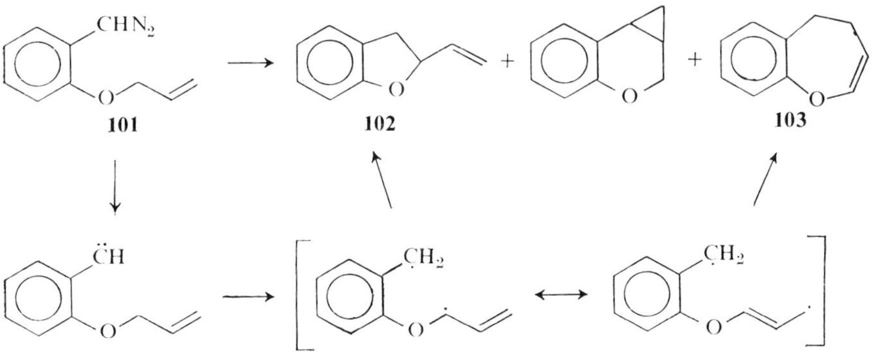

101 **102** **103**

decomposition of compounds such as **104**.[388]

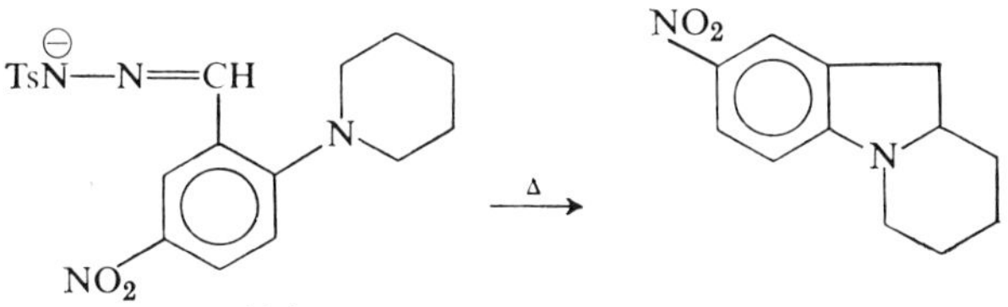

104

Aromatic cyclopropenes have been formed by intramolecular reaction,[389] but the more strained benzocyclobutadiene is avoided by the appropriate phenylcarbene.[318] A small amount of an unidentified yellow oil is reported, however.

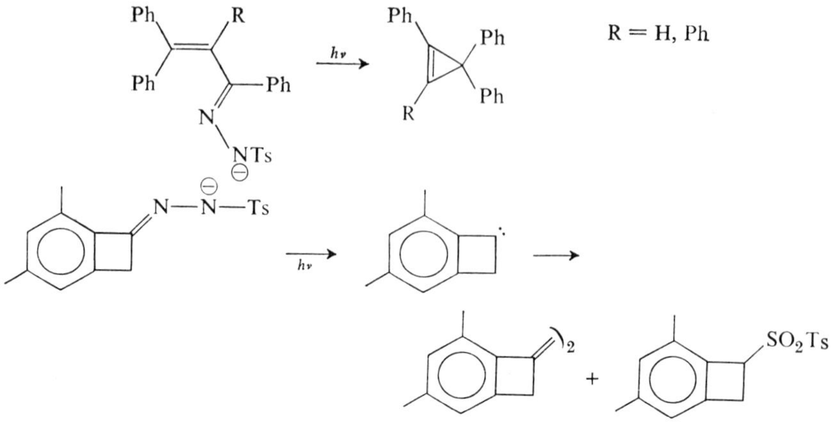

Ph R Ph Ph R = H, Ph
Ph Ph $\xrightarrow{h\nu}$ Ph
N R
NTs ⊖

⊖ N—N—Ts $\xrightarrow{h\nu}$ ⟶

)₂ + SO_2Ts

Naturally enough, aromatic carbenes have found use in studies of migratory aptitudes. Landgrebe and Kirk[390] showed that migration of substituted phenyl groups to the divalent carbon followed a Hammett σ^+ plot with

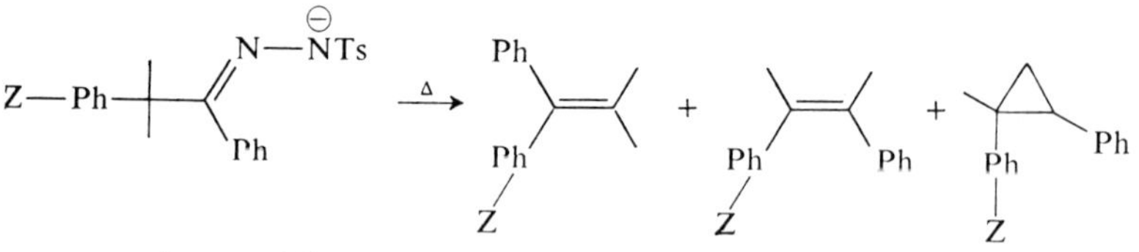

Z—Ph— N—NTs ⊖ Ph $\xrightarrow{\Delta}$ Ph + + Ph
Ph Ph Ph
Z Z Z

Average Migratory Aptitude

H	1.00	(72.8)	(27.2)	—
p-CH_3	1.90	(83.6)	(16.4)	—
m-Cl	0.67	(64.2)	(35.8)	—

$\rho = -0.68$. The effects are smaller than for migration to carbonium ions, but in the same direction, since electron releasing groups increase the ease of migration to an electron-deficient center. Zimmerman and Munch[391] similarly found that the migratory aptitude of anisyl relative to phenyl ranged, depending on solvent, from 1.5 to 2.3. In related work Robson

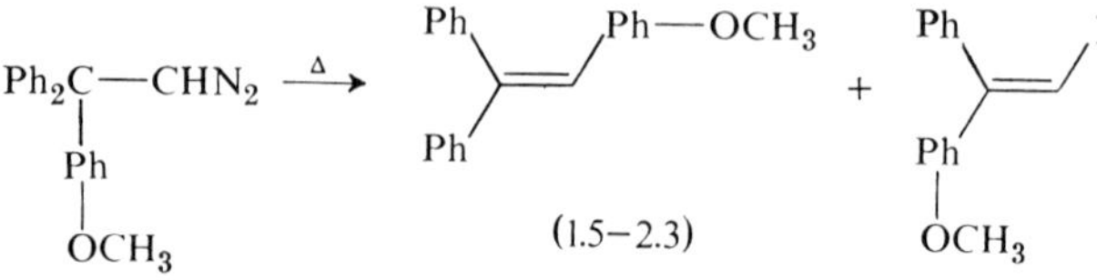

Ph_2C—CHN_2 $\xrightarrow{\Delta}$ Ph, Ph—OCH_3 + Ph, Ph
Ph Ph Ph
OCH_3 (1.5–2.3) OCH_3 (1)

and Shechter[392] found that hydrogen migration was preferred over heteroatom migration in every case save that of sulfur, which was supposed to avail itself of 3*d* orbitals in an ylid-like transition state (**105**).

$$PhC(=N_2)CH_2X \xrightarrow{\Delta} PhCH{=}CHX \qquad X = OCH_3,\ OPh,\ NH_2$$

$$PhC(=N_2)CH_2SR \xrightarrow{\Delta} PhCH{=}CHSR + Ph(RS)C{=}CH_2$$

	R=Et	R=Ph
PhCH=CHSR	(9–15)	(0–8)
Ph(RS)C=CH₂	(85–91)	(92–100)

105

B. Diphenylcarbene

Diarylcarbenes surpass even the monoarylcarbenes in their capacity for strange reactions. Again the base upon which difficulties rest is a duality of spin state, which complicates what is at best a series of reactions difficult to interpret.

The simplest diarylcarbene has been the source of much of the trouble. A very early report by Etter, Skovronek, and Skell[393] claimed that the addition of diphenylcarbene to *cis*- and *trans*-2-butene was nonstereospecific. A later correction appeared by Closs and Closs,[12,394] to the effect that addition was very largely stereospecific, and, more important, that the predominant products of the reaction were not cyclopropanes at all but olefins formed by an abstraction-recombination process. The cyclopropanes formed no more than 10% of the products, and the stereochemistry of addition was predominantly cis! The values given are independent of the addition of oxygen or dilution with cyclohexane but become somewhat more stereospecific on dilution with hexafluorobenzene. The thesis of P. W. Humer[395] agrees fairly well with these values in the cis case, but disagrees in the trans. Note that in Humer's work the cis/trans ratio is 38/62, not 4/96, and the cyclopropanes make up 25% of the products, not less than 10%. Since it is generally more likely that too much rather than too little of a product will be found by gas-chromatographic analysis, Closs' values appear to be more acceptable.

Ph$_2$C:

	% Ph Ph	% Ph Ph	Total % cyclo-propanes	Total % olefins	Reference
→	(77)	(23)	(10)	(90)	395a
→	(67)	(33)	(10)	(90)	395
→	(38)	(62)	(25)	(75)	395
→	(4)	(96)	(10)	(90)	12

Furthermore, it would be surprising to find a greater lack of specificity in the reaction with the trans olefin than in that with the cis. Regardless of the details, we are left with a species which preferentially reacts with *cis*-2-butene by hydrogen abstraction and is somewhat nonstereospecific.

This meager amount of information was augmented when it was discovered that the ratio of hydrogen-abstraction to cycloaddition was highly dependent on the substitution pattern of the olefin (Table 7).[396] Others[397–400] have

TABLE 7

Reaction of Diphenylcarbene with Olefins

Olefin	Cyclopropane	Olefin products
2,3-Dimethyl-2-butene	0	100
2-Methyl-2-butene	≤ 8	≥ 92
trans-2-Butene	22	78
3-Methyl-1-butene	52	48
1-Butene	74	26
Methylenecyclohexane	87	13
Isobutylene	100	0
Propylene	100	0

studied the formation of cyclopropanes from irradiation of diphenyldiazomethane in olefins but generally were not in a position to comment on the abstraction/cycloaddition controversy. The rapid addition of diphenylcarbene to dienes was previously commented upon[393] and has received some further attention,[401] and we shall return soon to the subject.

Reference

$Ph_2C{=}CH_2 \xrightarrow{h\nu}$ 1,1,2,2-tetraphenylcyclopropane (Ph, Ph / Ph Ph) — 397

$(CH_3)_3Si{-}(CH_2)_n{-}CH{=}CH_2 \xrightarrow{\Delta}$ 1,1-diphenyl-2-$(CH_2)_nSi(CH_3)_3$-cyclopropane, $n = 0\text{–}2$ — 398

$Ph_2CN_2 + (CH_3)_3Si{-}CH{=}CH_2 \xrightarrow{h\nu}$ 1,1-diphenyl-2-$Si(CH_3)_3$-cyclopropane — 399

1-heptene $\xrightarrow{h\nu}$ 1,1-diphenyl-2-pentylcyclopropane — 399, 400

n-Bu—O—CH=CH$_2$ $\xrightarrow{h\nu}$ 1,1-diphenyl-2-(O—n-Bu)cyclopropane — 400

Now one is faced with two questions. First, why are abstraction-recombination products formed, and second, what does one make of the stereochemistry of addition found? Although we must largely beg the first question, a few comments are in order. From the first,[393] it has been assumed that diphenylcarbene was reacting in the triplet state. This assumption was bolstered by several observations of triplet ESR spectra.[402–413] Early indications were that diphenylcarbene was bent but roughly planar.[405,407] More refined measurements[412] show that this is not so, and diphenylcarbene is both bent and twisted. Whatever the spin state, the dependence of mode of reaction on substitution of the olefin[396] seems consistent with a species that experiences steric problems as it approaches the olefin. The steric difficulties may tell little about geometry, but the necessarily planar fluorenylidene (*vide infra*) always forms mainly cyclopropanes in its reaction with olefins. This tells us that a flat species can penetrate to the double bond. If diphenylcarbene were flat, or nearly so, we would not expect large *steric* differences between it and fluorenylidene to appear. It has been pointed out, however, that electronic differences also exist. Diphenylcarbene is odd, alternant, whereas fluorenylidene is not.[12] In addition, in a picture of the cyclopropane-forming reaction in which we imagine the transition state as a donor-acceptor charge transfer complex, it is fluorenylidene that should form the stronger complex and therefore might be the more prone to cyclopropane formation.[414]

Apparently dimesitylcarbene is so sterically hindered that even abstraction from solvent cannot occur easily.[415] Decomposition of dimesityldiazomethane led only to dimers, a product of intramolecular insertion and dimesitylketone. No product of intermolecular abstraction or azine could be detected. Dimers often appear in reactions involving carbenes, but it is generally accepted[416] that an actual dimerization of two carbenes is less likely than a reaction

$$R_2C: + \overset{\oplus}{N_2}-\overset{\ominus}{C}R_2 \longrightarrow R_2\overset{\ominus}{C}-CR_2(\overset{\oplus}{N_2}) \longrightarrow R_2C{=}CR_2$$

between the carbene and diazo compound. Dimesitylcarbene may be so hindered that a population of carbene great enough to permit dimerization builds up. The approach necessary for dimerization[417,418] might well be less sterically demanding than that for either cyclopropane formation or hydrogen abstraction. The absence of azine was thought to be a consequence of the triplet nature of dimesitylcarbene.[415] We return to the question of cycloaddition versus abstraction in discussing other diaryl carbenes.

Now we turn to the stereochemistry of addition. If a triplet is reacting, why do we not get complete stereochemical scrambling? Other triplets such as methylene, fluorenylidene,[419] and a variety of carboalkoxycarbenes (see Section V) apparently achieve full equilibration. Why should diphenylcarbene be an exception? This problem has been recognized by many and a lucid explanation put forth by Closs.[12] It is postulated that the two spin states are in thermal equilibrium with intersystem crossing (k_1) and its reverse (k_{-1}) being fast. The ratio of singlet and triplet reaction with the olefin will be determined by the ratio k_1/k_{-1} and by the relative rates of singlet and triplet addition:

$$Ph_2C:^1 \underset{k_{-1}}{\overset{k_1}{\rightleftharpoons}} Ph_2C:^3$$

The triplet is apparently the ground state,[402–413] so k_1 must be greater than k_{-1}. One need not, however, postulate a much greater rate of addition of singlet carbene to olefins than triplet carbene, since the triplet may well spend most of its time abstracting hydrogen in the butene system. Thus the product of addition, the cyclopropane, would owe its existence primarily to the singlet state even though relatively little of that state exists. Obviously the critical experiment involves a test of the stereochemistry of addition in a case where little or no abstraction occurs, say propene or styrene. In the latter case, diphenylcarbene gives roughly 65% cis addition and 35% trans.[420] If the crude assumptions are made that the rates of addition of the singlet and triplet are the same, and that the singlet state is stereospecific, a 65/35 ratio implies a singlet/triplet mixture of ≈30/70.

$$Ph_2CN_2 \xrightarrow[h\nu]{PhCH=CHD} \text{cis product (65)} + \text{trans product (35)}$$

Ph D Ph Ph D Ph Ph Ph Ph D

(65) (35)

The reaction of thermally generated diphenylcarbene with alcohols has been studied.[421] The kinetics favor the reasonable mechanism shown:

$$Ph_2CN_2 \xrightarrow{85^\circ} Ph_2C: \xrightarrow{ROH} Ph_2\overset{\ominus}{C}-\overset{\oplus}{O}(R)(H) \longrightarrow Ph_2CH-OR$$

The reaction of diphenylcarbene with isopropanol in the presence of oxygen depends on alcohol concentration in a manner which is consistent with a reversible interconversion of singlet and triplet.[422] The product of the reaction of diphenylcarbene with oxygen is apparently the Criegee zwitterion, although a different species may be formed first.[423] The zwitterion can dimerize to give benzophenone diperoxide,[424] oxidize a hydrocarbon solvent,[423] or in the presence of aldehydes, give ozonides.[425]

Acetylenes, as well as olefins, are attacked in two steps. Monosubstituted acetylenes apparently give a diradical intermediate which is internally trapped by one benzene ring to ultimately yield indenes.[426] Dimethylacetylene gives mainly cyclopropene, and a steric explanation is advanced to explain the dependence on substitution.

A variety of heterocyclic and acyclic compounds can be isolated from the formal addition of diphenylcarbene to carbon-heteroatom multiple bonds. A few examples follow. In most such cases little certainty as to mechanism exists, and carbenes may not always be involved.

Reference

Ph_2CN_2 +

$Ph-N=S=O \xrightarrow{h\nu}$ [three-membered ring: N(Ph)–S(=O)–CPh₂] $\longrightarrow Ph-N=CPh_2$ — 427

$Ph-N=C=O \xrightarrow{h\nu}$ [three-membered ring: N(Ph)–C(=O)–CPh₂] $\longrightarrow$ 2,2-diphenyl-indoxyl (N, Ph, Ph, O) — 428

$(CF_3)_2C=S \xrightarrow{-78^\circ}$ thiirane with Ph, Ph, CF_3, CF_3 — 429

RO–C(=S)–S–C(=S)–OR $\xrightarrow{\Delta}$ thiirane with Ph, Ph, OR, and S–C(=S)–OR — 430

$$(CH_3O_2)P(=O)—C(=O)—CH_3 \xrightarrow{\Delta} \text{Ph(Ph)C—(O)—C(CH}_3\text{)P(=O)(OCH}_3)_2 \qquad 431$$

$$Ph_2CN_2 + CO_2 \xrightarrow[-78]{h\nu} \left[\text{Ph(Ph)C—O—C=O} \right] \longrightarrow \text{Polymer} \qquad 432$$

Kirmse, Horner, and Hoffmann[319] studied the decomposition of diphenyldiazomethane in saturated hydrocarbons and found strong evidence for an abstraction-recombination mechanism. The amount of tetrarylethane produced depended upon the ability of the solvent (RH) to lose hydrogen. In benzene, more azine than radical dimer was found, whereas in cyclohexane, toluene, and isopropyl ether the ethane was the major product. Although no

$$Ph_2CN_2 \xrightarrow{h\nu} Ph_2C: \xrightarrow{RH} Ph_2\dot{C}H + R\cdot$$

$$Ph_2C: \xrightarrow{Ph_2CN_2} Ph_2C{=}N{-}N{=}CPh_2 \qquad Ph_2\dot{C}H \longrightarrow Ph_2CH{-}CHPh_2$$

systematic studies of the reaction of diphenylcarbene with the carbon-hydrogen bond have been published, it seems probable that such a reaction occurs primarily by abstraction. Only the singlet state should be able to undergo an efficient insertion reaction,[433] and since diphenylcarbene is certainly predominantly in the triplet state, an abstraction process seems likely. Only in the event of a singlet-triplet equilibrium coupled with a large difference in reactivity favoring the singlet should insertion be the main reaction.

It has been reported[434] that the photolysis of tetraphenylmethane gives diphenylcarbene. This may well be so, but the claimed observation of the singlet state seems to be unwarranted. The claim is based on the observation of "insertion" into the carbon-hydrogen bond of cyclohexane, a reaction not observed in an earlier photolysis of diphenyldiazomethane.[319] When this earlier work is repeated at concentrations of the modern work, the product appears.[435] In addition, an "insertion" into the carbon-hydrogen bond of cyclopentane has recently been reported.[436]

A formal sulfur-sulfur insertion is known,[437] although the reaction conditions seem to favor an ionic or ylid process rather than a direct insertion.

Another paper that reports the formation of **107** from the irradiation of diphenyldiazomethane in the presence of **106** describes the reaction in terms of a carbon-iron insertion reaction.[438] This seems optimistic, since no evidence is quoted which *requires* a *bona fide* insertion reaction.

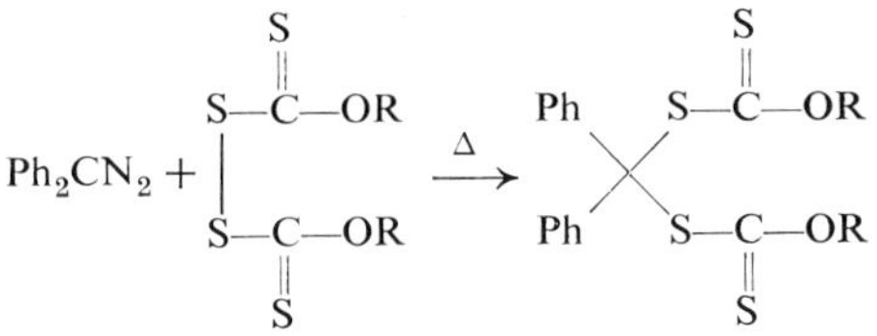

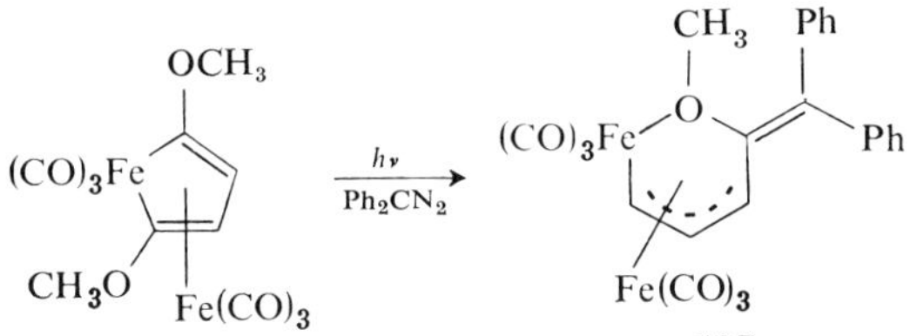

C. Fluorenylidene

Having examined one extreme of the diarylcarbene series, let us now look at the other before discussing the intermediate cases. Diphenylcarbene models a compound bridged by an infinite chain. Fluorenylidene has the two aryl groups connected by a zero bridge.

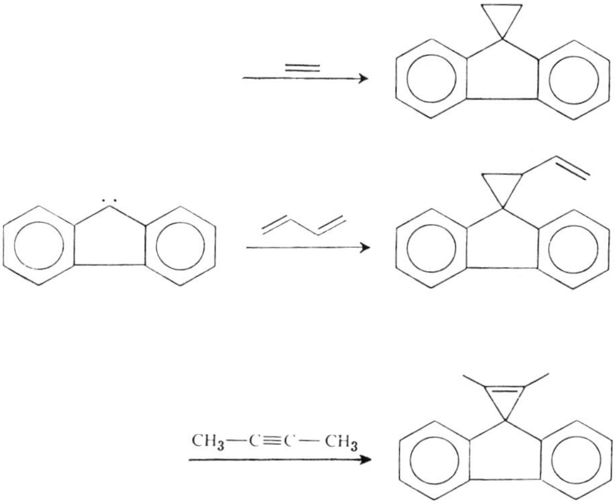

The first systematic studies of the addition of fluorenylidene to carbon-carbon multiple bonds never received broad publication.[439] Suffice to say it that fluorenylidene, unlike diphenylcarbene, adds to π-systems without the complication of large amounts of hydrogen-abstraction. Fluorenylidene seems to be an unexceptional carbene (unlike diphenylcarbene) in that it adds easily to a variety of olefins.[419,440] The stereochemistry of the addition to the butenes showed predominant but by no means complete retention of stereochemistry.[419] In contrast, either maleic or fumaric ester gave very predominantly the trans isomer.[441,442] 2,7-Dibromofluorenylidene behaves in a manner

	cis-Cyclopropane	*trans*-Cyclopropane
→	(~67)	(~33)
→	(~0)	(~100)
CH_3OOC $COOCH_3$ →	2%	65%
CH_3OOC $COOCH_3$ →	0	85%

indistinguishable from the parent substance.[442] Here, as in the case of diphenylcarbene, we are tempted to postulate a mixture of spin states, which would account for the predominant retention observed in the reaction with the butenes and the 4-methyl-2-pentenes. Presumably, the far less specific results of the Japanese workers[441,442] reflect a reluctance of the singlet state to react with the electron-deficient double bond. This would either allow a triplet to play a more dominant role in the addition reaction or allow time for further intersystem crossing from the singlet to the triplet.

The latter explanation gains credibility from the work of Jones and Rettig,[419] who borrowed a technique long used in the gas phase and showed that the stereochemistry of the addition of fluorenylidene to *cis*-2-butene and *cis*-4-methyl-2-pentene was dependent on the amount of inert moderator present. The more moderator, the less specific the cycloaddition. The inert material chosen was hexafluorobenzene. As previously noted in Section III, hexafluorobenzene is not universally inert.[218,328] Presumably hexafluorobenzene owes its reluctance to react to a combination of properties. The carbon-fluorine bond apparently has never been found to take part in an intermolecular insertion reaction.[122,148,149] The benzene carbon-carbon double bond should be fairly reluctant to participate in the cycloaddition reaction, since aromaticity would be lost and any addition reaction would initially place two fluorines on a three-membered ring, a place they dislike even more than a double bond.[329] A fluorocarbon would presumably be an even more inert solvent than hexafluorobenzene, but fluorinated hydrocarbons are poor solvents for "ordinary" compounds and the possibility of micelle formation seems great.

This experiment indicates that the triplet state is the ground state of fluorenylidene, a fact borne out by spectroscopic measurements,[403,405,407,409,410,413,442–444] and that under ordinary conditions a mixture of spin states is produced by photolysis of diazofluorene. Whether the original mixture arises via intersystem crossing in the carbene itself or in the diazo compound precursor cannot be decided with certainty. The properties of triplet fluorenylidene can be examined at moderator concentrations of $>50\%$, but, as is not usually the case, the singlet state is difficult to observe. It was thought[419] that it might be possible to selectively filter out the triplet state by allowing it to react with a triplet trap. Oxygen functions as such a trap and experiments in air show a greater specificity than runs under nitrogen. Presumably, oxygen is reacting preferentially with the triplet state of the carbene, giving a mixture richer in the singlet state. The effect is small but apparently real.

A more easy-to-use trap is butadiene. Triplets seem to react especially fast with dienes,[393] as would be expected if a diradical intermediate were involved.

$R_2C{:}\uparrow\uparrow$ + (cis-2-butene) ⟶ R–Ċ(R)–CH(CH₃)–ĊH–CH₃

$R_2C{:}\uparrow\uparrow$ + (butadiene) ⟶ R₂Ċ–CH₂–ĊH–CH=CH₂ (·)

As can be seen from Table 8, the more butadiene added, the more stereospecific the addition. Of course the singlet state is also reacting with the diene,

TABLE 8

Reaction of Fluorenylidene with *cis*-2-Butene in the Presence of 1,3-Butadiene

Butadiene (mmoles)	*cis*-2-Butene (mmoles)	cis/trans Ratio of cyclopropanes
0	10.0	2.1
3.0	6.0	10.1
11.5	3.0	>49

and so at high butadiene concentration very little product is observed from reaction with the butene, and precise numbers become most difficult to obtain. Nevertheless, it appears as if singlet fluorenylidene is stereospecific or very nearly so.

An *estimate* of the composition of singlet and triplet fluorenylidene formed on irradiation of the diazo compound can be gleaned from the unpublished relative rate data of Walton (see Table 9).[445] These numbers must be regarded

TABLE 9

Relative Rates of Addition of Fluorenylidene[445]

Olefin	K_{rel}	
	90% Hexafluorobenzene	Pure olefin
2,3-Dimethyl-2-butene	0.37	0.66
2-Methyl-2-butene	1.00	1.00
1-Pentene	1.70	0.47
3,3-Dimethyl-1-butene	0.86	0.38
cis-4-Methyl-2-pentene	0.40	0.43
trans-4-Methyl-2-pentene	0.69	0.56
2,3-Dimethyl-1,3-butadiene	9.0	3.5

as preliminary, but the indication is that in the absence of moderator a mixture of *approximately* 50% singlet and 50% triplet is formed.

Neither fluorenylidene nor diphenylcarbene can be induced to add in 1,4 fashion to dienes.[446] Occasional accounts of 1,2 additions to acyclic[401] and cyclic[447,448] dienes have appeared, but even a deliberate search for the 1,4 reaction using the various dienes shown below failed to reveal it.[446]

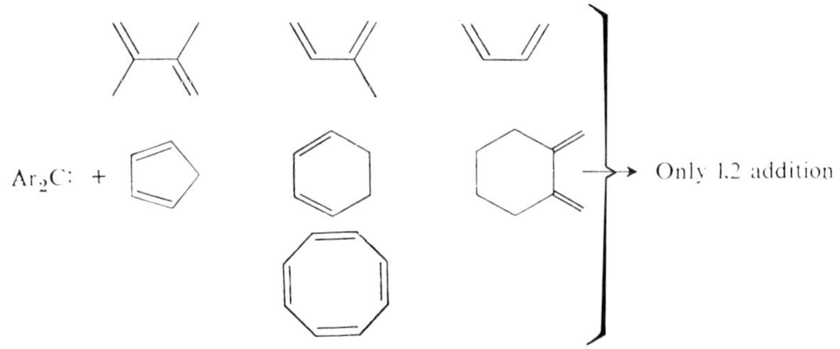

$Ar_2C:$ + [dienes] → Only 1,2 addition

Benzene is attacked to give the equilibrating norcaradiene-cycloheptatriene system **107a**.[449]

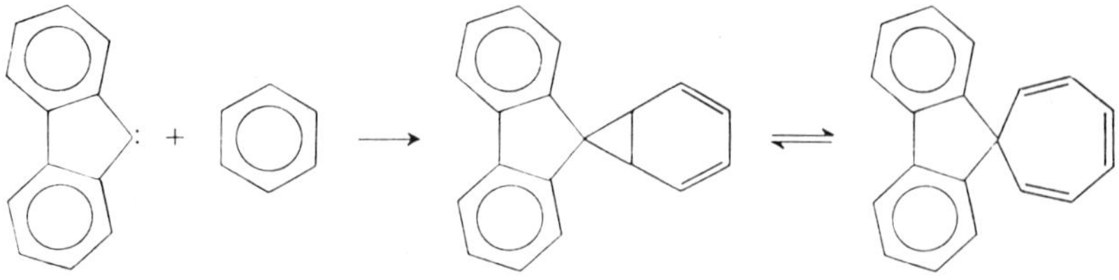

107a

No systematic study of the reaction with the carbon-hydrogen bond exists, but Kirmse, Horner, and Hoffman[319] examined the photolysis of diazofluorene in cyclohexane and decided that the 9-cyclohexylfluorene formed was produced primarily by direct insertion. This seems a tenuous assertion, since a large amount of radical dimer was formed. In later work small amounts of fluorene were found along with a somewhat larger amount of 9-cyclohexylfluorene. It was also found that fluorenylidene reacted with isobutane to give only the product of insertion into the tertiary carbon-hydrogen bond, 9-*t*-butylfluorene.[450] Of course the fact that a mixture of spin states is probably present makes the data very difficult to interpret. A systematic, detailed study of carbon-hydrogen insertion is wanting.

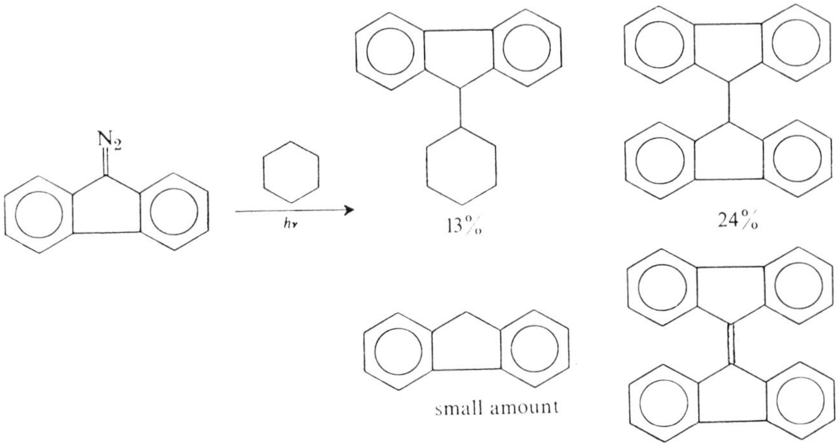

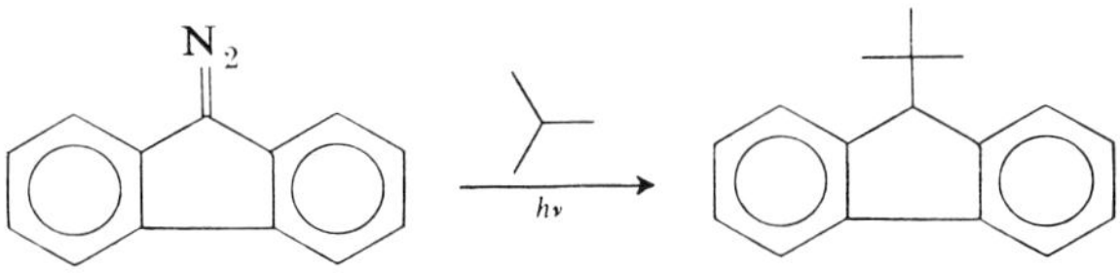

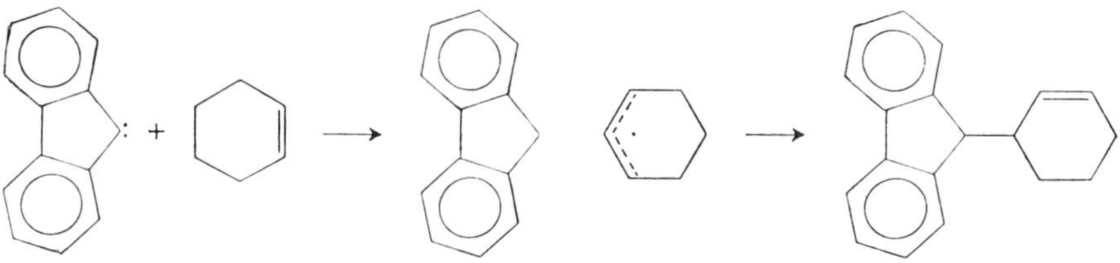

Baldwin and Andrist[451] have examined the product of allylic "insertion" of fluorenylidene into cyclohexene. By means of deuterium labeling experiments it was determined that the product is formed by abstraction of the allylic hydrogen followed by recombination. No mention is made in this paper of the other major product of the reaction, which is the normal addition product **107b**. In view of the variation in properties of the diarylcarbenes with regard

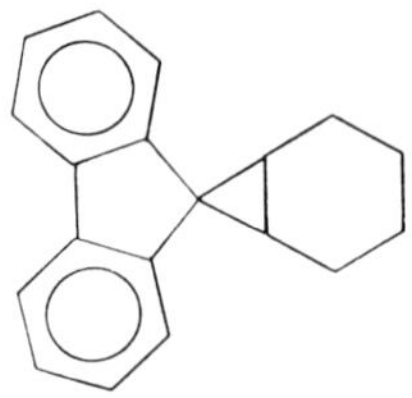

107b

to addition and abstraction, it seems worthwhile to point out that **107b** is formed.[450]

D. Other Diarylcarbenes

A small number of carbenes in which the bridge connecting the two rings is between zero and infinity has been studied. The hope was that some light might be shed upon the differences in the $n = 0$ compound (fluorenylidene) and the $n = \infty$ compound (diphenylcarbene) by carbenes **108**, **109**, and **110**. Far from being the case, however, the few studies that have been made have only added to the confusion.

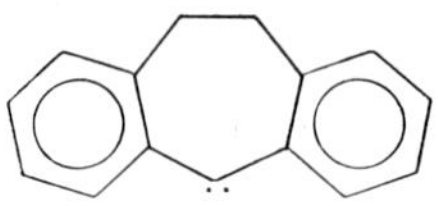

108 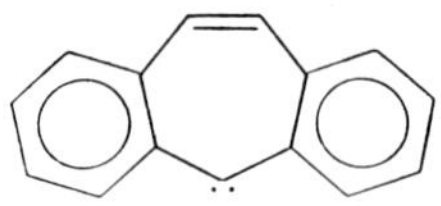 **109** 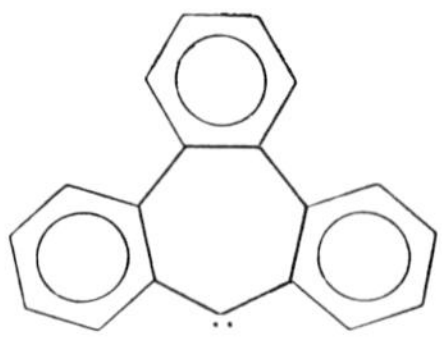 **110**

Carbene **108** seems to mirror diphenylcarbene in its properties. Moritani et al.[452] found that cyclopropanes were not formed by attack of **108** on the

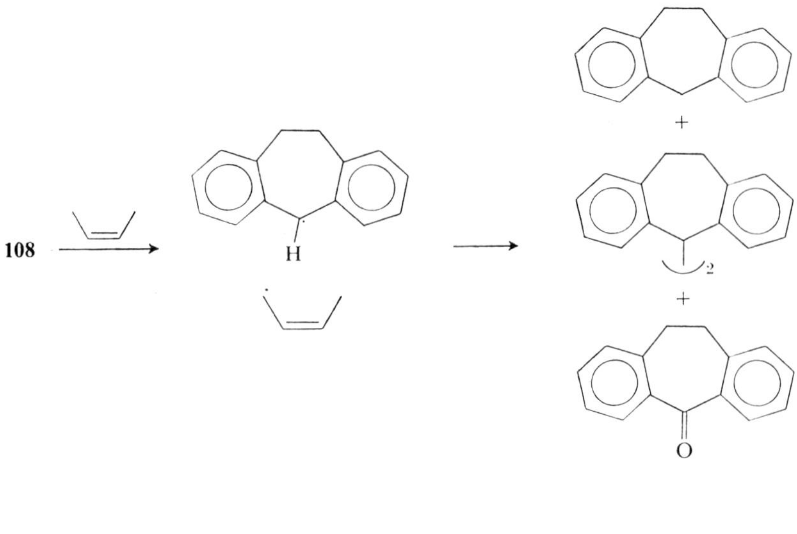

108 → H → + ₂ + O

2-butenes but that some products of abstraction-recombination could be found. Greene[453] had previously found low yields of the products of recombination of the radicals formed in the initial abstraction. Little difference

108

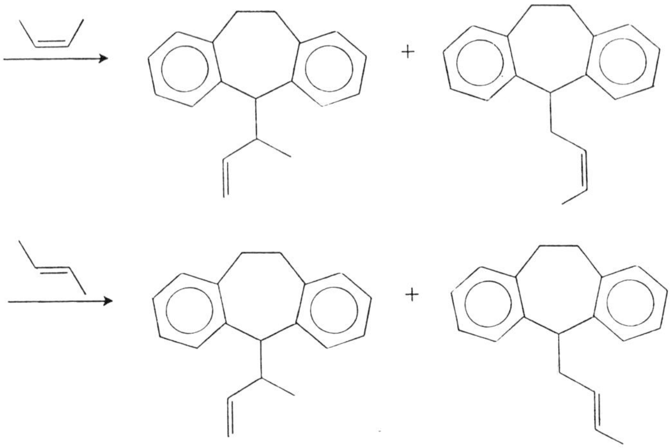

+

+

between carbene **108** and diphenylcarbene can be seen. At first glance this may seem odd since **108** appears flat, at least on paper, and diphenylcarbene has been shown, certainly in rigid medium[412] and probably in solution,[396] to be bent and twisted. Inspection of models reveals, however, that **108** can achieve a structure necessarily bent and easily twisted. Perhaps by the time $n = 2$ we are already at a good approximation of the $n = \infty$ case.

Flash photolysis of the precursor to **108** at 77° K[454] allowed the recording of an ultraviolet spectrum attributed to **108**. Further spectra recorded at room temperature in liquid alkanes revealed the same spectrum observed at 77° K. A second spectrum also appeared which was apparently that of the radical formed by hydrogen abstraction from solvent.[455] The lifetime of the carbene was estimated at 1–50 microseconds.

Carbenes **109** and **110** are less flexible than **108**, however, and differences in properties appear. Both **109** and **110** give products of addition to olefins.[456,457] Remarkably, the cyclopropanes are formed in a stereospecific fashion. Carbene **109** gives a spectrum of products which seems to implicate a mixture of spin states. Both abstraction and stereospecific cycloadditions are occurring. This seems to indicate mixture of a singlet, which adds stereospecifically to olefins (no existing evidence points to a stereospecific triplet) and a triplet, which like diphenylcarbene and **108** abstracts hydrogen. Neither dilution studies using cyclohexane as inert(?) diluent nor benzophenone-photosensitized decomposition reveals any change.[456] Were an equilibrium between singlet and triplet reached, dilution and photosensitization would have no effect. This seems the likely explanation for the behavior of **109**, although it

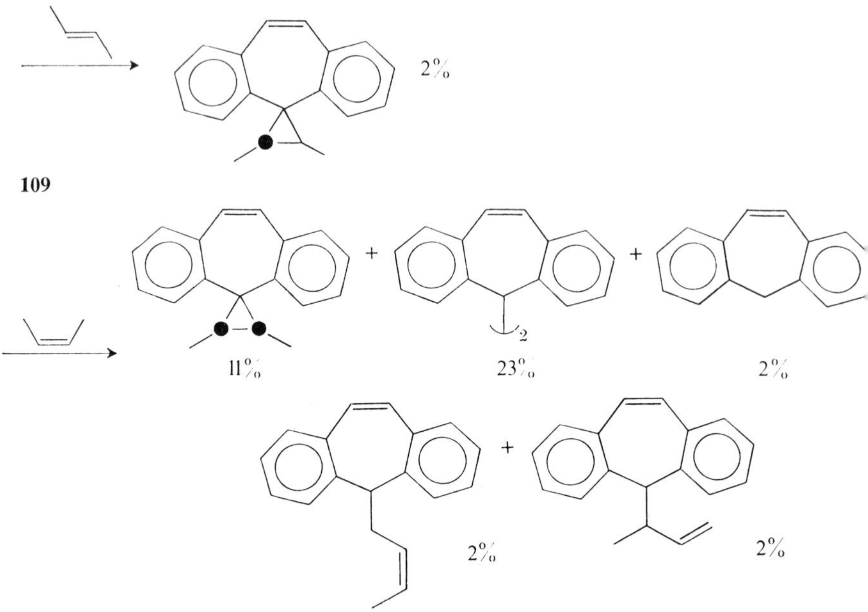

2%

109

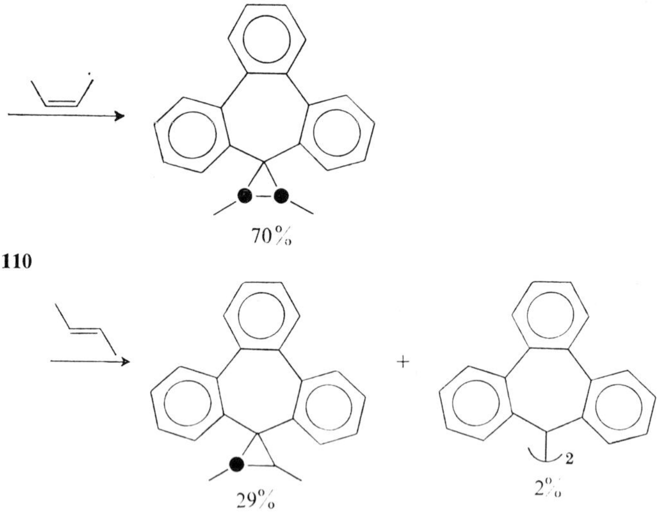

+ +

2

11% 23% 2%

+

2% 2%

70%

110

+

2

29% 2%

has been suggested[456] that a charge transfer complex plays a crucial role in producing stereospecific addition from triplet carbenes **109** and **110**.

Carbene **110** poses similar problems. Its properties appear to be those of a singlet! Large amounts of cyclopropane are formed, and addition is stereospecific. Only small amounts of products formed by abstraction are found. Carbene **110** apparently has a triplet ground state,[458] since the triplet spectrum persists for hours at 77° K. Is it possible that a thermally populated triplet was being observed? This is probably not the case, since kT at 77° K is very small. Thus the properties of **110** are somewhat puzzling. Again it has been suggested that the extended conjugation in **110** allows for formation of a stable charge transfer complex, which can lead to stereospecific addition.[456]

The ultraviolet absorption spectra of diphenylcarbene and of **109** can be obtained at both 77° K and room temperature in saturated hydrocarbon solvent. In olefinic solvents or matrices, however, the spectrum of **109** could not be obtained. This seems to indicate that even at 77° K reaction of **109** with olefins is extremely rapid.[456,459]

The thermal decomposition of diazoanthrone in olefins bearing electron-withdrawing substituents gives cyclopropanes, and occasionally nitrogen-containing products formed through 1,3-dipolar addition reactions.[460,461] Photochemical generation of **111** in benzene led to products of hydrogen abstraction.

Irradiation of diazoanthrone in cyclohexene again led to products of hydrogen abstraction, with no cyclopropane being formed.[462] Olefins not,

111

containing allylic hydrogen did form adducts, however.[463] No careful examination of the stereochemistry of addition has been made, although the stilbenes both give primarily *trans*-cyclopropane. Isomerization of the stilbenes must be facile under the reaction conditions, however.

	Product	Yield (%)
$CH_2{=}CH_2$	*a*	39
$PhCH{=}CH_2$	*b*	93
$Ph_2C{=}CH_2$	*c*	91
111 + *trans*-PhCH=CHPh	*d*	19
cis-PhCH=CHPh	*d*	19
$Ph_2C{=}CHPh$		0
$Ph_2C{=}CPh_2$		0

(*a*) $R_1, R_2, R_3 = H$
(*b*) $R_1, R_2 = H, R_3 = Ph$
(*c*) $R_1 = H, R_2, R_3 = Ph$
(*d*) $R_2 = H, R_1, R_3 = Ph$

A comparison of the deoxygenation of pyridine *N*-oxide with abstraction of hydrogen from either solvent benzene or added cyclohexene has been used to infer the existence of an equilibrium between singlet and triplet **111** and of intersystem crossing from singlet diazoanthrone to the triplet.[464]

The ratio of addition to abstraction from solvent has been studied as a function of olefin substitution.[465] The data given in Table 10 were thought

TABLE 10
Reaction of **111** with Olefins

Olefin	Addition	Abstraction
α-Methyl styrene	91	1
Methyl vinyl ketone	70	9
Isopropenyl methyl ketone	60	10
Ethyl fumarate	58	13
2,3,3-Trimethyl-1-butene	58	29
Mesityl oxide	0	76
Ethyl maleate	0	41
trans-4-Methyl-2-pentene	0	81
cis-4-Methyl-2-pentene	0	83
2-Methyl-2-pentene	0	85

consistent with steric control of the reaction. The absence of cyclopropanes from *cis*- and *trans*-4-methyl-2-pentene in which the double bond is substituted with electron-donating groups was used as evidence that only the triplet underwent the addition reaction.[465] Singlet addition is apparently less fast than abstraction of the tertiary allylic hydrogen.

The products of reaction of **111** with various heterocyclic compounds are also consistent with addition of the triplet.[466–468] Such results have also been obtained for biscarbomethoxycarbene (see Section V) and diphenylcarbene.[469]

$Ph_2C:$ + O → O Ph Ph → O Ph Ph

Since it appears that **108** (where $n = 2$) is already a good model for diphenylcarbene (where $n = \infty$), it would be very helpful to know the properties of **112** ($n = 1$), unfortunately an unknown species. Carbene **112** should be nearly planar and if, as has been postulated, the abstraction/cycloaddition problem can be resolved on steric grounds,[396] it should form cyclopropanes on reaction with olefins, or at least be intermediate in properties between diphenylcarbene and fluorenylidene. Carbene **111** is not an ideal model for

112 because electronic interaction between the two rings exists in the $n = 1$ bridge, but does not in **112**.

112

E. Miscellaneous Aryl- and Diarylcarbenes

This section is more of a compendium than a discussion. One is always faced with a number of compounds that do not seem to fit logically into one category or another. We are taking the simple expedient of combining these and of presenting only the most interesting.

Diazoindane[470] has been synthesized, and although little beyond the conversion to indene and indanone on irradiation in ether/pentene has been reported, it would seem a good substrate for a study of the effects of photosensitized decomposition on an aromatic carbene.

N_2 $\xrightarrow{h\nu}$ + O + Azine

10% 5% 50%

Ferrocenylphenylcarbene and ferrocenylmethylcarbene have been found to mimic diphenylcarbene in their reactions with olefins.[471–474] With 1,1-diphenylethylene, cyclopropanes were formed in modest yields.[472]

$\xrightarrow[80°]{}$ Fc—CH_2—PH + Fc—CH(Ph)—CH(Ph)—Fc + Azine

27% 1.5% 30%

Fc, Ph, N—N—Ts, ⊖

$\xrightarrow{Ph_2C{=}CH_2}$ Ph, Ph, Fc Ph + Azine

Fc = ferrocenyl

14% 33%

The bridged ferrocenylcarbene **112a** has also been generated.[475] Competition experiments using decene and 1,1-diphenylethylene gave only the product

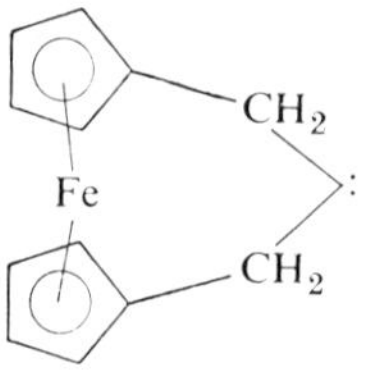

112a

of addition to the phenyl-substituted double bond. This certainly seems to implicate a triplet carbene, and, in turn, a remarkable effect of the remote ferrocene on the multiplicity of the alkylcarbene.

9-Anthrylcarbene has been found to initiate the polymerization of methyl methacrylate[476,477] and to react with protonic solvents.[478] This carbene has also figured prominently in a spectroscopic study which demonstrated that arylcarbenes are bent.

α- and β-Naphthylcarbene were found to show two triplet ESR spectra.[479] This might be due to two bent structures as shown, or to a flat carbene plus an out-of-plane species. 9-Anthrylcarbene, however, would show two triplets

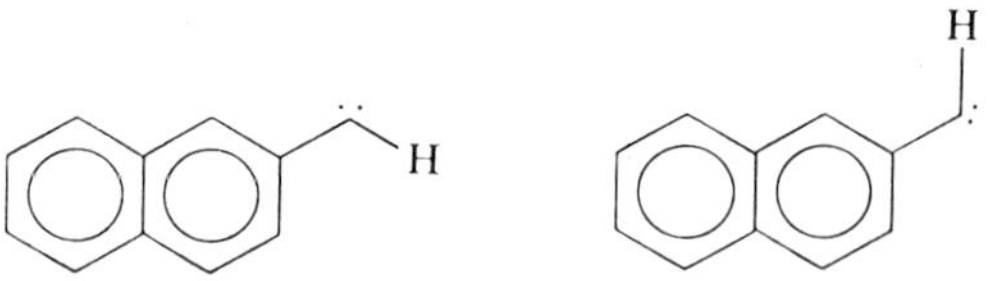

only in the latter event. Since it has only one, the reasonable assumption was made that two in-plane, bent triplets were formed from the α- and β-naphthyl species.

In contrast, 9,9′-dianthrylcarbene has been found to have a linear ground state triplet.[480] When generated in 9,9′-dianthryl ketone as host, the ESR signal of the triplet could be observed at temperatures as high as 200°. In various glassy matrices, however, thawing of the matrix resulted in irreversible loss of signal.

The furfurylidene **112b** can be trapped via either insertion or addition reactions, but its main path involves a ring opening as shown.[481]

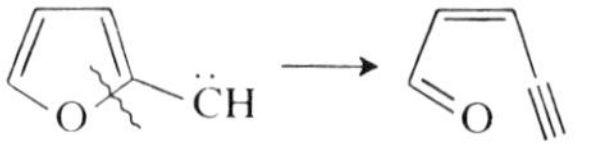

112b

A few di- and polycarbenes are known in the aromatic series, but these interesting compounds have been studied mainly by spectroscopic techniques.

In 1963, dicarbene **113** was generated at 77° K by the irradiation of the corresponding bisdiazo compound.[482] Evidence for a ground state species was found, and it was noted that the spectrum was that of the triplet **114**, not the quintet state. Carbenes **115** and **116** would be incapable of forming a stable triplet and it was indeed found by two groups[483–485] that they possessed quintet ground states. It was explicitly noted that the possibility of the observed quintet states being a few small calories above unobserved ground singlet states existed. The observed chemistry of these interesting species is

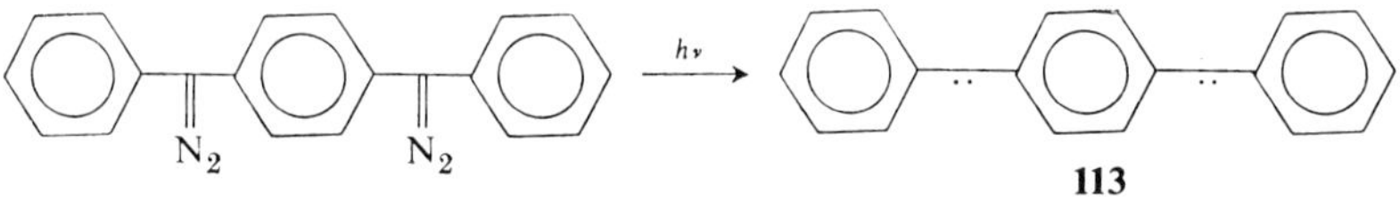

N_2 N_2 $h\nu$ **113**

114

HC̤ C̤H

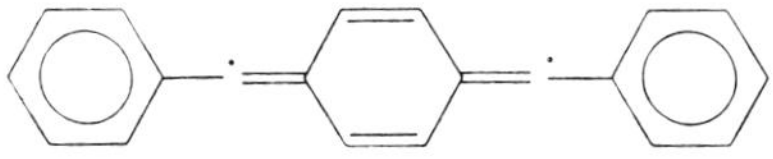

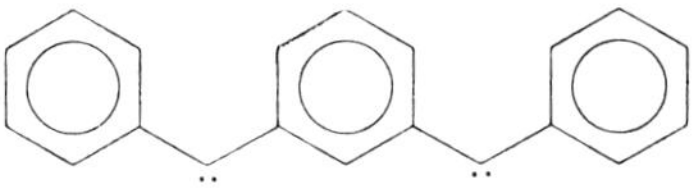

115 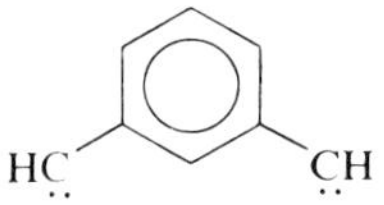 **116**

limited to the formation of **117** on generation of **113** in toluene,[482] and the addition of **116a** to benzene.[486]

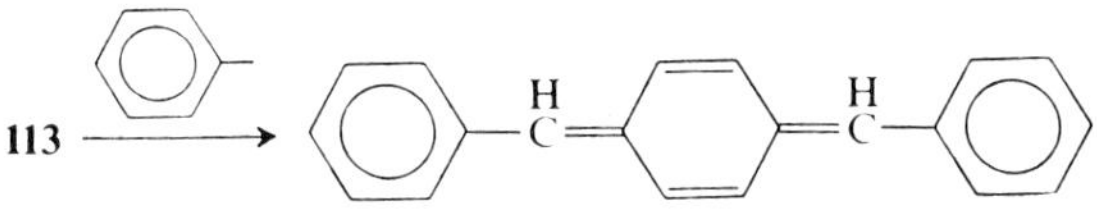

113 → H C H C **117**

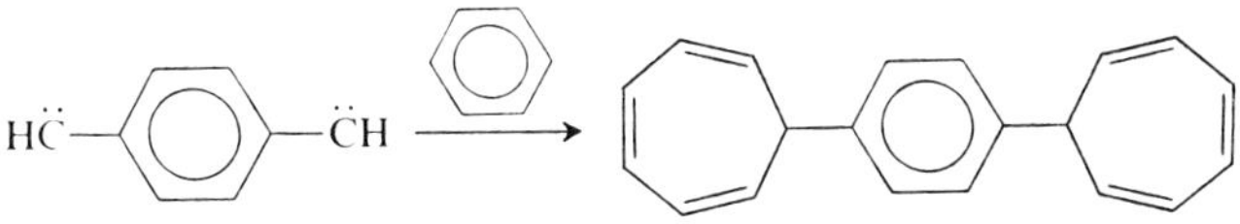

HC̈—C̈H **116a**

A large number of aryldiazo compounds, many including heteroatoms, has been made,[487–490] but thus far little attention has been paid to their reactions.

Finally, the phosphorus-substituted phenylcarbene **118** made by Regitz[491,491a] is worthy of mention because it adds to benzene to give not the usual cycloheptatriene but the stable norcaradiene **119**.

F. Intramolecular Reactions

Beyond simple intramolecular insertions or additions, the question arises as to what ways a carbene will find out of its misery when easy routes are

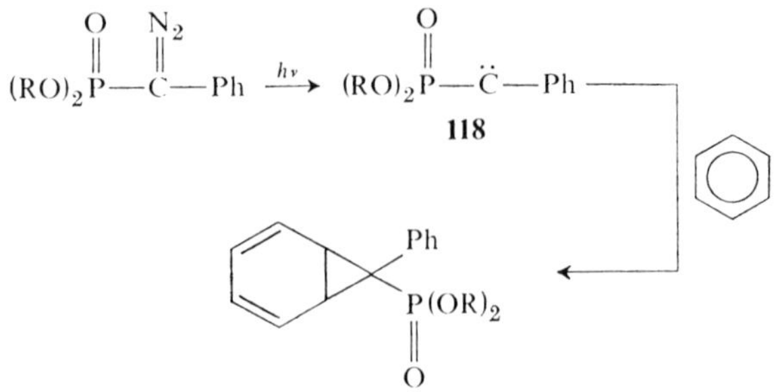

$$(RO)_2P(=O)-C(=N_2)-Ph \xrightarrow{h\nu} (RO)_2P(=O)-\ddot{C}-Ph$$

118

119

removed. That strange reactions might occur could have been foreseen from the 1913 work of Staudinger and Endle,[492] who found that the gas-phase pyrolysis of diphenylketene yielded fluorene in some 35% yield. Staudinger and co-workers also noted that shortly before exploding, a blue flash could be observed from diphenyldiazomethane while diazofluorene showed a momentary reddish flamelet.[493] These interesting observations apparently have received no further attention beyond informal attempts at repetition at Yale University during the 1960s.[494] The failures of the Yale workers should not deter other investigators.

Fluorene was confirmed as a product of diphenylcarbene by Harrison and Lossing in 1960[495] and Rice and Michaelsen in 1962[496] and a mechanism proposed,[495] which even at the time must have seemed hardly probable. A similar observation was made by Franzen and Joschek,[497] who found **120** on generation of di-α-naphthylcarbene.

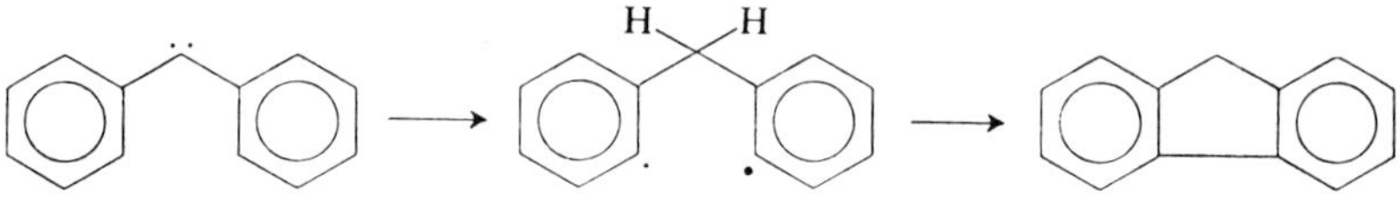

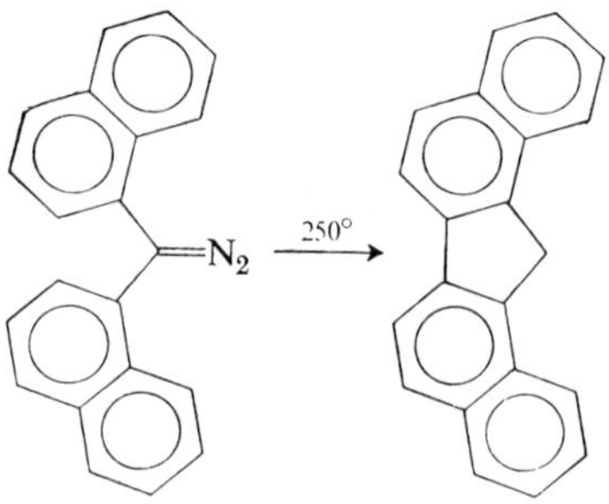

250°

120

The plot thickened somewhat when Crow and Wentrup found that apparent sources of α-pyridylcarbene gave products similar to those found from phenylnitrene.[498] The following example is typical, although more detailed studies have been made:

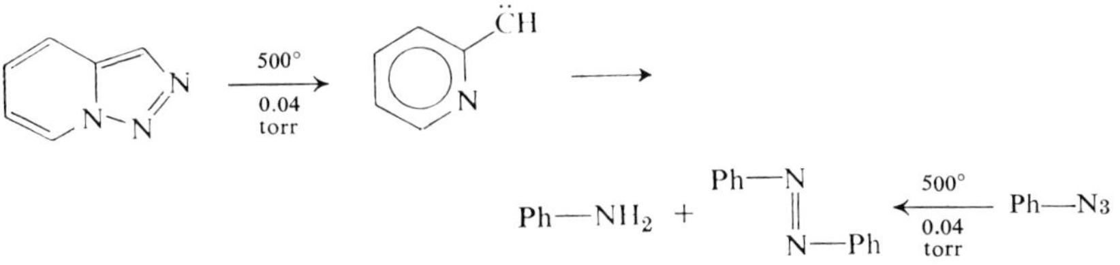

Many varied systems have been studied, and it is clear that the nitrene-carbene and nitrene-nitrene interconversions are common.[499] It must be stressed that these interconversions occur mainly in the gas phase and only rarely in solution.[500]

One would have expected a concerted search for the all-carbon counterpart, and in 1969 one part of the interconversion was discovered, as it was found that phenylcarbene rearranged to cycloheptatrienylidene.[501–502] A possible

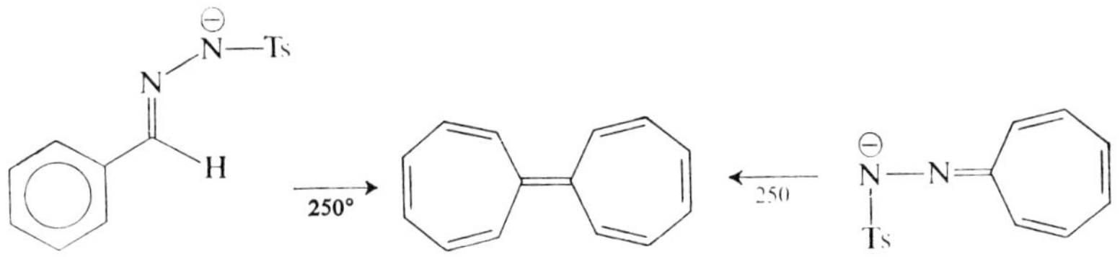

reversal had been noted previously; that is, ferrocenylcycloheptatrienylidene generated from the ferrocenyltropilium ion gave products attributable to a phenylcarbene.[503] The authors specifically note that such a mechanism is not required, but it does seem possible.

Further work which implicates (but does not require) the interconversion of phenylcarbene and cycloheptatrienylidene has involved the gas-phase interconversion of *ortho*-, *meta*-, and *para*-tolylcarbene and their further conversion to phenylmethylcarbene.[504–506] Pyrolysis of *o*-, *m*-, or *p*-tolyldiazomethane gives benzocyclobutene and styrene as products. The *ortho* compound gives somewhat more benzocyclobutene than the others. Perhaps the *ortho*-methyl group forces the carbene to be born in an orientation especially favorable for insertion, or perhaps some direct insertion process is occurring from the diazo compound.[504] At even higher temperatures, phenylcarbene gives, in addition to the dimers mentioned previously,[501] fulvenallene (**121**) and cyclopentadienylacetylene.[507] It is now clear from carbon-13 labeling studies that the intermediate responsible for the formation of fulvenallene is not involved in the equilibration of the tolylcarbenes.[508]

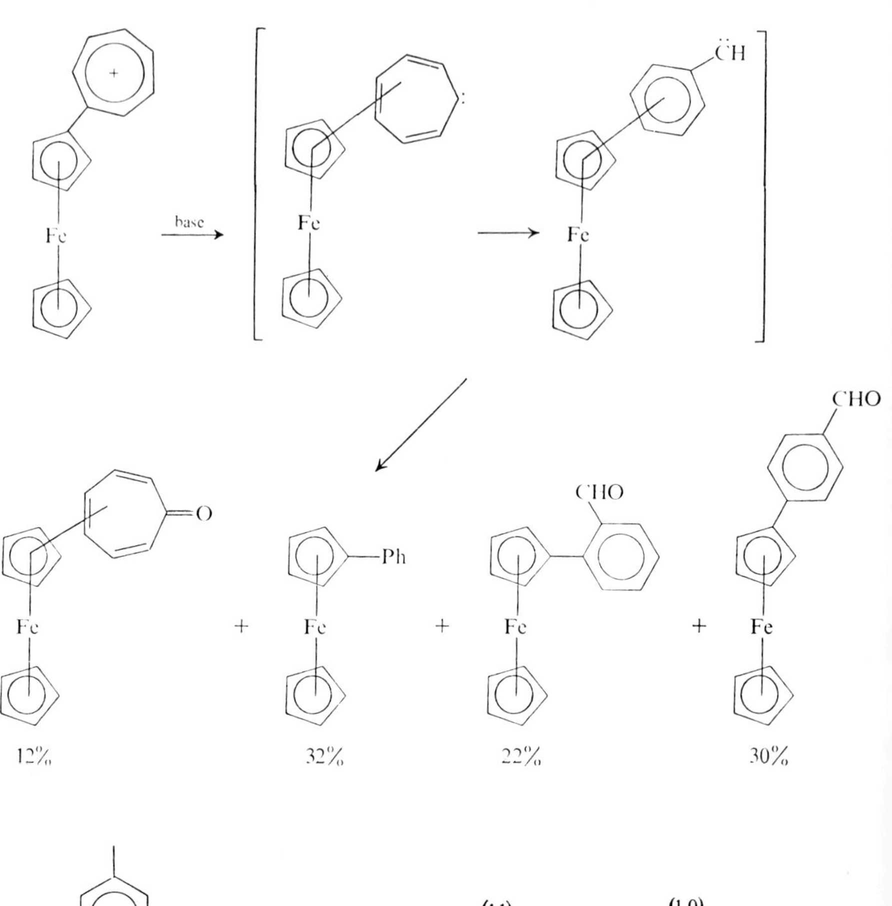

base
Fe
ĊH
CHO
O
Ph
12%
32%
22%
30%

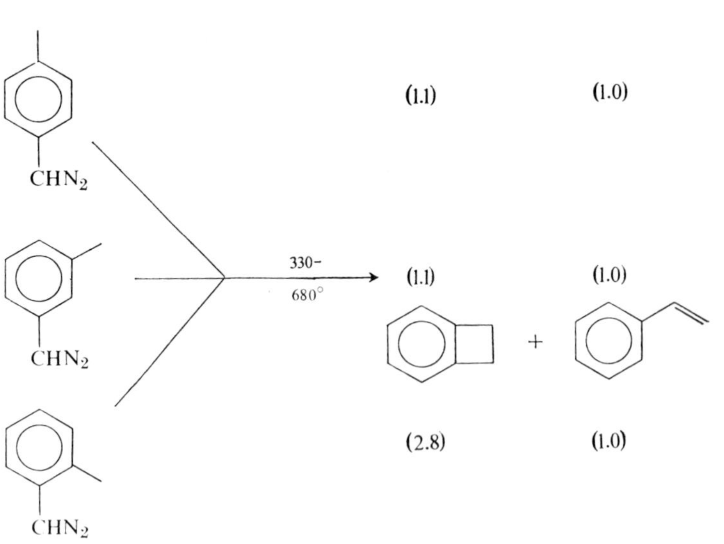

CHN_2
CHN_2
CHN_2
330–
680°
(1.1)
(1.0)
(1.1)
(1.0)
+
(2.8)
(1.0)

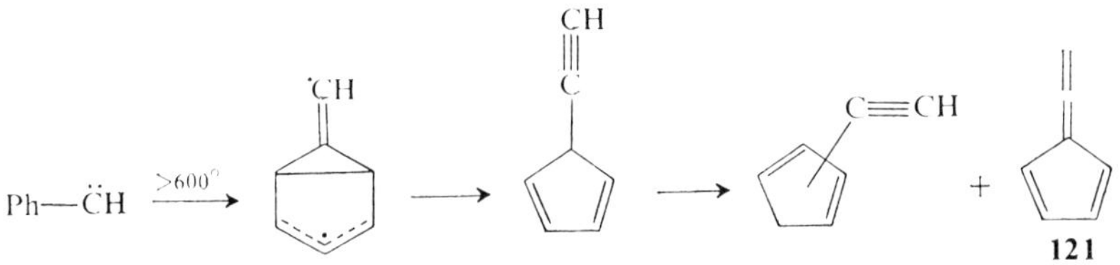

Some of the very early results can be reinterpreted in terms of this later work. For instance, it has now been shown[502,509] that fluorene is being produced through a rearrangement similar to these recently discussed, and not by the old mechanism.[495]

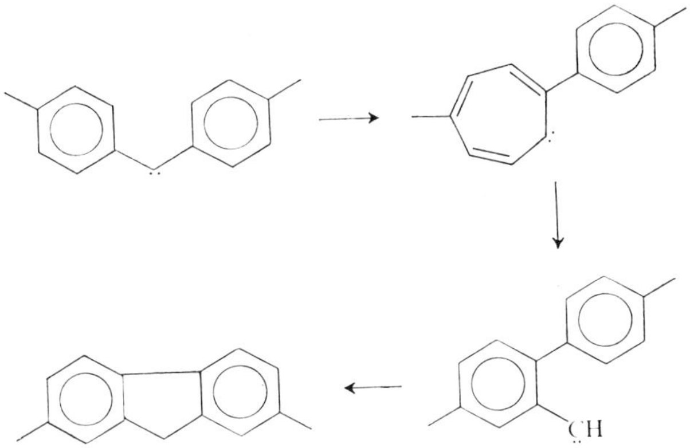

V. KETO- AND CARBOALKOXYCARBENES

These two groups are considered together here, since most of their differences are of degree rather than kind. Problems similar to those encountered in the discussion of alkylcarbenes are found as intramolecular reactions typically complicate keto- and carboalkoxy diazo compound decompositions. In this case such reactions have a special name, the Wolff rearrangement,[18a,510] and it is now recognized that this process is common to both keto- and carboalkoxycarbenes. We discuss the Wolff rearrangement in a separate section. Questions of spin state also arise, and we deal with them as we go along.

A. Carboalkoxycarbenes

Both insertion and cycloaddition reactions have been intensively studied, and we start by discussing the former. The early work of Doering and Knox,[511] which showed that carboethoxycarbene and biscarboalkoxycarbenes were quite selective, has been extended in several ways. Doering and Basinski[320] showed that carboethoxycarbene reacted with diethyl ether to give insertion into the α and β positions in the ratio 4/1, in addition to a product **122** probably formed from an ylid.

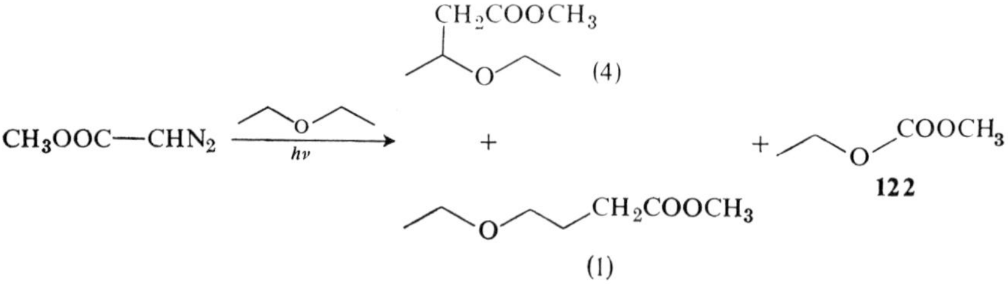
CH_2COOCH_3
O
(4)
$CH_3OOC—CHN_2$
O
hν
+
+
O
$COOCH_3$
122
O
CH_2COOCH_3
(1)

The reactivity of the bridgehead hydrogen in bicyclic systems is quite low, as one would expect. Willcott has supplied data on reactions of carboethoxycarbene and biscarboethoxycarbene with bicyclic hydrocarbons,[88] and Sauers and Kiesel have investigated the reaction of nortricyclane with carboethoxycarbene[512] (Table 11). Both groups conclude, as did Doering and

TABLE 11

Relative Rates of Reaction with Bicyclic Hydrocarbons

		1-Position	All others	3°/2°
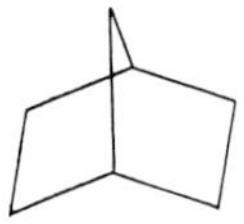	:CHCOOR :C(COOR)$_2$	(1.0) (1.0)	(6.7) (27.3)	(0.7) (0.2)
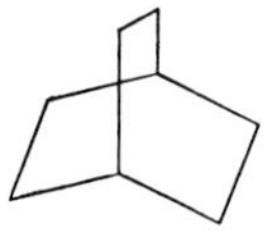	:CHCOOR :(COOR)$_2$	(1.0) (1.0)	(6.0) (6.7)	(1.0) (0.7)

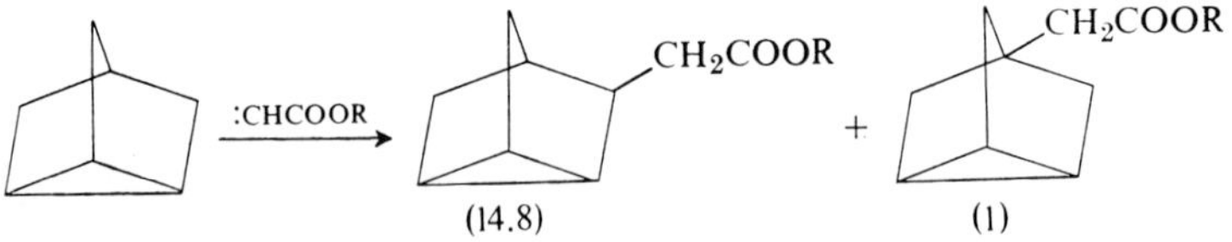
:CHCOOR
CH_2COOR
+
CH_2COOR
(14.8)
(1)

Knox,[511] that the transition state for insertion at the bridgehead is destabilized by the relative lack of importance of resonance structures such as **123**.

The insertion reaction has been shown to proceed with retention,[513] which contrasts with the report of Franzen that methylene inserts with inversion.[86] Optical activity is maintained in the reaction with the silicon-hydrogen bond.[514,515]

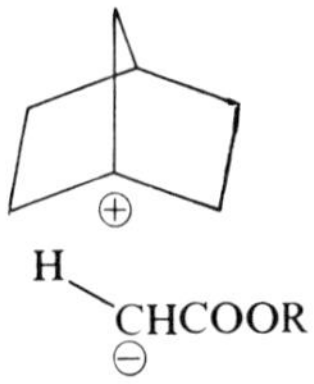
H
CHCOOR

123

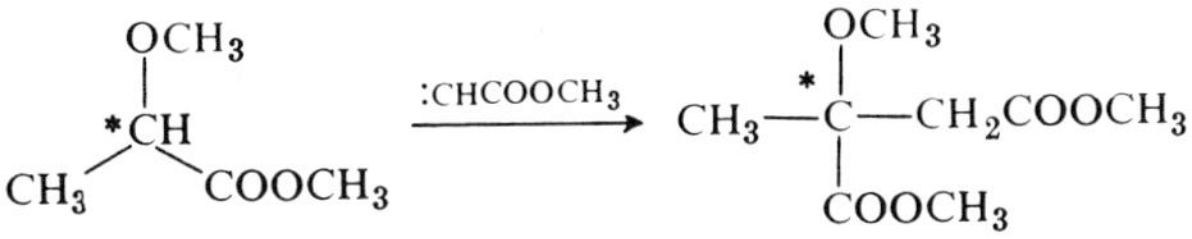
OCH3
*CH
CH3
COOCH3
:CHCOOCH3
OCH3
*
CH3—C—CH2COOCH3
COOCH3

retention

Intramolecular insertion reactions have been found and can compete with the intermolecular variety.[516–519]

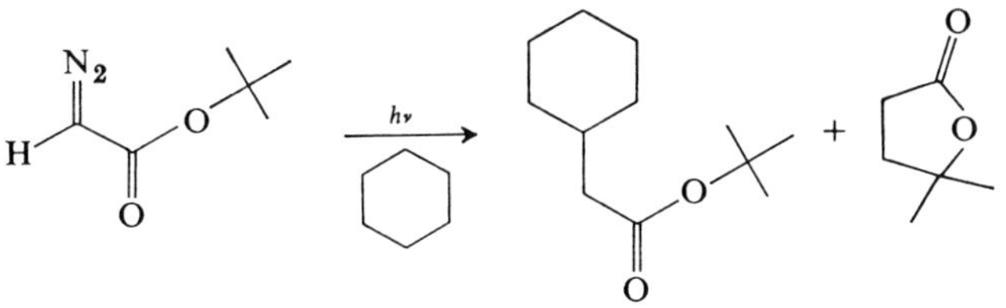
N2
H
O
O
hν
O
O
+
O
O

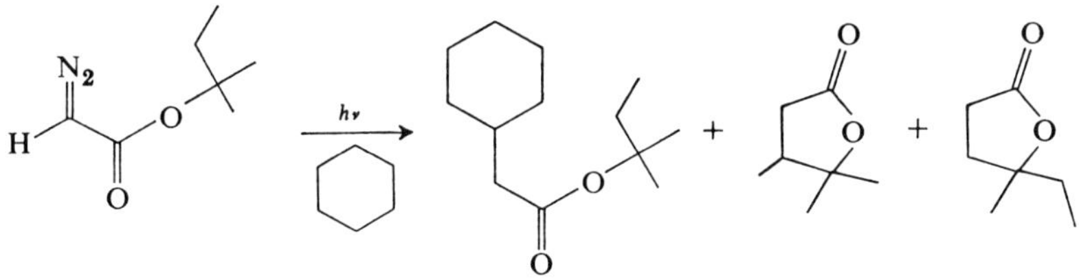
N2
H
O
O
hν
O
O
+
O
O
+
O
O

Decomposition of diazosuccinic ester apparently goes through the carbene even in solvents such as acetic acid.[520] Deuterium is not incorporated in deuterio acetic acid, although DCl/D_2O does give 50% D incorporation.

$$\mathrm{EtOOC{-}\overset{\overset{\textstyle N_2}{\|}}{C}{-}CH_2{-}COOEt} \xrightarrow[\Delta]{\mathrm{DOAc}} \mathrm{Et_2OOC{-}CH{=}CH{-}COOEt}$$

Little seems to have been done on the activity of allylic carbon-hydrogen bonds, but during a study of cyclopropene formation Lindt and Deutschmann[521] found that the propargyl carbon-hydrogen bond was not especially reactive.

$H\ddot{C}COOEt + Pr{-}C{\equiv}C{-}Pr \longrightarrow$

Pr = *n*-propyl

CH_2COOEt

Pr Pr

+ Pr—C≡C—CH_2—CH_2—CH_3

COOEt

Experiment: 14 15 10

Statistical: 4 4 6

Ethyldiazoacetate apparently reacts with carbon tetrachloride via a radical chain reaction,[522] as does diazomethane.

N,*N*-diethyldiazoacetamide undergoes intramolecular self-insertion in the positions α and β to the nitrogen. Polar solvents favor α insertion because, it was thought, of an especially polar transition state for the formation of this particular product.[523]

H O N N_2 α β $h\nu$ O N + O N

Addition of carboethoxycarbene to olefins has long been known[524] and has been found to be stereospecific in two cases.[525] Addition to cyclic or asymmetric double bonds gives as the major product the less hindered exo adduct.[526,527]

+ :CHCOOR ⟶ COOR (5) + COOR (2)

+ :CHCOOR ⟶ COOR 16% + ROOC 10% + CH_2COOR 21%

+ :CHCOOR ⟶ COOR (3) + ROOC (1)

There is little recent work to add to these older examples and to others contained in Kirmse's book.[1] What there is concerns the induction of optical activity by the use of asymmetric copper catalysts[528,529] and intramolecular reactions. Decomposition of ethyldiazoacetate using an optically active copper chelate gives optically active products in both the addition and carbon-hydrogen insertion reactions and in ylid ring expansion. The enantiomeric copper chelate gives enantiomeric products. This and other

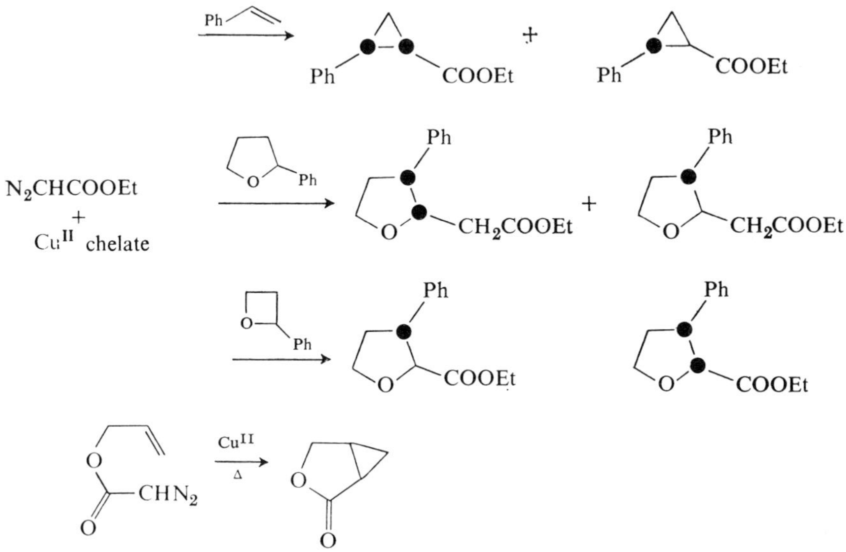
Ph
Ph
COOEt
+
Ph
COOEt
$N_2CHCOOEt$
+
Cu^{II} chelate
O
Ph
Ph
O
CH_2COOEt
+
Ph
O
CH_2COOEt
O
Ph
Ph
O
COOEt
Ph
O
COOEt
O
O
CHN_2
Cu^{II}
Δ
O
O

124 **125**

work[530,531] has led to the postulate of a copper-carbene-olefin complex as the active species. It is worth noting that photolysis of **124** does not give any lactone **125**. A copper catalyst is required.[531]

Biscarbomethoxycarbene has been studied in some detail, and the singlet and triplet states have been compared.[446,532,533] The addition is only 90% stereospecific, a fact that leads to the type of questions posed in Section III about the stereochemistry of addition of 4,4-dimethylcyclohexadienylidene. Here, though, the benzophenone-photosensitized decomposition of diazomalonic ester is apparently successful in producing the triplet. Unlike the original experiment of Kopecky, Hammond, and Leermakers on the photosensitized decomposition of diazomethane,[73] products are formed in good yield.

The effect of hexafluorobenzene (presumably an inert solvent) on the stereochemistry of addition is unusual. Whereas at very high (greater than 98%) concentrations of hexafluorobenzene more *trans*-cyclopropane is formed, at lower concentrations more *cis*-cyclopropane appears.[446] Although

$N_2C(COOR)_2$, $h\nu$

Olefin	ROOC–cyclopropane–COOR (cis)	ROOC–cyclopropane–COOR (trans)	Yield (%)
cis	(92)	(8)	39.8
trans	(10)	(90)	24.3
cis, Ph_2CO	(14)	(86)	43.0
trans, Ph_2CO	(15)	(85)	—

one cannot make much of the high-concentration end of the curve, it is interesting to speculate on the initial part. If the 8–10% of the "wrong" stereoisomer were due to intersystem crossing from the singlet carbene to a ground state triplet,[534] one would expect hexafluorobenzene to allow time for further crossing. This it clearly does not do. It is difficult to speculate beyond this, however, since the paths available are too numerous and the effect of hexafluorobenzene on the rates too unpredictable. The remaining

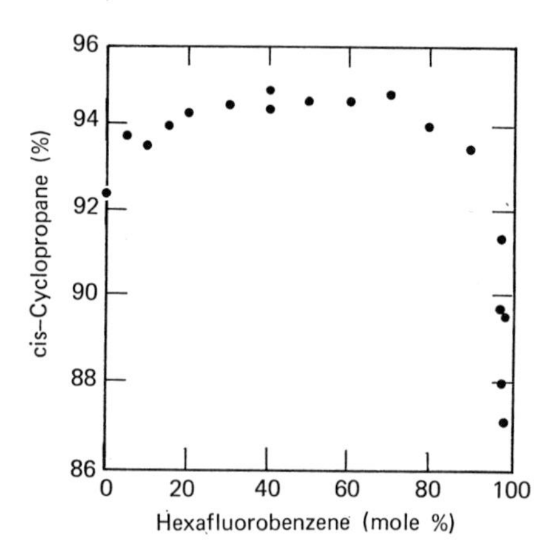

possibilities include intersystem crossing in the diazo compound, an equilibrium at reaction temperatures between singlet and triplet, and direct

$N_2C(COOR)_2$ + 2,3-dimethylbutane $\xrightarrow{h\nu}$ or $\xrightarrow{h\nu,\ Ph_2CO}$

	$-CH(COOR)_2$ (tertiary) + $-CH(COOR)_2$ (primary)	$[(ROOC)_2CH—]_2$	$(ROOC)_2CH_2$
$h\nu$	46% ($3^\circ/1^\circ = 13.1$)	trace	trace
$h\nu, Ph_2CO$	13% ($3^\circ/1^\circ = 20$)	38%	23%

reaction of singlet and triplet diazo compound to give cyclopropane. The last possibility seems attractive, since an increase in the concentration of hexafluorobenzene should remove chances for triplet diazo compound to react and thus increase the amount of overall cis addition.

Spin state also affects the carbon-hydrogen insertion reaction, which clearly goes through a process of abstraction-recombination in the triplet case. The selectivity of the triplet is slightly greater than that of the singlet.[446]

One particular cycloaddition of biscarbomethoxycarbene which seems worthy of mention is the route to 7,7-disubstituted cycloheptatrienes used by Berson et al.:[535]

$+ (ROOC)_2C: \longrightarrow$ COOR COOR $\longrightarrow \longrightarrow$ COOR COOR

Reactions of carboalkoxycarbenes with aromatic compounds lead to several products. Doering et al.[536] worked out the structures of the compounds (Büchner esters) formed from the copper-catalyzed decomposition of ethyldiazoacetate in benzene. More recently, the corresponding photochemical[537] and thermal[538,539] reactions have been examined. In the latter case carboethoxycarbene was compared to carboethoxynitrene. Both species were found to be electrophilic (ρ carbene $= -0.38$, ρ nitrene $= -1.32$), but the carbene was far less discriminating. A concerted addition was indicated, as a stepwise process should have had a greater ρ.

Addition to naphthalene gives the norcaradiene **126** and the exo form is preferred.[540–542] Numerous additions to substituted naphthalenes and heteroaromatic compounds are described by Kirmse.[1]

:CHCOOR $\longrightarrow$ COOR

126

Thermal decomposition of diazoacetic ester in anthracene is reported[543–545] to give only **127**, although the copper-catalyzed decomposition also gave small amounts of the unusual product of 1,4-addition, **128**. The melting point of the acid derived from it matches that reported by Meinwald and Miller,[546] who synthesized it by another route. The possibility remains, of course, that **128** is formed by rearrangement of some other product.

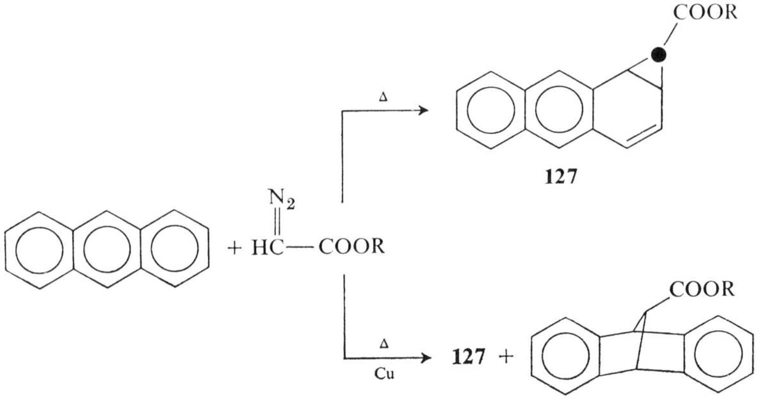

COOR

Δ

127

N_2

+ HC—COOR

COOR

Δ

Cu

127 +

128

Phenanthrene is also attacked and, although originally only the exo and endo products of addition to the 9,10 double bond (**129** and **130**) were identified,[547] several others have recently come to light.[543] The source of compound **131** is claimed to be a two-step addition followed by hydrogen shift. Direct insertion seems a strong possibility, however.

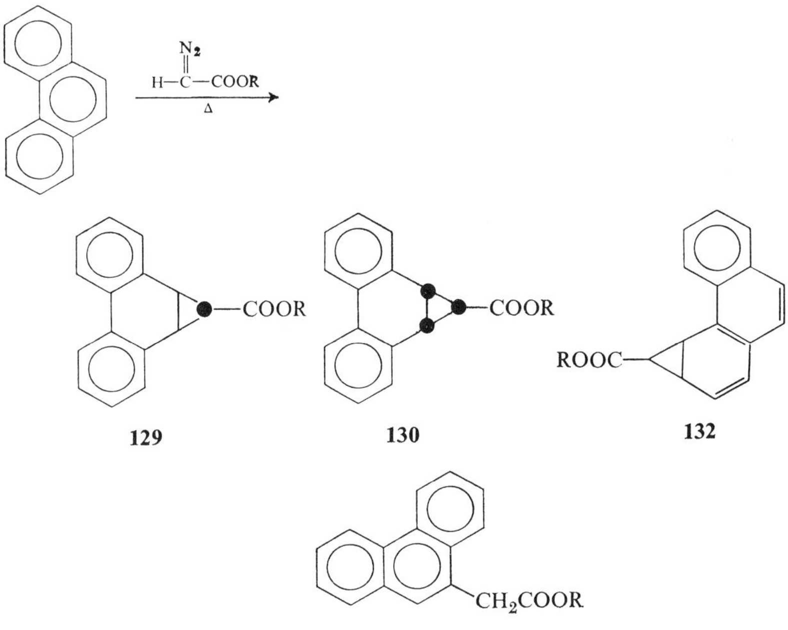

N_2

H—C—COOR

Δ

—COOR

—COOR

ROOC—

129

130

132

CH_2COOR

131

Biphenylene and carboethoxycarbene yield the remarkable fluorene **133**. The mechanism proposed[548] is shown here. Contrary to original reports,[549]

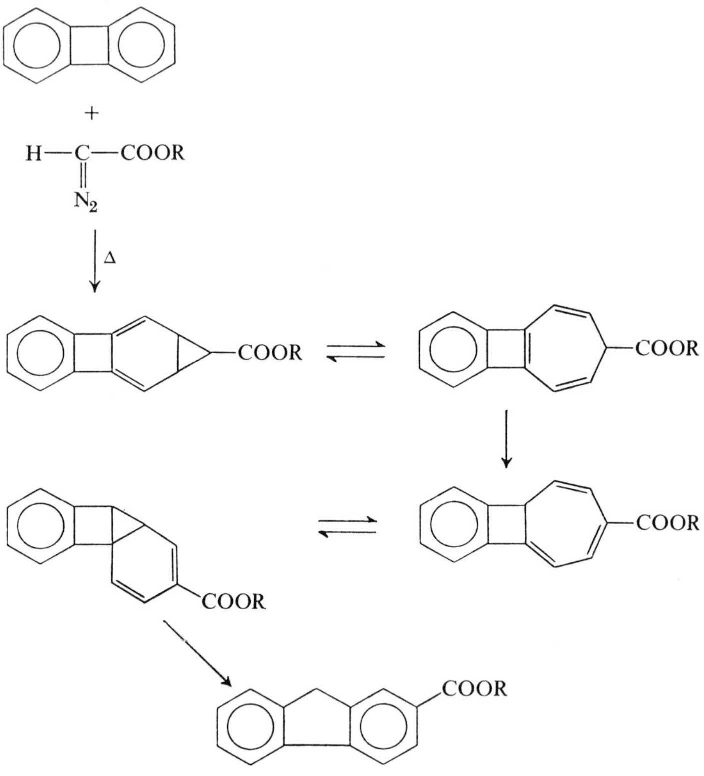

H—C—COOR
‖
N_2
↓ Δ

133

biscarbomethoxycarbene also adds to benzene to give, initially, 7,7-dicarbomethoxycycloheptatriene, which easily rearranges on workup.[446]

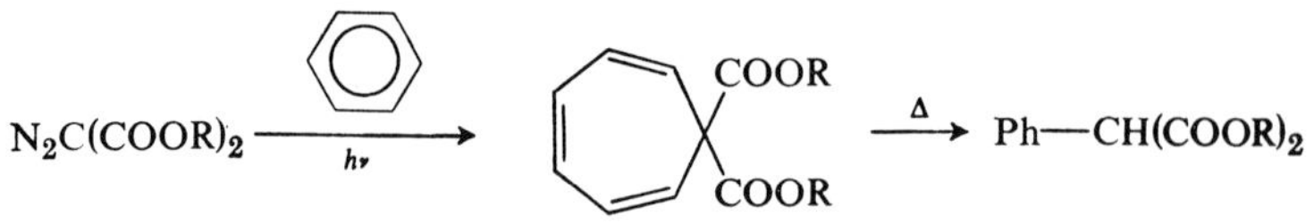

$N_2C(COOR)_2 \xrightarrow[h\nu]{} \quad \xrightarrow{\Delta}$ Ph—CH(COOR)$_2$

Alkyl acetylenes are attacked to give cyclopropenes.[550] Phenylacetylenes give both cyclopropenes and furans, although the furans are formed only when copper catalysts are present.[551–553]

$$\ddot{H}CCOOR \xrightarrow{CH_3-C\equiv C-CH_3} \text{(1,2-dimethylcyclopropene-3-COOR)}$$

$$\ddot{H}CCOOR \xrightarrow{Ph-C\equiv C-CH_3} \text{(1-phenyl-2-methylcyclopropene-3-COOR)}$$

$$\ddot{H}CCOOR \xrightarrow{Ph-C\equiv C-Ph} \text{(1,2-diphenylcyclopropene-3-COOR)}$$

Furans also appear when triplet biscarbomethoxycarbene is generated.[554] A typical example is shown below along with the mechanism proposed.

$$CH_3-C\equiv C-CH_3 + (CH_3OOC)_2CN_2 \xrightarrow[Ph_2CO]{h\nu} \text{(diradical intermediate: } CH_3OOC, OCH_3) \longrightarrow \text{(furan: } CH_3OOC, CH_3O)$$

Schöllkopf and his collaborators have synthesized and examined a variety of metal- and halo-substituted carboalkoxycarbenes, often cleverly using the long-known[555] compound **134** as starting material. The tin- and silicon-substituted carbenes related to **135** and **136** seem to be normal.[556,557] The

$$\textbf{134}\ Hg\left(\overset{N_2}{\overset{\|}{C}}-COOR\right)_2 \xrightarrow{(CH_3)_3SnCl} ROOC-\overset{N_2}{\overset{\|}{C}}-Sn(Ph)_3\ (\textbf{135}) \xrightarrow[h\nu]{\text{isobutylene}} \text{ROOC, Sn(Ph)}_3\text{-2,2-dimethylcyclopropane}\quad 49\%$$

$$\textbf{134} \xrightarrow{(CH_3)_3SiCl} ROOC-\overset{N_2}{\overset{\|}{C}}-Si(CH_3)_3\ (\textbf{136}) \xrightarrow[h\nu]{\text{isobutylene}} \text{ROOC, Si(CH}_3)_3\text{-2,2-dimethylcyclopropane}\quad 35\%$$

silver-substituted diazo compound is not stable but serves as a source of alkyl and iododiazoacetic esters.[557–559] Iodo-, bromo-, and chlorodiazo compounds

$$N_2CHCOOR \xrightarrow[0°]{Ag_2O} \left[Ag{-}\overset{\overset{\large N_2}{\|}}{C}{-}COOR \right] \begin{cases} \xrightarrow{I_2} I{-}\overset{\overset{\large N_2}{\|}}{C}{-}COOR \\ \xrightarrow{RX} R{-}\overset{\overset{\large N_2}{\|}}{C}{-}COOR \end{cases}$$

can be made from **134**[560] and yield carbenes which add to olefins to give cyclopropanes.[560,561] Addition to olefins is stereospecific[562] and no insertion products are found.

$$\mathbf{134} \xrightarrow[X = Cl,\ Br,\ I]{X_2} X{-}\overset{\overset{\large N_2}{\|}}{C}{-}COOR \xrightarrow[h\nu]{\text{olefin}} \text{cyclopropane (X, COOR)}$$

Ethyl lithiodiazoacetate, prepared either from **134** or by direct lithiation of diazoacetic ester, can be used as the source of many substituted diazocompounds.[563]

$$\mathbf{134} \xrightarrow{RLi} \text{ or } H{-}\overset{\overset{\large N_2}{\|}}{C}{-}COOET \xrightarrow{RLi} \;\; Li{-}\overset{\overset{\large N_2}{\|}}{C}{-}COOET \xrightarrow{(CH_3)_3C{-}SiCl} (CH_3)_3CSi{-}\overset{\overset{\large N_2}{\|}}{C}{-}COOET$$

Compound **134** has been directly decomposed by Strausz and his coworkers[564,565] to give a carbyne and a mercury-containing carbene. A mechanism involving initial hydrogen abstraction from cyclohexene by the carbyne followed by addition is eliminated by the endo/exo ratio of 31/17. Carboethoxycarbene reacts with cyclohexene to give endo and exo product in the ratio 1/2.[527]

$$\mathbf{134} \longrightarrow ROOC-\ddot{C}-Hg-\overset{N_2}{\overset{\|}{C}}-COOR \longrightarrow ROOC-\ddot{C}$$

31% 17% 38%

B. Formylcarbenes

Very little is known about such species, although several precursors are now available.[566–568] One would expect facile ketene formation from the direct photolysis of α-diazoaldehydes, and photosensitized decomposition in olefins might be more likely to produce cyclopropyl aldehydes. Copper-

$$N_2CH-CHO + \text{(CH}_3)_2C{=}C(CH_3)_2 \xrightarrow[\Delta]{Cu} \text{tetramethylcyclopropyl}-CHO \quad 25\%$$

catalyzed decomposition of diazoacetaldehyde in 2,3-dimethyl-butene-2 gives up to 25% cycloaddition.[567,568]

C. Ketocarbenes

Ketocarbenes are most prone to Wolff rearrangement, and relatively few data on their properties exist. External reactions are facilitated by copper catalysis (probably by the formation of an olefin-carbene-copper complex), and therefore little work has been done on the uncatalyzed decomposition. We reserve much of the material on the Wolff rearrangement for a later

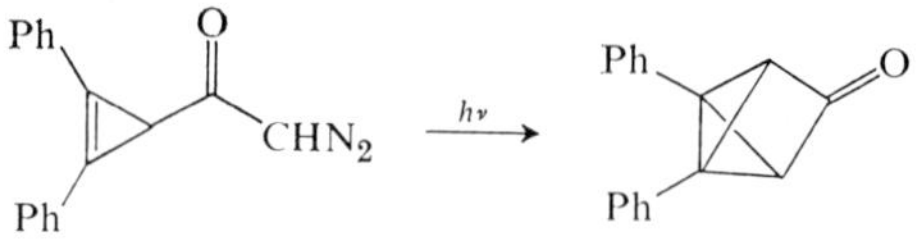

Ph Ph O CHN$_2$ $h\nu$ Ph Ph O

137 **138**

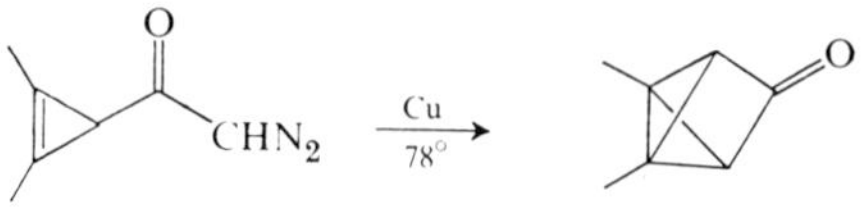

O CHN$_2$ Cu 78° O

section and give only representative examples of the copper-catalyzed decompositions.

The tricyclic ketone **138** is formed on photolysis of **137**[569] and a copper-catalyzed version of this reaction also exists.[570]

Freeman and Kuper observed a similar reaction in **139** and also isolated **140**, the product of the rearrangement of **141**, itself formed by Wolff rearrangement[571,572] of **139** (or the related carbene).

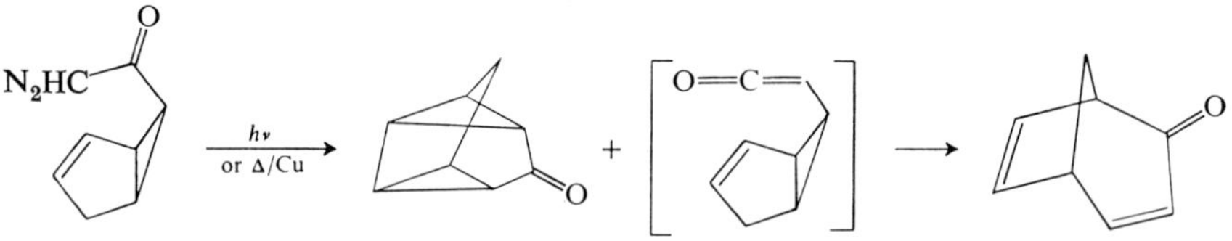

O N$_2$HC $h\nu$ or Δ/Cu O + O=C= O

139 **141** **140**

The mysterious product **143** reported[573] to be formed from the diazoketone **142** is incorrect; the correct structure[574] is formed from the Wolff rearranged compound as shown:

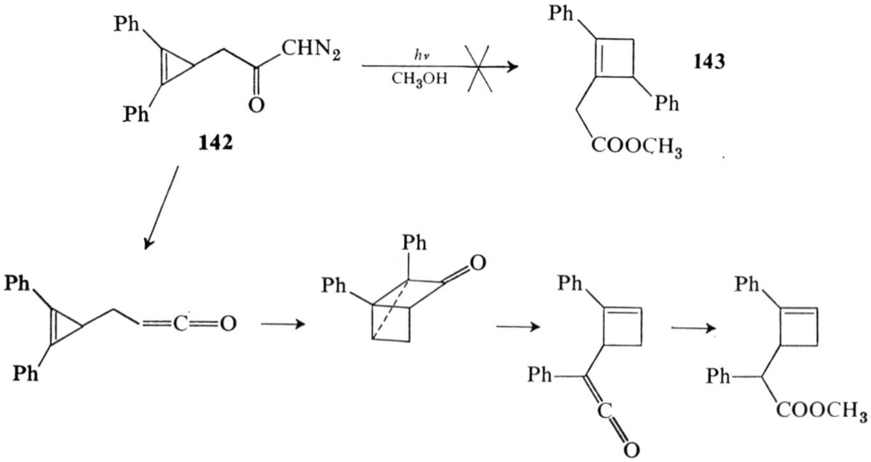

Ph Ph CHN$_2$ O $h\nu$ CH_3OH Ph Ph $COOCH_3$ **143**

142

Ph Ph =C=O Ph Ph O Ph Ph C O Ph Ph $COOCH_3$

Photolysis of diazoketone **144** in cyclohexene or 1,1-diphenylethylene gives adducts, presumably because the mobility of an amide in the Wolff rearrangement is low,[575] and imine **145** gives a carbene which adds to cyclooctene.[576]

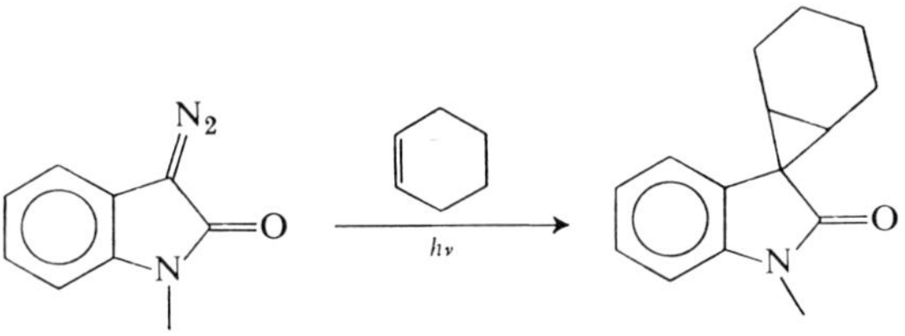

144

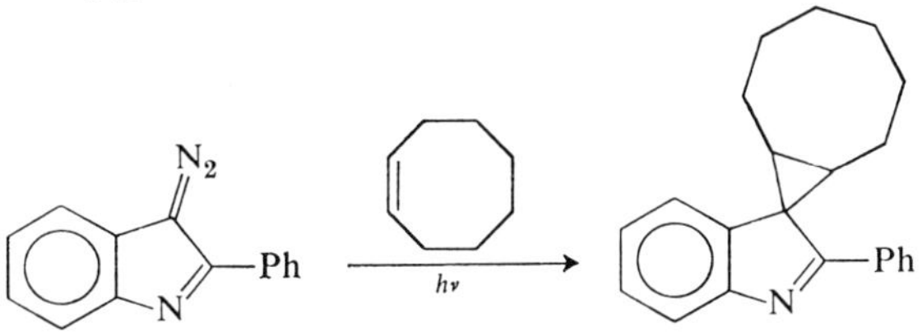

145

Additions are most common when a copper catalyst is employed, but occasionally uncatalyzed additions are successful. These are often suspect, however. For instance, addition of **145a** to olefins was reported on thermal

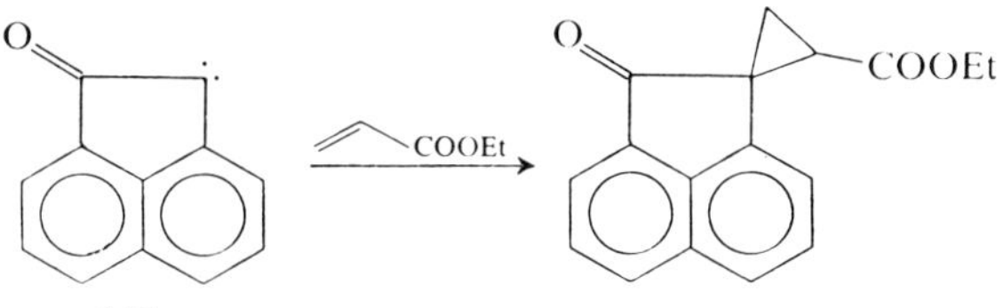

145a

decomposition of diazoacenapthenenone.[577] Olefins not bearing electron-withdrawing groups do not form cyclopropanes, however, and the intermediary of pyrazolines seems quite possible. Indeed, two pyrazolines were isolated. A few examples of catalyzed additions follow.

Doering et al.[578] used the reaction to make barbaralone (**146**), a precursor of bullvalene, and had previously[578,579] used more saturated systems to make a series of tricyclic ketones. Typically, photolysis in the absence of copper gave a complex mixture of products.

Both acyclic and cyclic cases were examined by Fawzi and Gutsche,[580] who came to the reasonable conclusion that "proximity of the olefin to the diazoalkyl group is an important factor."

N_2CH O Cu Δ O **146**

$R—CH{=}CH—(CH_2)_n—\overset{O}{\overset{||}{C}}—CHN_2 \xrightarrow{CuSO_4}$ R— $(CH_2)_n$ O

	n	Yield (%)		n	Yield (%)
R = H	2	59	R = Ph	2	59
	3	37		3	30
	4	3		4	very low

Mori and Matsui[581] achieved a 59% yield of epimers **147** and **147a** from the copper/copper sulfate-catalyzed decomposition of **148**, and the pre-benzvalene **149** was made from the copper-catalyzed decomposition of **149a** by Monahan.[582] Similar reactions were performed by House and Blankley and here we show an example in a cyclic system:[583]

CHN_2 $CuSO_4$ O O **148** **147, 147a**

Ph Ph O Ph CHN_2 **149a** Cu Δ Ph Ph O Ph **149**

CHN_2 O O Δ Cu O O

The reaction has been applied in an intermolecular sense using benzoyldiazomethane:[584]

$Ph—\overset{O}{\overset{\|}{C}}—CHN_2$ → ($CH_2=CH$–OAc, $Cu(AcAc)_2$) AcO–cyclopropyl–C(=O)–Ph (1) ; AcO–cyclopropyl–C(=O)–Ph (6)

$Ph—\overset{O}{\overset{\|}{C}}—CHN_2$ → (cyclohexene, $Cu(AcAc)_2$) bicyclo[4.1.0]heptyl–C(=O)–Ph

Additions of ketocarbenes to aromatic systems have generally failed, even when copper-catalyzed,[1] but it has recently been demonstrated that both intermolecular[585] and intramolecular[586,587] reactions are possible.

NO_2 ... CHN_2, O → (benzene, CuCl, Δ) NO_2 ... O

N_2 = HC, O → (Cu, Δ) [=O] → O

N_2 = HC, O → (Cu, Δ) O + O + O

Some insertion reactions are known, although they are rare. Wolff rearrangement is resisted by carboethoxytrifluoroacetylcarbene, and a 26%

$CF_3—\overset{O}{\overset{\|}{C}}—\underset{N_2}{\underset{\|}{C}}—\overset{O}{\overset{\|}{C}}—OEt$ → (cyclohexene, $h\nu$) O, COOEt, CF_3 + O, CF_3, COOEt

yield of insertion into cyclohexane has been realized,[588] as well as a mixture of cyclopropane and insertion product from cyclohexene.[589]

An intramolecular insertion involves the irradiation of **150**, which prefers insertion into the benzylic carbon-hydrogen bond to Wolff rearrangement.[590] Similarly, the copper sulfate-catalyzed decomposition of **151** yields **152**.[591]

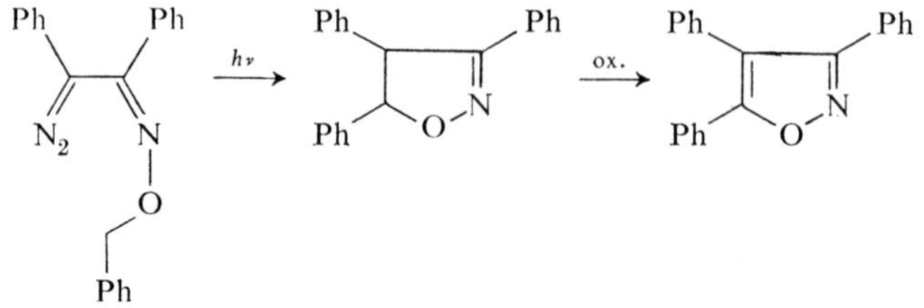

Ph Ph N_2 N O Ph **150** $\xrightarrow{h\nu}$ Ph Ph Ph N O $\xrightarrow{\text{ox.}}$ Ph Ph Ph N O

151 C=O CHN_2 $\xrightarrow[\text{CuSO}_4]{\Delta}$ =O **152**

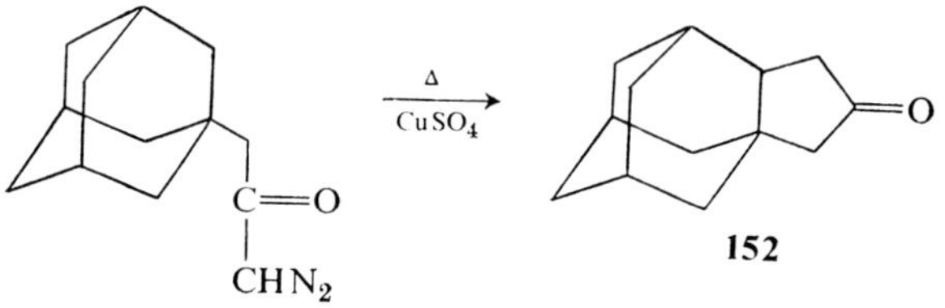

1,3-Dipolar additions are common and are often catalyzed by copper. We supplement the discussion in reference 592 with a few uncatalyzed examples. Although the mechanism is not discussed, a 1,3-dipolar addition is probably responsible for the formation of **153**:[593]

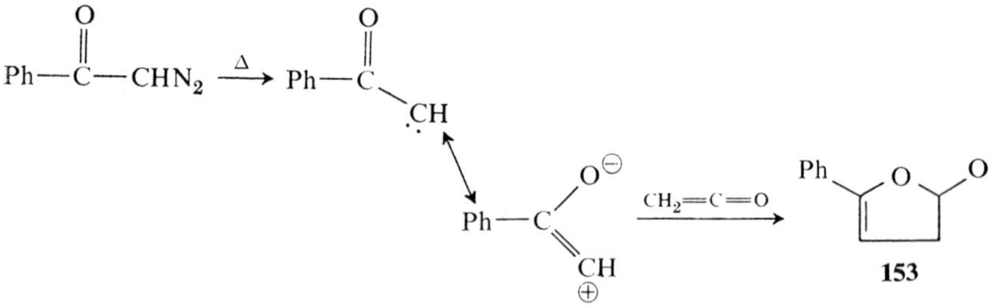

$$\text{Ph—C(=O)—CHN}_2 \xrightarrow{\Delta} \text{Ph—C(=O)—}\ddot{\text{C}}\text{H} \longleftrightarrow \text{Ph—C(O}^{\ominus}\text{)=CH}^{\oplus} \xrightarrow{\text{CH}_2=\text{C}=\text{O}}$$

Ph O O **153**

Addition of **154** and carboethoxycarbene to benzonitrile and phenylketene has been achieved by Huisgen and co-workers:[594,595]

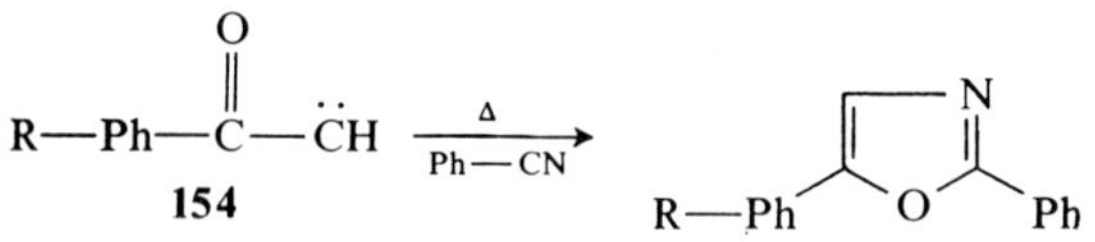

$$\text{R—Ph—C(=O)—}\ddot{\text{C}}\text{H} \xrightarrow[\text{Ph—CN}]{\Delta}$$

154

R—Ph N O Ph

More recently, Dworschak and Weygand have trapped carbomethoxytrifluoroacetylcarbene as a 1,3-dipole:[596,597]

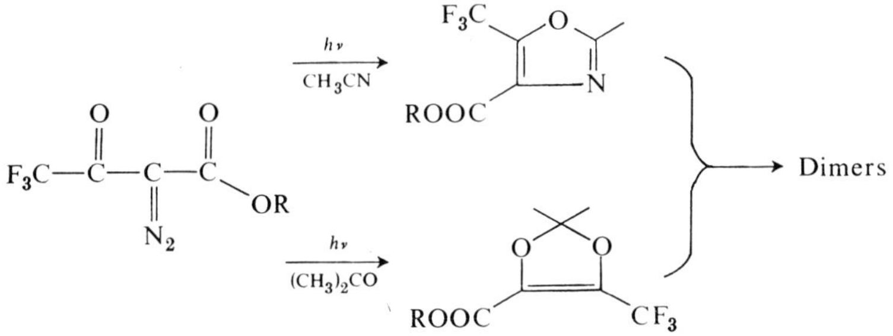

Carbene **155** also may be trapped by olefins in such fashion, and addition to the stilbenes is nearly stereospecific.[598] Interception of **155** can also be

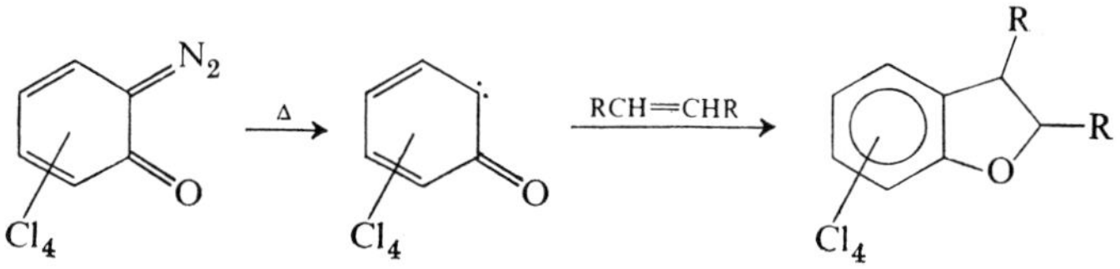

155

Olefin	Yield (%)
Ph—CH=CH$_2$	21
trans-PH—CH=CH—Ph	18
cis-Ph—CH=CH—Ph	8
trans-Ph—CH=CH—COOR	14
trans-ROOC—CH=CH—COOR	57
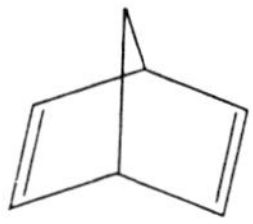 (norbornadiene)	9

achieved by acetylenes,[599] carbon disulfide,[600] phenylisothiocyanate,[600] and chlorobenzene.[599]

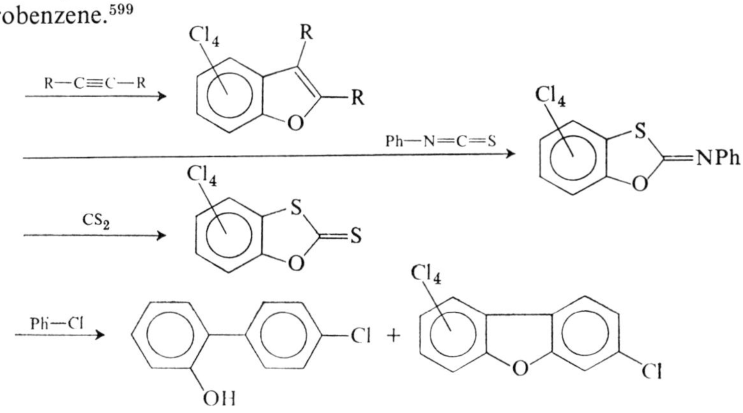

Whereas "ordinary" diazoketones yield carbenes that are trapped in tetrachloro-*o*-benzoquinone as shown, the carbene **156**, which cannot easily undergo Wolff rearrangement, is trapped in 1,4 fashion.[601]

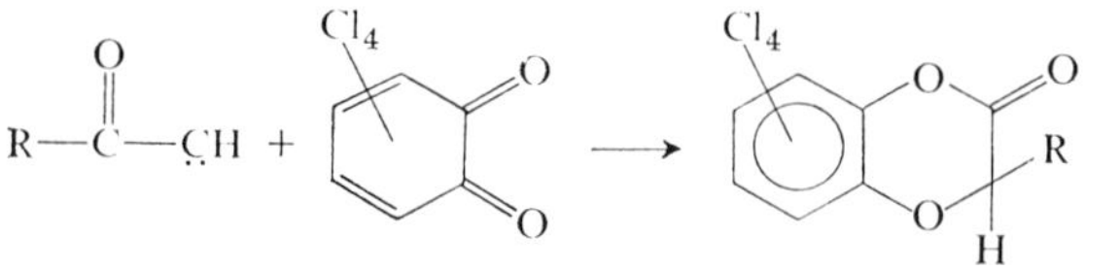

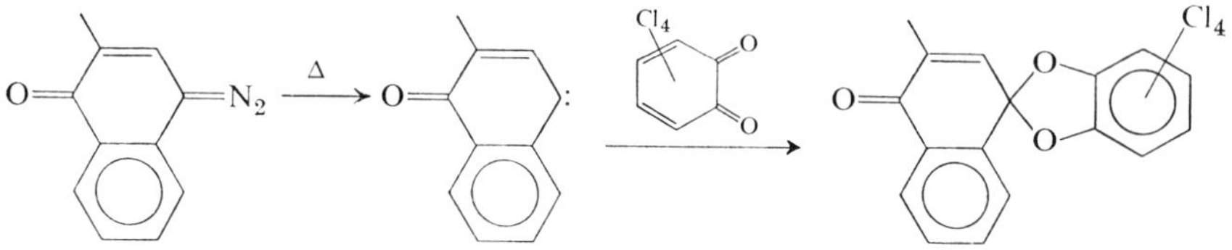

$$R{-}\overset{O}{\overset{\|}{C}}{-}\ddot{C}H + (Cl_4\text{-}o\text{-benzoquinone}) \longrightarrow (Cl_4\text{-benzodioxanone, R, H})$$

$$O{=}(\text{naphthalenone}){=}N_2 \xrightarrow{\Delta} O{=}(\text{naphthalenone}){:} \xrightarrow{Cl_4\text{-}o\text{-benzoquinone}} O{=}(\text{spiro-benzodioxole, } Cl_4)$$

156

No Wolff rearrangement is observed from **157**, which is trapped in aromatic solvents via an electrophilic attack.[602]

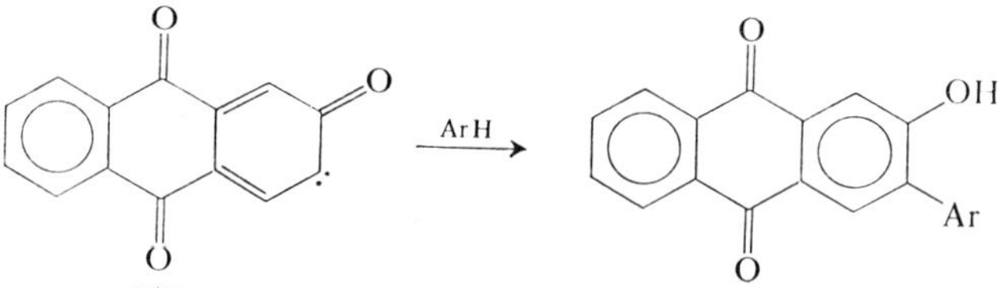

$$\textbf{157} \xrightarrow{ArH} (\text{anthraquinone with OH, Ar})$$

157

D. Ylid Formation

Although few stable ylids have been made from keto- or carboalkoxycarbenes, many authors have postulated such species as intermediates. Stable ylids are formed from carbenes and alkyl sulfides[603] and from isoquinoline and carboethoxycarbene.[604] As one might expect, the ylid is not formed from the

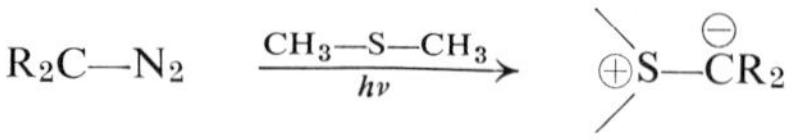

$$R_2C{-}N_2 \xrightarrow[h\nu]{CH_3{-}S{-}CH_3} \oplus S{-}\overset{\ominus}{C}R_2$$

$R = COOCH_3$
$COOCH_2CH_3$
$COCH_3$

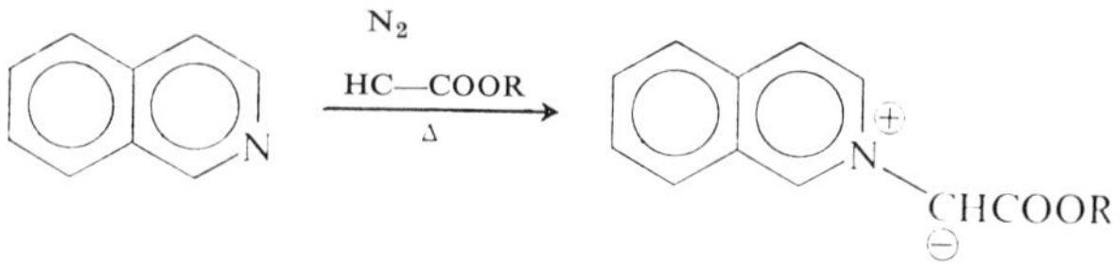

$$\text{isoquinoline} \xrightarrow[\Delta]{N_2HC{-}COOR} \text{isoquinolinium } N^{\oplus}{-}\overset{\ominus}{C}HCOOR$$

triplet state. Benzophenone-sensitized decomposition of diazomalonic ester yielded no ylids.[603]

Unstable nitrogen and especially oxygen ylids have long been thought to explain the products of carboalkoxycarbenes and ketones and ethers.[605] Decomposition of ethyldiazoacetate by either thermal or photochemical means in styrene oxide gave products well-rationalized by ring expansion and fragmentation reactions of the ylid **158**.[606,607]

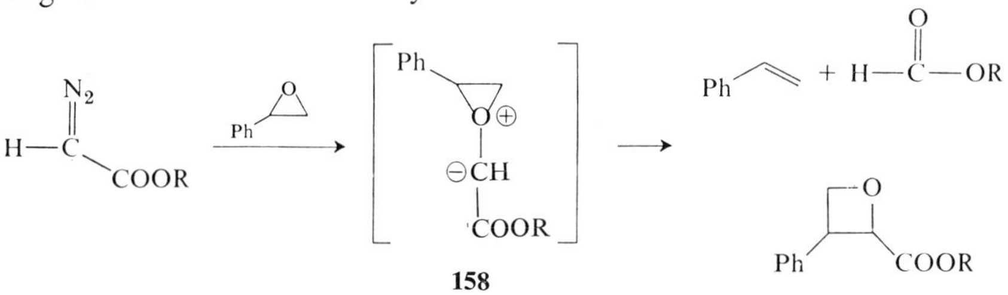

158

2-Phenyloxetane gave ring expansion but no deoxygenation.[607] Ring expansion has also been noticed in **159**.[608]

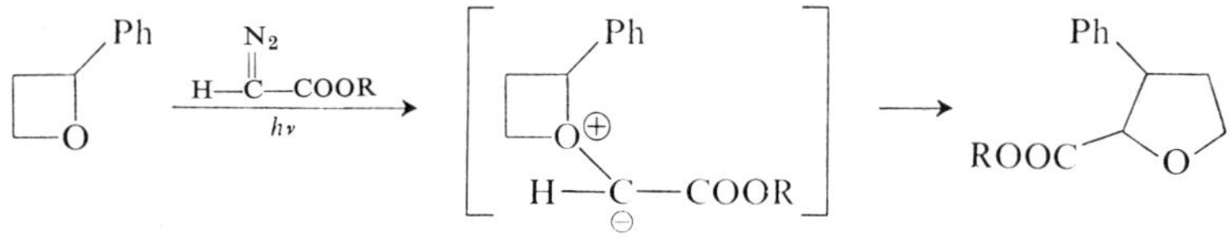

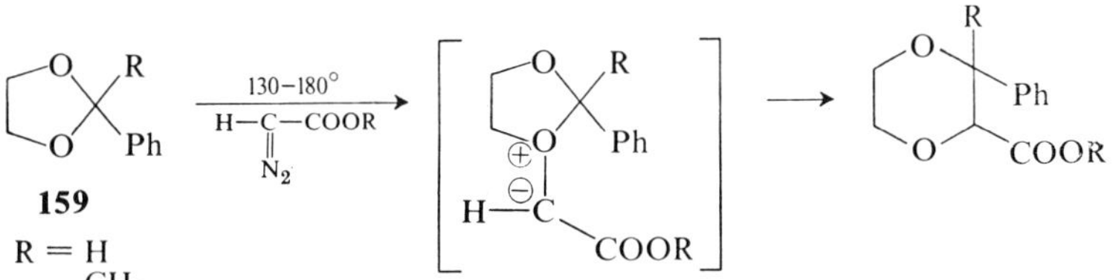

159

R = H
CH_3

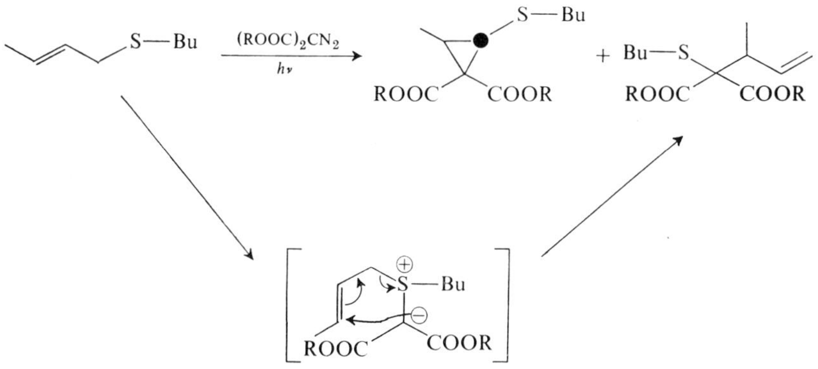

Ando followed up his isolation of stable sulfur ylids with work that strongly suggests the intervention of sulfur ylids in which one group bound to sulfur is unsaturated. Here the ylid cannot be isolated but rearranges *in situ*.[609] Very similar results were reported for carboethoxycarbene.[610] A different mechanism was suggested in the benzophenone-sensitized photolysis of diazomalonic ester.[609] It was postulated that rearrangement took place through the "Skell" intermediate **160**, as shown, but it seems clear that a complete mechanism must include an intersystem crossing step somewhere in the process.

$$\text{CH}_3\text{CH}{=}\text{CHCH}_2\text{SR} \xrightarrow[(\text{ROOC})_2\text{CN}_2]{h\nu,\ \text{Ph}_2\text{C}{=}\text{O}} \text{cyclopropane (RS, ROOC, COOR)} + \text{CH}_2{=}\text{CHCH(CH}_3)\text{C(SR)(COOR)}_2$$

$$\text{R—S—CH}_2\text{—}\dot{\text{C}}\text{H—CH—CH}_3 \text{ with } \dot{\text{C}}(\text{COOR})_2 \quad \mathbf{160}$$

Similar experiments indicate the presence of chloro ylids.[611] Again, rearranged product is reduced to traces by use of benzophenone as photosensitizing agent.

$$\text{CH}_3\text{CH}{=}\text{CHCH}_2\text{Cl} \xrightarrow[(\text{ROOC})_2\text{CN}_2]{h\nu} \text{cyclopropane (CH}_2\text{Cl, ROOC, COOR)} + \text{CH}_2{=}\text{CHCH(CH}_3)\text{CCl(COOR)}_2$$

(via ylid: $(\text{ROOC})_2\text{C}^{\ominus}\text{—Cl}^{\oplus}$)

Use of either methylene halides or cyclohexane as diluent decreased the ratio of "insertion" to addition in the reaction of biscarbomethoxycarbene with allyl chloride (Table 12).[612] This was interpreted as indicating collision-induced conversion of singlet to triplet. However, in the heavily diluted runs the yields are severely reduced, and it is not certain that intersystem crossing is being induced. Perhaps singlet is merely being scavenged by methylene halide, leaving a mix of carbene richer in the triplet. Regardless, these results do implicate the presence of the triplet.[612]

The formation of **161** could occur through an ylid process but, as noted, dipolar addition followed by rearrangement is also a possibility.[613]

$$\text{Ph(ROOC)C}{=}\text{N}_2 + (\text{CH}_3)_3\text{C—N}{\equiv}\text{C} \xrightarrow{\Delta} \text{Ph(ROOC)C}{=}\text{C}{=}\text{N—C(CH}_3)_3 \quad \mathbf{161}$$

TABLE 12

Reaction of Biscarbomethoxycarbene with Allyl Chloride

Solvent	"Insertion"/"addition"	Yield (%)
None	2.35	76
Cyclohexane, 10%	3.39	64
Cyclohexane, 50%	2.52	26
Cyclohexane, 90%	1.95	5
Methylene bromide, 10%	2.19	36
Methylene bromide, 50%	1.60	14
Methylene bromide, 90%	0.06	4

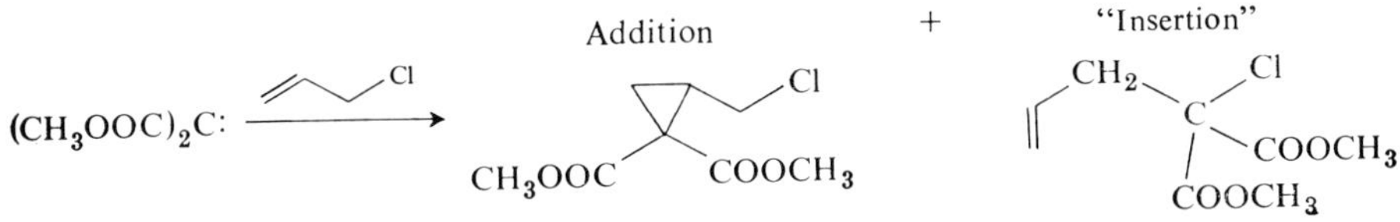

E. Wolff Rearrangement

The Wolff rearrangement, as the crucial step in the Arndt-Eistert synthesis of homologated acids or esters, has found wide use.[614,615] An example of the use of the photochemical Wolff rearrangement in chain lengthening follows.[616] When applied in cyclic systems, the Wolff rearrangement becomes a ring contraction and the photochemical reaction in particular has been most useful in synthesis of strained small ring compounds,[617–619] although the thermal reaction has been used as well.[620]

Reference

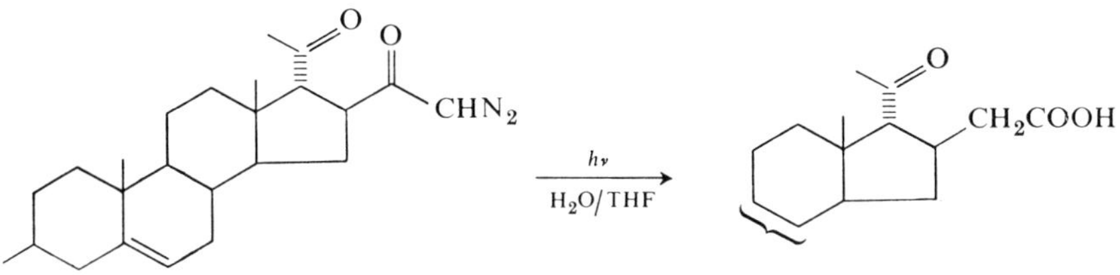

616

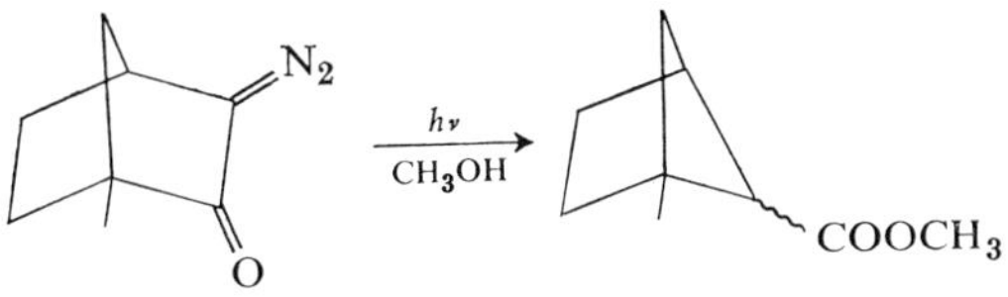

617

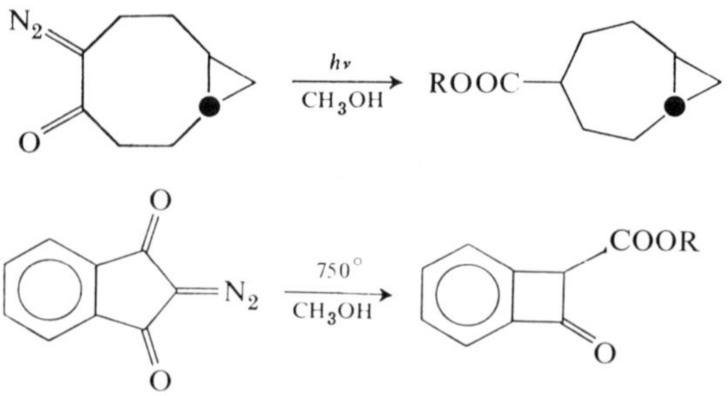

$\xrightarrow[CH_3OH]{h\nu}$ ROOC— 618 619

$\xrightarrow[CH_3OH]{750°}$ COOR 620

The crucial question related to the mechanism of the uncatalyzed Wolff rearrangement involves the presence or absence of the nitrogen. Is the reactive intermediate a carbene or the diazoketone itself? The related question can be asked of the Curtius rearrangement. At least in the decomposition of pivaloylazide, the nitrene is not involved in the Curtius rearrangement.[621] Some are of the opinion that the Wolff rearrangement follows a similar path, with migration accompanying loss of nitrogen. A study using NMR spectroscopy revealed the equilibrium between the *cis*- and *trans*-diazoketones **162** and **163**.[622] It was suggested that the highly preferred cis form **162** was responsible for the Wolff rearrangement, and indeed it is surely only **162** which enjoys an appropriate arrangement of migrating and leaving group. In support of this idea it was noted that the di-*t*-butylcompound **164**,

163 R—C(N_2)—C(=O)—R

⇌

162 R—C(N_2)—C(=O)—R → $R_2C{=}C{=}O$

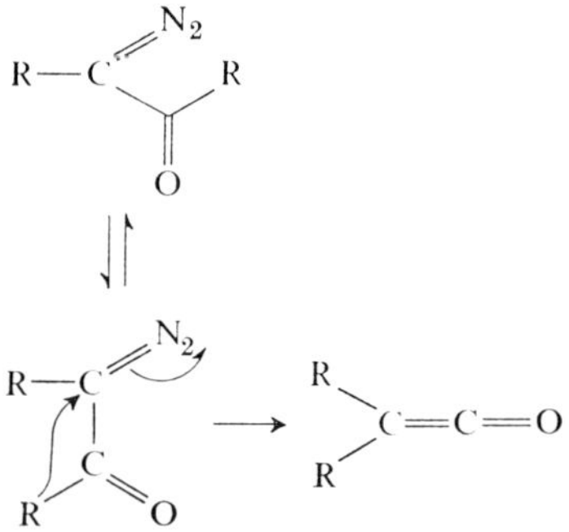

in which the cis form should be less favorable, does not undergo the Wolff rearrangement,[623] and it is supposed that a carbene would not show such a geometric preference.[622] The reluctance of **164** to undergo the Wolff

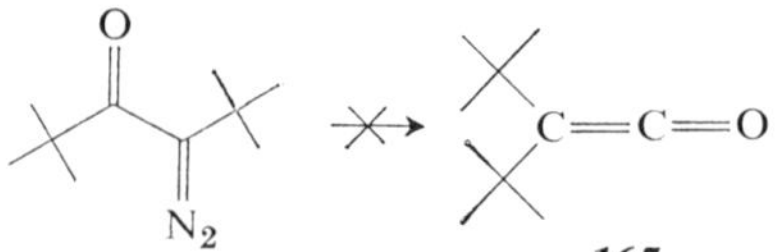

164 ↛ **165** C=C=O

rearrangement is a strong argument, but it might just be that **165** resists formation in any event and control of the course of the reaction lies in the stability of the products.

Kinson and Trost[624] postulated concerted rearrangement under electron-impact conditions in **166** and **167**. The frequent coincidence of characteristics between photochemical and electron-impact conditions makes this especially relevant to the photochemical Wolff rearrangement.

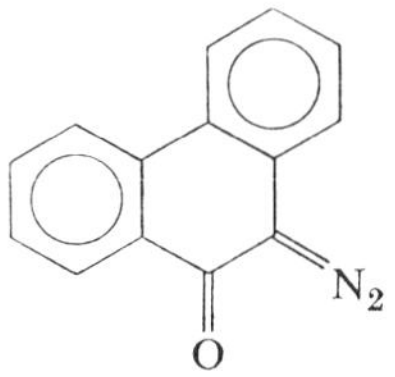

N_2 O **166**

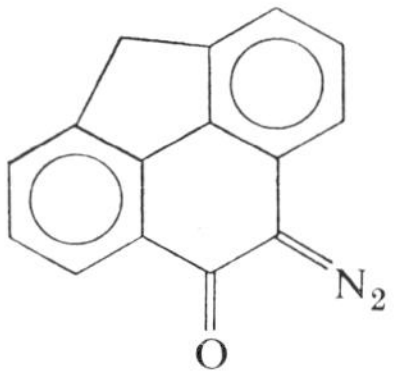

N_2 O **167**

There is evidence that under certain conditions some diazo compounds are protonated and that therefore Wolff rearrangement takes place in the diazonium ion. Thus, although Jugelt and Schmidt[625] found no strong effect of added aniline on the rate of nitrogen evolution of **168**, Bartz and Regitz[626] did find such effects in some cases. Nitrogen evolution in phenylbenzoyldiazomethane (**169**), diazocyclohexanone (**170**), and diazocyclododecanone (**171**)

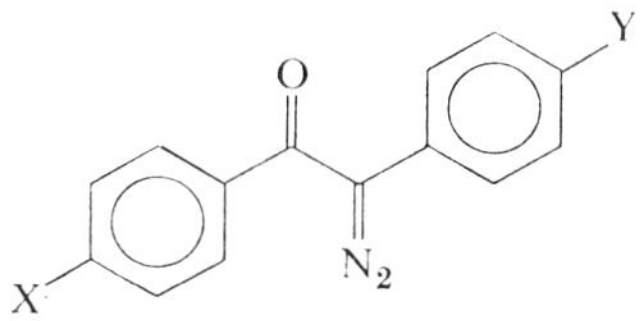

X Y O N_2 $\rho = 0.75$ $\rho = -1.49$ **168**

Ph N_2 Ph O **169**

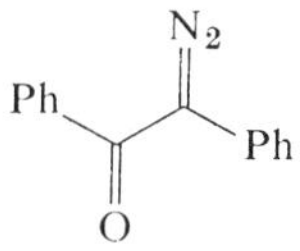

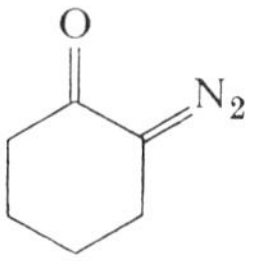

O N_2 **170**

O N_2 12 **171**

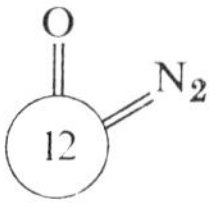

was sensitive to the presence of alcohols and amines. With these compounds a linear increase in N_2 evolution with the nucleophile concentration was observed, indicating that the nucleophilic additives were acting as proton donors toward the diazo compound.

It had previously been proposed by Padwa and Layton[627] that benzoyldiazomethane in alcohol solvent was hydrogen bonded and decomposed to a hydrogen bonded singlet carbene, which then rearranged.

Wilds et al.[628] deemed a protonated diazoketone responsible for the abnormal properties found in the thermal but not photochemical Wolff rearrangement of **172**.

Others have not agreed and have postulated a carbene. The rate of nitrogen evolution from **173** is not dependent on the concentration of the Schiff-base acceptor.[629] This does not seem to preclude a concerted rearrangement to diphenylketene. The factor of 8000 is going from **174** to **175** was thought to argue against a concerted loss of nitrogen.[630]

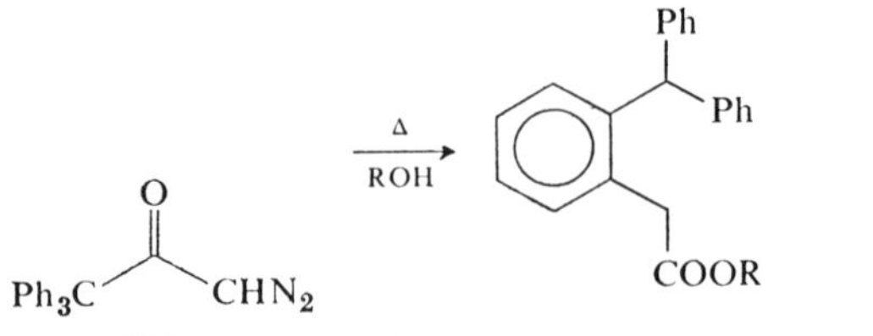

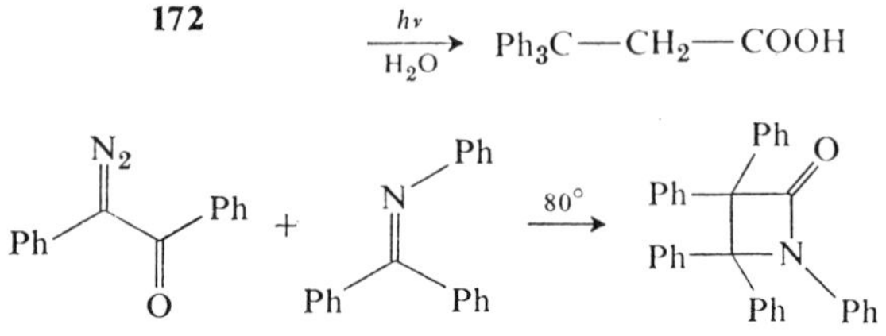

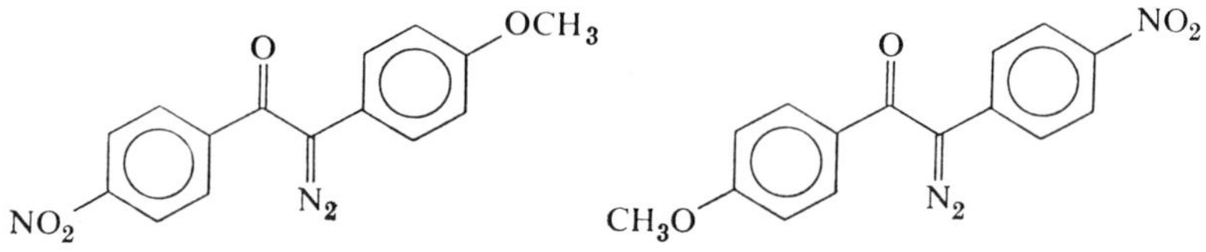

174 $K_{rel} = 12000$

175 $K_{rel} = 1.4$

There seems little on either side which absolutely requires one mechanism or another. On one thing we can agree, however, and that is that singlets must be involved. Work by Padwa and Layton[627] and Cowan et al.[631] on benzoyl-diazomethane indicated that the triplet species would not undergo Wolff rearrangement. Jones and Ando[632] showed that aliphatic diazoketones such as diazocyclohexanone underwent Wolff rearrangement on unsensitized irradiation. The corresponding carbenes could not be trapped with olefins.

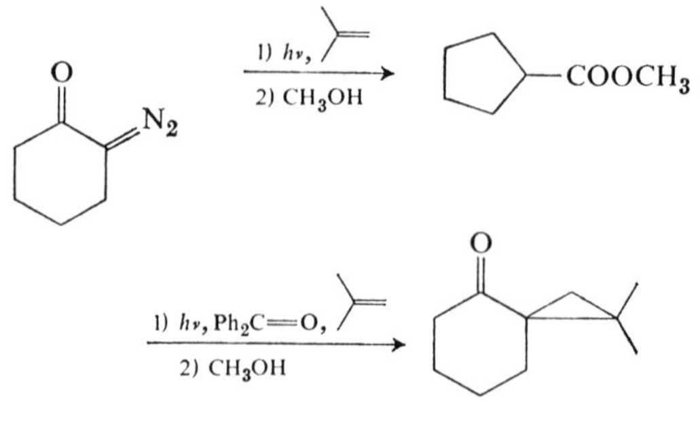

Photosensitized decomposition gave an intermediate that did not Wolff rearrange and was trapped by olefins. It is not certain, however, that triplet carbenes were produced here. Wolff rearrangement might be resisted by a triplet diazo compound, which could add to the olefins. The authors feel this is unlikely, but it is possible. That triplet carbene will not undergo the rearrangement was shown by Trozzolo and Fahrenholtz,[633] who allowed phenylbenzoylcarbene generated in a glass at 77° K to warm. Less than 3% of diphenylketene was formed. Irradiation of **173** at room temperature gave 85–90% of the ketene.

hν, room temp → $Ph_2C{=}C{=}O$

173

hν, 77°K

The importance of oxirenes in the Wolff rearrangement has been made clear. For instance, in contrast to an earlier report,[634] azibenzil has now been shown by a carbon-13 labeling study to rearrange to diphenylketene by two paths, one of which (54%) involves diphenyloxirene.[635] Similarly, other diazoketones and esters have been shown to rearrange via oxirenes.[636,637] A test for oxirene participation which uses as a diagnostic tool not a labeling experiment but formation of two α,β-unsaturated ketones has recently appeared.[638]

hν

Until rather recently Wolff rearrangement of carboalkoxycarbenes was a neglected reaction. In 1965 it was noticed in enzymatic systems that Wolff rearrangement occurred[639] and in 1968 detailed studies began to appear on simpler systems.

Deuteriocarboethoxycarbene is trapped in two ways by CD_3OD,[640] one of them involving Wolff rearrangement. Product of Wolff rearrangement was

$$D\ddot{C}COOR \longrightarrow ROOC{-}CD_2{-}O{-}CD_3 + RO{-}CD_2{-}COOCD_3$$

intercepted also by isopropyl alcohol, and **176** was isolated in 29% yield.[641]

$$ROOC{-}C(=N_2)H \xrightarrow[(CH_3)_2CHOH]{h\nu} RO{-}CH_2{-}C(=O){-}O{-}CH(CH_3)_2 + (CH_3)_2CH{-}O{-}CH_2{-}C(=O){-}O{-}CH(CH_3)_2$$

29% **176** ; 12% **177**

The ketenes formed by Wolff rearrangement can also be trapped in cycloaddition reactions.[642] In the gas phase, apparently, further fragmentation to ethoxycarbene occurs, and the end-products are derived from it:

$$ROOC{-}C(=N_2){-}H \xrightarrow{h\nu} RO(H)C{=}C{=}O \xrightarrow{\text{cis-2-butene}} \text{cyclobutanone (OR)}$$

$$CH_3CH_2{-}O{-}C(=O){-}CHN_2 \xrightarrow[\text{gas-phase}]{h\nu} [CH_3CH_2{-}O(H)C{=}C{=}O] \longrightarrow CH_3CH_2O{-}\ddot{C}H$$

$$CH_3CH_2O{-}\ddot{C}H \longrightarrow CH_3CH_2{-}CH \quad ; \quad CH_3CH_2O{-}\ddot{C}H \longrightarrow [\text{methyloxetane}]^{**}$$

$$CH_3CHO + CH_3CH_2CHO + CH_3COCH_3 + \text{methyloxetane} \longleftarrow [\text{methyloxetane}]^{**}$$

In benzene solution further reactions occur, the end-product being **178**:[643]

$$RO{-}C(=O){-}CHN_2 \xrightarrow[\text{solution}]{h\nu} RO{-}C(=O){-}\ddot{C}H \longrightarrow [RO(H)C{=}C{=}O] \longrightarrow [\text{RO, COOR cyclopropanone}] \longrightarrow \mathbf{178}$$

178: 1,4-dioxane ring bearing RO, =CHCOOR, ROOCCH=, OR substituents

The odd product **177** was accounted for in 1969 by invoking the reaction of isopropyl alcohol with the ion pair **179**:[644]

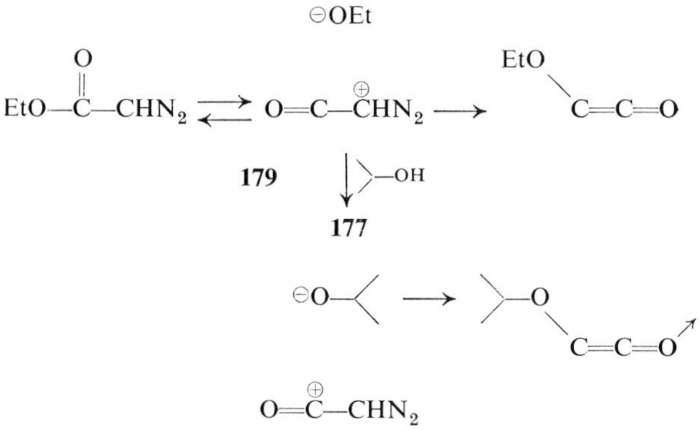

$\ominus$OEt

EtO—C(=O)—CHN$_2$ ⇄ O=C—$\overset{\oplus}{C}$HN$_2$ ⟶ EtO—C=C=O

179 ↓ (CH$_3$)$_2$CH—OH

177

$\ominus$O—CH(CH$_3$)$_2$ ⟶ (CH$_3$)$_2$CH—O—C=C=O

O=$\overset{\oplus}{C}$—CHN$_2$

Labeling experiments demonstrate that there are two sources of the ethoxyketene formed from the gas-phase decomposition of ethyldiazoacetate:[636,637]

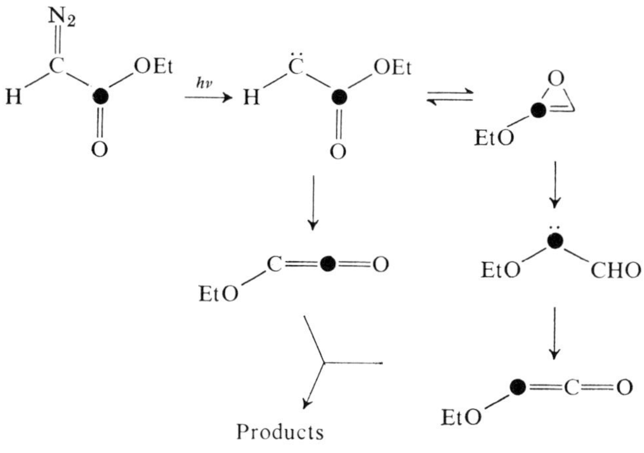

H—C(=N$_2$)—C(=O)—OEt $\xrightarrow{h\nu}$ H—$\ddot{C}$—C(=O)—OEt ⇌ EtO-oxirene

↓ EtO—C=C=O ; ↓ EtO—$\ddot{C}$—CHO

↓ Products ; ↓ EtO—C=C=O

The similarity of the properties of the intermediates formed from **180** and di-*t*-butylacetylene and *m*-chloroperbenzoic acid has led to the positing of an oxirene formed from each.[645] The lack of similarity of products in other systems was attributed to conformational effects.

The Wolff rearrangement of methyldiazomalonate has also recently been observed. In contrast to the preceding examples, this is a thermal reaction. The following sequence adequately explains the products:[646]

Carbomethoxymethoxycarbene has been generated from the appropriate

COOOH

Cl

Δ

N_2

(7)

(6)

(93)

(94)

180

CH_3OOC, CH_3OOC $C{=}N_2$ $\xrightarrow[\text{gas phase}]{\Delta}$ [CH_3OOC, CH_3OOC C:] $\longrightarrow$ [CH_3OOC, CH_3O $C{=}C{=}O$]

[CH_3OOC (β-lactone, O, O)] $\longrightarrow$ CH_3OOC (vinyl)

CH_3OOC, CH_3 C=O

[CH_3OOC, CH_3O C:]

[CH_3O, CH_3O $C{=}C{=}O$]

[CH_3O (β-lactone, O, O)] $\longrightarrow$ CH_3O (vinyl)

$CH_3{-}C({=}O){-}OCH_3$ $\longleftarrow$ [CH_3O, CH_3O C:]

tosylhydrazone salt,[647] and does not undergo Wolff rearrangement in solution at 140°. Only dimers are isolated.

CH_3OOC, CH_3O $C{=}N{-}\overset{\ominus}{N}{-}Ts$ $\xrightarrow{\Delta}$ CH_3OOC, CH_3O C: $\longrightarrow$ Dimers

VI. MISCELLANY

A. Halocarbenes

The vast bulk of work in this area is not concerned with carbenes generated from diazo compounds but rather with what have become known as carbenoids. However, in 1965 Closs and Coyle[648] generated chloro- and bromo-diazomethane (**181**, **182**) by the action of *t*-butylhypochlorite or *t*-butylhypobromite on diazomethane. Pyrolysis of **181** gave an intermediate that was less selective than its carbenoid counterpart, as Table 13 demonstrates.

TABLE 13

Relative Rates of Addition

Olefin	CH_2Cl_2[649] + RLi	$ClCHN_2$ (**181**)	$BrCHN_2$ (**182**)
2,3-Dimethyl-2-butene	2.81	1.20	1.18–1.21
2-Methyl-2-butene	1.78	1.18	—
trans-2-Butene	0.45	1.09	1.10
cis-2-Butene	0.91	0.99	1.02
Isobutene	1.00	1.00	1.00
1-Butene	0.23	0.74–0.77	0.75

Addition to the butenes was stereospecific, as is carbenoid addition, but the syn/anti ratios in various olefins were markedly different (Table 14). Unlike

TABLE 14

Syn/Anti Ratios

Olefin	CH_2Cl_2[649] + RLi	$ClCHN_2$	$BrCHN_2$
2-Methyl-2-butene	1.6	1.0	—
Cyclohexene	3.2	1.0	1.0
cis-2-Butene	5.5	1.0	1.0
1-Butene	3.4	1.0	1.0

most carbenoids, photochemically or thermally generated monohalocarbenes undergo the insertion reaction easily.

$ClCHN_2$, Δ or $h\nu$ → CH_2Cl + CH_2Cl + CH_2Cl

Yield = 9–11%, branched/unbranched = 20

All the foregoing data are consistent with the picture of a reactive, free halocarbene, very probably in a ground singlet state.

Mitsch[650] has produced difluorocarbene by irradiation or pyrolysis of difluorodiazirine. 1,1-Difluorocyclopropanes are formed and addition is stereospecific. Again a ground singlet state is indicated.

B. Sulfonylcarbenes

This subject has been reviewed recently,[18] thus a summary suffices here. The diazo compounds are synthesized by the action of weak bases on the appropriate nitroso compounds[651] or by use of the diazo transfer reaction.[652,653]

$$RSO_2\text{—}CH_2\text{—}N(N{=}O)\text{—}C({=}O)\text{—}OEt \xrightarrow[Et_2O]{Al_2O_3} RSO_2\text{—}CHN_2$$

Addition reactions are common and upon direct irradiation addition is cis.[654] With cyclohexene (Ar = Ph) an exo/endo ratio of 2.1 is found.[655]

CH_3O—C_6H_4—SO_2—CHN_2 $\xrightarrow{h\nu}$ (trans-2-butene) → trans-2,3-dimethylcyclopropyl SO_2—C_6H_4—OCH_3

$\xrightarrow{h\nu}$ (cis-2-butene) → cis-2,3-dimethylcyclopropyl SO_2—C_6H_4—OCH_3 (36%) + SO_2—C_6H_4—OCH_3 isomer (47%)

Addition to acetylenes is also very easy and a 36% yield of **183** can be achieved,[18,656] but benzene gives only the apparent insertion product **184**.[655]

$$CH_3O\text{—}C_6H_4\text{—}SO_2CHN_2 \xrightarrow[CH_3\text{—}C\equiv C\text{—}CH_3]{h\nu} CH_3O\text{—}C_6H_4\text{—}SO_2\text{—}(2,3\text{-dimethylcyclopropenyl})$$

183

$$C_6H_5\text{—}SO_2CHN_2 \xrightarrow[h\nu]{C_6H_6} C_6H_5\text{—}SO_2\text{—}CH_2\text{—}C_6H_5$$

184

It seems possible that **184** is formed by rearrangement of an initially produced norcaradiene, although this is by no means certain.

Insertion into the carbon-hydrogen bond of cyclohexane has been observed for phenylsulfonylcarbene,[655] and formal insertion into the oxygen-hydrogen bond of methanol is also known.[657] An ylid mechanism has been proposed[18] for the latter reaction and it seems certain that ylids are involved in the reaction of sulfonylcarbenes with vinyl ethers.[656] Stable sulfur ylids were isolated by Diekmann[658] from a bissulfonylcarbene and butyl- and dibutyl-sulfide.

$$C_6H_5—SO_2—\ddot{C}H + CH_2{=}CH—O—CH_2CH_3 \longrightarrow [\text{ylide}] \longrightarrow Ph—SO_2—CH_2—O—CH{=}CH_2 + CH_2{=}CH_2$$

A rearrangement analogous to the Wolff rearrangement also occurs,[657] although a competition experiment showed the Wolff rearrangement to be the easier process.[659]

$$p\text{-}CH_3C_6H_4—SO_2—C(=N_2)—C(=O)—CH_3 \xrightarrow{h\nu} p\text{-}CH_3C_6H_4—SO_2—\ddot{C}—C(=O)—CH_3 \longrightarrow p\text{-}CH_3C_6H_4—SO_2—C(CH_3){=}C{=}O$$

C. Mercapto- and Alkoxycarbenes

Relatively little is known of mercaptocarbenes; simple olefins such as cyclohexene are not attacked[660,661] but more nucleophilic species such as α-morpholinostyrene, ketene diethyl acetal, and propyl propenyl ether are.[661] Cyclic species **185** fragments to carbon disulfide and ethylene.[660]

$$(CH_3S)_2C{=}N—\overset{\ominus}{N}—Ts \xrightarrow[\Delta]{\alpha\text{-morpholinostyrene}} \text{1-morpholino-1-phenyl-2,2-bis(methylthio)cyclopropane}$$

$$\mathbf{185}\ \text{(1,3-dithiolan-2-ylidene)} \longrightarrow CH_2{=}CH_2 + CS_2$$

The related oxycarbenes have also evaded intensive study. The tosylhydrazone salt of methyl formate is not stable at room temperature and decomposes spontaneously to a large number of products. Neither *cis*- nor *trans*-dimethoxyethylene was formed,[662] and there seems no compelling evidence for the formation of methoxycarbene.

Ethyl acetate tosylhydrazone also gives a variety of products, some of which may reasonably be attributed to ethoxymethylcarbene.[662] Efforts with methylbenzoate tosylhydrazone salt led mainly to azine,[662] an event confirmed by Crawford and Raap.[663] These authors were able to achieve additional

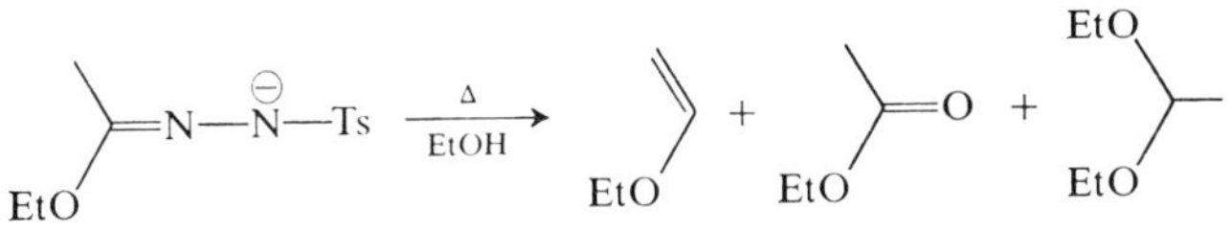

reactions of the presumed phenylmethoxycarbene, however.[663] Diethylfumarate gave cyclopropane in 62% yield while 1-decene gave only 22%. This is in line with a nucleophilic character for the carbene, but the large yield in the fumarate case makes one suspicious of pyrazoline formation, or of some other

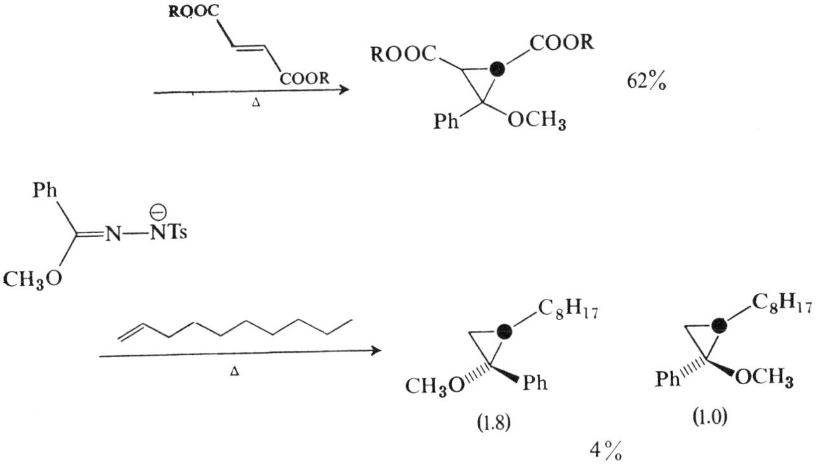

reaction of a diazo compound. An intramolecular trapping was also carried out as shown:[663]

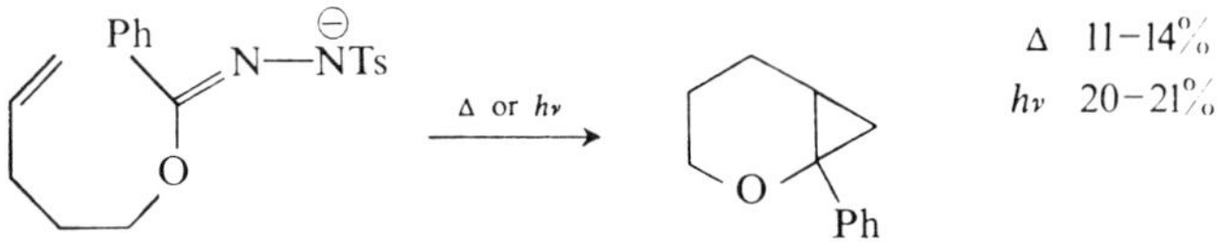

Diethoxycarbene could not be trapped with a variety of olefins and apparently decomposes to carbon monoxide and ethoxy and ethyl radicals.[664]

Tetraethoxyethylene was not isolated, although thermal decomposition of concentrated solutions of the norbornadiene ketal **186** does give carbene dimer.[665] Under dilute conditions trimethylorthoformate appeared. When the pyrolysis was run in the gas phase, methyl acetate could be isolated.

$$(EtO)_2C{=}N{-}\overset{\ominus}{N}{-}Ts \xrightarrow{\Delta} CO + EtO\cdot + Et\cdot$$

$$\mathbf{186} \xrightarrow[\text{dilute}]{\text{concentrated}\ \Delta} (CH_3O)_2C{=}C(OCH_3)_2,\ (CH_3O)_3CH$$

$$\xrightarrow[\text{phase}]{\text{gas}} CH_3O{-}\overset{O}{\overset{\|}{C}}{-}CH_3$$

186

Agosta and Foster have decomposed tosylhydrazone salt **186a** and isolated, in addition to the product of hydrogen shift in **180b**, 2,2-dimethylcyclobutanone.[666] Interestingly, photolysis of the cyclobutanone has been suggested to give oxocarbene **186b**.[667]

$$\mathbf{186a} \xrightarrow{310^\circ} \mathbf{186b} \longrightarrow (30) + (70)$$

186a **186b** (30) (70)

D. Cyanocarbenes

We have previously alluded to the ESR studies of cyanocarbenes which showed them to have triplet ground states. Their chemistry, although not voluminous, is very lively.

Addition of dicyanocarbene to olefins is largely but not completely stereospecific.[668] Addition of cyclohexane as diluent evidently induces intersystem crossing, as the stereochemistry of addition becomes independent of the starting butene at 100/1 cyclohexane/olefin (Table 15). Addition to acetylenes gives cyclopropenes, and the carbon-hydrogen insertion reaction is common ($3^\circ/2^\circ/1^\circ = 12/4.6/1$).[668]

It is the addition to benzene which first brought wide attention to dicyanocarbene.[669] Remarkably, the adduct is a norcaradiene, not a cycloheptatriene.

$$C_6H_6 + (CN)_2C{=}N_2 \xrightarrow{80^\circ} \text{7,7-dicyanonorcaradiene}$$

It has previously been mentioned that bistrifluoromethylcarbene adds to benzene to give a cycloheptatriene, and so it is not surprising that cyanotrifluoromethylcarbene adds to benzene to give a species in which the position

TABLE 15

"Stereochemistry of Dicyanocarbene Addition"

Addition/insertion	Olefin	Dilution	*cis*-Cyclopropane (%)	*trans*-Cyclopropane (%)
92/8	*cis*-2-butene	neat	92	8
94/6	*cis*-2-butene	1:10	60	40
>98/2	*cis*-2-butene	1:100	30	70
88/12	*trans*-2-butene	neat	6	94
94/6	*trans*-2-butene	1:10	22	28
>98/2	*trans*-2-butene	1:100	30	70

of the norcaradiene-cycloheptatriene equilibrium is intermediate between the two forms.[670]

NC, F_3C $=N_2$ —(benzene, Δ)→ cycloheptatriene with CF_3, CN ⇌ norcaradiene with CF_3, CN

A further remarkable reaction of dicyanocarbene is 1,4-addition to cyclooctatetraene. Apparently only the triplet state is capable of this process, since the 1,4-adduct increases as inert solvent is added to the reaction medium.[671]

cyclooctatetraene —:C(CN)$_2$→ 1,2-adduct (CN, CN) + 1,4-adduct (NC, CN)

Aminocyanocarbene was the presumed product of the thermal or photochemical decomposition of tosylhydrazone salt **187**,[672] but no addition

NC, H_2N $=N—\overset{\ominus}{N}—Ts$ —(Δ or hν)→ NC, CN, H_2N, NH_2 (olefin)

187

products with ordinary olefins could be found. In view of the probable nucleophilic nature of the carbene, one might have hoped for success were the trapping agent something like a fumarate or maleate, but the experiment has not been reported. An absorption spectrum is observed at 77° K which disappears on warming and is attributed to the carbene. The assignment seems tenuous, as does the key role claimed for the carbene in the hydrogen cyanide polymerization.

It is claimed that cyanovinylcarbenes **188** and **189** formed via the diazo compounds **190** and **191** add to olefins in addition to forming cyclopropenes.[673] This seems correct, although the intermediacy of a pyrazoline is not disproved.

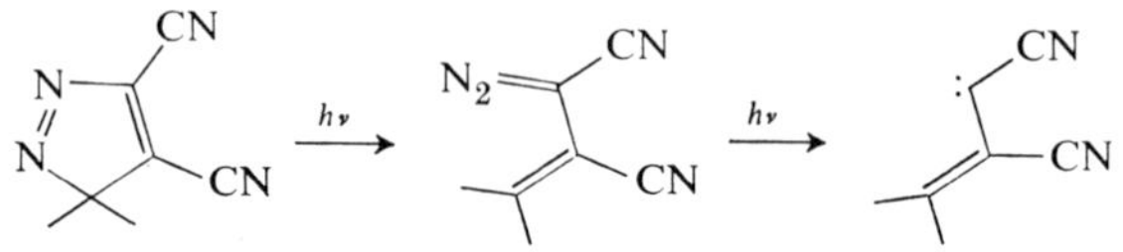

191 **189**

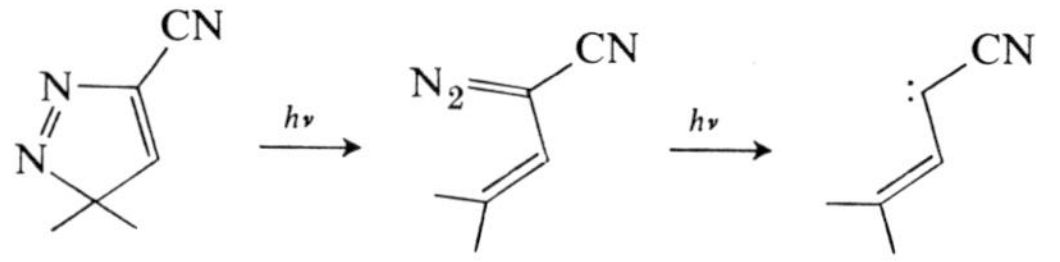

190 **188**

Furan is attacked by **188** to give an adduct and benzene gives the norcaradiene **192** on reaction with **189**.

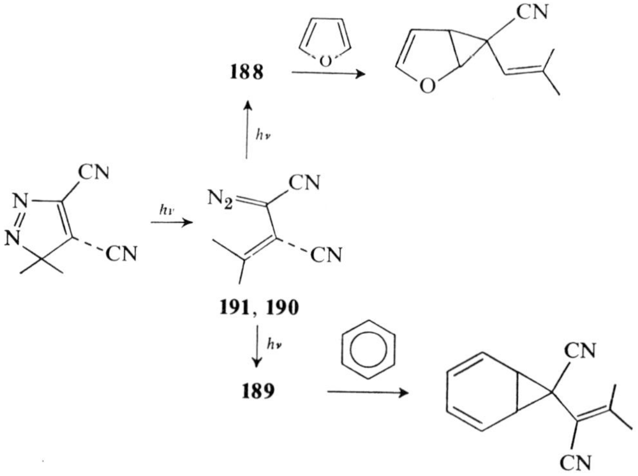

188

191, 190

189

192

E. Other Nitrogen-containing Carbenes

Other aminocarbenes have not been made from diazo compounds, but both mono- and dinitrodiazomethane are known, as are a variety of substituted nitrodiazomethanes.[674–676] Reactions of nitrocarbenes have not been reported.

F. Phosphorus-containing Carbenes

Several diazo compounds of this variety are known,[677–682] and copper-catalyzed additions to olefins and 1,3-dipolar additions have been reported.[678–680]

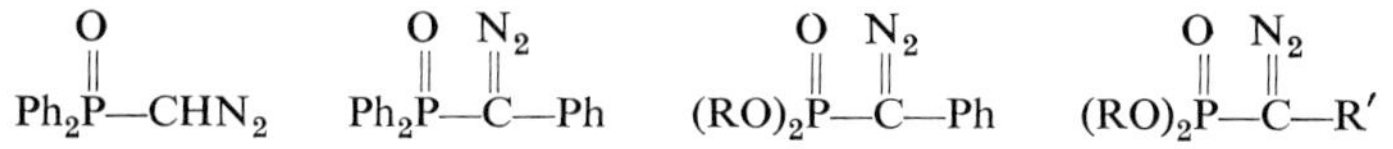

$$\mathrm{Ph_2\overset{\overset{O}{\|}}{P}{-}CHN_2} \qquad \mathrm{Ph_2\overset{\overset{O}{\|}}{P}{-}\overset{\overset{N_2}{\|}}{C}{-}Ph} \qquad \mathrm{(RO)_2\overset{\overset{O}{\|}}{P}{-}\overset{\overset{N_2}{\|}}{C}{-}Ph} \qquad \mathrm{(RO)_2\overset{\overset{O}{\|}}{P}{-}\overset{\overset{N_2}{\|}}{C}{-}R'}$$

Reference: 678 677 679 682

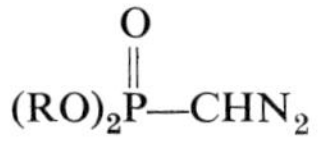

$$\mathrm{(RO)_2\overset{\overset{O}{\|}}{P}{-}CHN_2}$$

680
681

$R' = —Ph, —COCH_3$
$—COPh, —COOR$
$—P(OR)_2$ (with P=O)

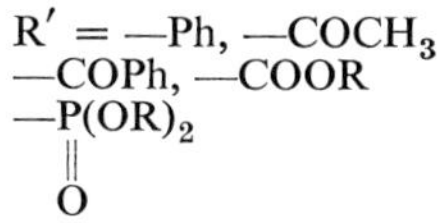

More recently photolysis in olefins and benzene has led to cyclopropanes, insertion products, and the norcaradiene previously mentioned.[491,491a]

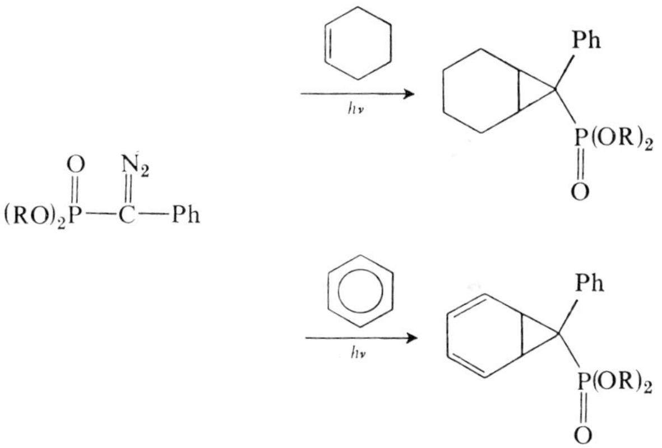

$$\mathrm{(RO)_2\overset{\overset{O}{\|}}{P}{-}\overset{\overset{N_2}{\|}}{C}{-}Ph}$$

→ (cyclohexene, $h\nu$) → 7-Ph-7-$P(O)(OR)_2$-norcarane

→ (benzene, $h\nu$) → 7-Ph-7-$P(O)(OR)_2$-norcaradiene

The relative ease of the Wolff-type rearrangement and insertion into the oxygen-hydrogen bond of methanol has been found to vary with the group attached to the divalent carbon.[491]

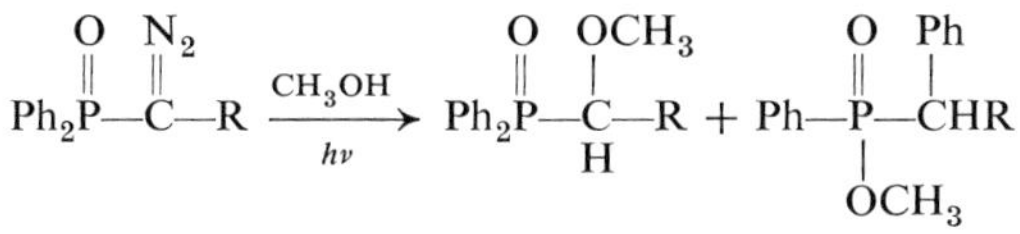

$$\mathrm{Ph_2\overset{\overset{O}{\|}}{P}{-}\overset{\overset{N_2}{\|}}{C}{-}R \xrightarrow[h\nu]{CH_3OH} Ph_2\overset{\overset{O}{\|}}{P}{-}\underset{H}{\overset{OCH_3}{C}}{-}R + Ph{-}\underset{OCH_3}{\overset{\overset{O}{\|}}{P}}{-}\overset{Ph}{C}HR}$$

R	Wolff	Insertion
Ph	100	0
$\overset{\overset{O}{\|}}{C}{-}NH_2$	81	0
COOEt	22	61
H	—	61
COPh	—	77

G. Carbenes Containing Other Elements

Diazo compounds containing lithium, silver, mercury, silicon, cadmium, zinc, germanium, lead, and tin are known. We have previously dwelt upon the mercury-containing compound **134** and its synthetic utility, and the syntheses of a wide variety of related compounds have been described.[683–685]

$$R{-}\overset{\overset{\displaystyle O}{\|}}{C}{-}CHN_2 \xrightarrow{HgO} (R{-}\overset{\overset{\displaystyle O}{\|}}{C}{-}CN_2)_2Hg$$

R = $(CH_3)_3C$, Ph, CF_3, *sym*—$(CH_3)_3C_6H_2$, CN, COR, COOR

Tin- and silicon-containing diazo compounds have been reported several times,[686–690] and one report of the germanium, lead, zinc, and cadmium compounds has appeared very recently.[691] As with other compounds reported in this section, to date the synthetic efforts have not been followed by much

		$R_3Si{-}\overset{\overset{\displaystyle N_2}{\|}}{C}{-}R'$	$R_3Sn{-}\overset{\overset{\displaystyle N_2}{\|}}{C}{-}R'$
R = Ph	R′ = COOR		556
R = CH_3,	R′ = COOR	557	
R = CH_3,	R′ = H	686, 687, 689	686
R = Ph,	R′ = Ph	688	

chemistry. We have mentioned (Section V) that the tin and silicon-containing diazo esters and carbenes appear to be normal in their properties. To this we can add here only the report of Seyferth et al.[689] that trimethylsilyl diazomethane gives adducts to olefins on CuCl catalyzed decomposition.

$$\text{cyclohexene} + (CH_3)_3Si{-}CHN_2 \xrightarrow{CuCl} \text{7-}Si(CH_3)_3\text{-norcarane (65)} + \text{7-}(CH_3)_3Si\text{-norcarane (9)}$$

Lithiated diazo compounds, which are quite unstable, are nonetheless known.[563,691,692] Unlike lithiodiazomethane,[692,693] ethyllithiodiazoacetate exists in the C-lithiated form.[563]

Acknowledgment

Of the many persons who contributed to this review, two, Professor Peter Gaspar of Washington University and Robert W. Shaw of this department, must be singled out for special thanks. We are also grateful to the many others who contributed comments and details of unpublished work. Financial support for this work was provided by the National Science Foundation through grants GP-12759 and GP-30797X.

References

1. W. Kirmse, *Carbene Chemistry*, Academic Press, New York, 1964.
2. J. Hine, *Divalent Carbon*, Ronald Press, New York, 1964.
3. H. M. Frey, *Prog. React. Kinet.*, **2,** 131 (1964).
4. J. I. G. Cadogen and M. J. Perkins, in S. Patai, *The Chemistry of Alkenes*, Interscience, New York, 1964, Chapter 9.
5. W. B. DeMore and S. W. Benson, *Adv. Photochem.*, **2,** 219 (1964).
6. J. A. Bell, *Prog. Phys. Org. Chem.*, **2,** 1 (1964).
7. C. W. Rees and C. E. Smithen in A. R. Katritsky Ed., *Adv. Heterocycl. Compds.*, **3,** 57 (1964).
8. W. Kirmse, *Prog. Org. Chem.*, **6,** 164 (1964).
9. G. G. Rozantsev, A. A. Fainzil'berg, and S. S. Novikov, *Russ. Chem. Rev.*, **34,** 69 (1965).
10. B. S. Herold and P. P. Gaspar, *Fortschr. Chem. Forsch.*, **5,** 89 (1965).
11. G. Köbrich, *Angew. Chem. Int. Ed.*, **6,** 41 (1967).
12. G. L. Closs, *Top. Stereochem.*, **3,** 194 (1968).
13. T. L. Gilcrist and C. W. Rees, *Carbenes, Nitrenes and Arynes*, Appleton-Century-Crofts, New York, 1969.
14. W. Kirmse, *Chem. unserer Zeit*, **3,** 184 (1969).
15. W. Kirmse, *Carbene, Carbenoide und Carbenanaloge*, Verlag Chemie, GmbH, Weinheim/Bergstr., 1969.
16. R. A. Moss, *Chem. Eng. News*, **47,** June 16, 30 (1969).
17. D. Bethell, *Adv. Phys. Org. Chem.*, **7,** 153 (1969).
18. A. M. van Leusen and J. Strating, *Q. Rep. Sulfur Chem.*, **5,** 67 (1970).
18a. W. Kirmse, *Carbene Chemistry*, 2nd ed., Academic Press, New York, 1971.
19. G. Herzberg and J. Shoosmith, *Nature*, **183,** 1801 (1959).
20. G. Herzberg, *Proc. Roy. Soc. (London)*, **A262,** 291 (1961).
21. G. Herzberg and J. W. C. Johns, *Proc. Roy. Soc. (London)*, **A295,** 107 (1967).
22. J. F. Harrison and L. C. Allen, *J. Am. Chem. Soc.*, **91,** 807 (1969).
23. M. C. Goldberg and J. R. Riter, Jr., *J. Chem. Phys.* **50,** 547 (1969).
24. C. F. Bender and H. F. Schaefer, *J. Am. Chem. Soc.*, **92,** 4984 (1970).
25. E. Wasserman, V. J. Kuck, R. S. Hutton, and W. A. Yager, *J. Am. Chem. Soc.*, **92,** 7491 (1970).
26. E. Wasserman, W. A. Yager, and V. J. Kuck, *Chem. Phys. Lett.*, **7,** 409 (1970).
27. G. Herzberg and J. W. C. Johns, *J. Chem. Phys.*, **54,** 2276 (1971).
28. R. Hoffmann, G. D. Zeiss, and G. W. Van Dine, *J. Am. Chem. Soc.*, **90,** 1485 (1968).
29. W. A. Lathan, W. J. Hehre, and J. A. Pople, *J. Am. Chem. Soc.*, **93,** 808 (1971).
30. W. A. Yeranos, *Z. Naturforsch.*, **26a,** 1245 (1971).
31. R. A. Bernheim, H. W. Bernard, P. S. Wang, L. S. Wood, and P. S. Skell, *J. Chem. Phys.*, **54,** 3223 (1971).
32. S. V. O'Neill, H. F. Schaefer, III, and C. F. Bender, *J. Chem. Phys.*, **55,** 162 (1971).
33. W. A. Lathan, W. J. Hehre, L. A. Curtiss, and J. A. Pople, *J. Am. Chem. Soc.*, **93,** 6377 (1971).
34. J. F. Harrison, *J. Am. Chem. Soc.*, **93,** 4112 (1971).
35. R. A. Bernheim, H. W. Bernard, P. S. Wang, L. S. Wood, and P. S. Skell, *J. Chem. Phys.*, **53,** 1280 (1970).

36. L. C. Allen and D. R. Franceschetti, private communication with permission to cite.
37. R. W. Carr, Jr., T. W. Eder, and M. G. Topor, *J. Chem. Phys.*, **53,** 4716 (1970).
38. M. L. Halberstadt and J. R. McNesby, *J. Am. Chem. Soc.*, **89,** 3417 (1967).
38a. H. M. Frey, *Chem. Commun.*, 1024 (1972).
39. V. H. Dibeler and S. K. Liston, *J. Chem. Phys.*, **48,** 4765 (1968).
40. W. A. Chupka and C. Lifshitz, *J. Chem. Phys.*, **48,** 1109 (1968).
41. W. A. Chupka, J. Berkowitz, and K. M. A. Refaey, *J. Chem. Phys.*, **50,** 1938 (1969).
42. A. H. Laufer and H. Okabe, *J. Am. Chem. Soc.*, **93,** 4137 (1971).
43. E. W. Schlag and B. S. Rabinovitch, *J. Am. Chem. Soc.*, **82,** 5996 (1960).
44. B. S. Rabinovitch, E. W. Schlag, and K. B. Wiberg, *J. Chem. Phys.*, **28,** 504 (1958).
45. H. M. Frey, *Proc. Roy. Soc. (London)*, **A250,** 409 (1959).
46. H. M. Frey, *Proc. Roy. Soc. (London)*, **A251,** 575 (1959).
47. H. M. Frey, *Chem. Commun.*, **1965,** 260.
48. F. A. L. Anet, R. F. W. Bader, and A. M. van der Auwera, *J. Am. Chem. Soc.* **82,** 3217 (1960).
49. H. M. Frey, *J. Am. Chem. Soc.*, **82,** 5947 (1960).
50. R. F. W. Bader and J. I. Generosa, *Can. J. Chem.*, **43,** 1631 (1965).
51. F. J. Duncan and R. J. Cvetanović, *J. Am. Chem. Soc.*, **84,** 3593 (1962).
52. D. C. Montague and F. S. Rowland, *J. Phys. Chem.*, **72,** 3705 (1968).
53. H. M. Frey and R. Walsh, *Chem. Commun.*, **1969,** 158.
54. R. J. Cvetanović, H. E. Avery, and R. S. Irwin, *J. Chem. Phys.*, **46,** 1993 (1967).
55. J. A. Bell, *J. Am. Chem. Soc.*, **87,** 4966 (1965).
56. R. Hoffmann, *J. Am. Chem. Soc.*, **90,** 1475 (1968).
57. J. W. Simons and B. S. Rabinovitch, *J. Phys. Chem.*, **68,** 1322 (1964).
58. R. W. Carr and G. B. Kistiakowsky, *J. Phys. Chem.*, **70,** 118 (1966).
59. B. A. De Graff and G. B. Kistiakowsky, *J. Phys. Chem.*, **71,** 3984 (1967).
60. T. W. Eder and R. W. Carr, *J. Phys. Chem.*, **73,** 2074 (1969).
61. S.-Y. Ho and W. A. Noyes, Jr., *J. Am. Chem. Soc.*, **89,** 5091 (1967).
62. A. N. Strachen and D. E. Thornton, *J. Phys. Chem.*, **70,** 952 (1966).
63. D. E. Thornton and A. N. Strachen, *J. Phys. Chem.*, **71,** 4583 (1967).
64. G. W. Taylor and J. W. Simons, *Can. J. Chem.*, **48,** 1016 (1970).
65. B. S. Rabinovitch, K. W. Watkins, and D. R. Ring, *J. Am. Chem. Soc.*, **87,** 4960 (1965).
66. D. F. Ring and B. S. Rabinovitch, *J. Phys. Chem.*, **72,** 191 (1968).
67. D. R. Ring and B. S. Rabinovitch, *Int. J. Chem. Kinet.*, **1,** 11 (1969).
68. C. S. Elliot and H. M. Frey, *Trans. Faraday Soc.*, **64,** 2352 (1968).
69. D. H. White, P. B. Condit, and R. G. Bergman, *J. Amer. Chem. Soc.*, **94,** 1348 (1972).
70. C. McKnight and F. S. Rowland, *J. Am. Chem. Soc.*, **88,** 3179 (1966).
71. W. von E. Doering and P. LaFlamme, *J. Am. Chem. Soc.*, **78,** 5447 (1956).
72. R. C. Woodworth and P. S. Skell, *J. Am. Chem. Soc.*, **81,** 3383 (1959).
73. K. R. Kopecky, G. S. Hammond, and P. A. Leermakers, *J. Am. Chem. Soc.*, **84,** 1015 (1962).
74. F. Casas, J. A. Kerr, and A. F. Trotman-Dickenson, *J. Chem. Soc.*, 1141 (1965).
75. P. P. Gaspar and W. R. Roth, quoted in W. von E. Doering and W. R. Roth, *Tetrahedron*, **19,** 715 (1963).
76. H. E. Simmons and R. D. Smith, *J. Am. Chem. Soc.*, **81,** 4256 (1959).
77. J. W. Simmons and G. W. Taylor, *J. Phys. Chem.*, **73,** 1274 (1969).
78. G. W. Taylor and J. W. Simmons, *J. Phys. Chem.*, **74,** 464 (1970).
79. R. L. Johnson, W. L. Hase, and J. W. Simmons, *J. Chem. Phys.*, **52,** 3911 (1970).
80. W. L. Hase and J. W. Simmons, *J. Phys. Chem.*, **52,** 4004 (1970).

81. W. von E. Doering, J. C. Gilbert, and P. A. Leermakers, *Tetrahedron*, **24,** 6863 (1968).
82. H. Meerwein, H. Rathjen, and H. Werner, *Chem. Ber.*, **75,** 1610 (1942).
83. W. von E. Doering, R. G. Buttery, R. G. Laughlin, and N. Chaudhuri, *J. Am. Chem. Soc.*, **78,** 3224 (1956).
84. D. B. Richardson, M. C. Simmons, and I. Dvoretzky, *J. Am. Chem. Soc.*, **82,** 5001 (1960).
85. W. von E. Doering and H. Prinzbach, *Tetrahedron*, **6,** 24 (1959).
86. V. Franzen and R. Edens, *Ann.*, **729,** 33 (1969).
87. V. Franzen, *Abstracts of Papers*, 141st ACS Meeting, Washington, D.C., 1962, p. 23-0.
88. M. R. Willcott, Ph.D. thesis, Yale University, New Haven, Conn., 1963.
89. L. E. Helgen, Ph.D. thesis, Yale University, New Haven, Conn., 1965.
90. W. Kirmse and M. Buschoff, *Chem. Ber.*, **102,** 1098 (1969).
91. R. C. Dobson, D. M. Hayes, and R. Hoffmann, *J. Am. Chem. Soc.*, **93,** 6188 (1971).
92. S. W. Benson, *Adv. Photochem.*, **2,** 1 (1964).
93. R. F. W. Bader and R. A. Gangi, *J. Am. Chem. Soc.*, **93,** 1831 (1970).
94. D. F. Ring and B. S. Rabinovitch, *J. Am. Chem. Soc.*, **88,** 4285 (1966).
95. D. F. Ring and B. S. Rabinovitch, *Can. J. Chem.*, **46,** 2435 (1968).
96. C. McKnight, P. S. T. Lee, and F. S. Rowland, *J. Am. Chem. Soc.*, **89,** 6802 (1967).
97. B. M. Herzog and R. W. Carr, *J. Phys. Chem.*, **71,** 2688 (1967).
98. R. W. Carr, *J. Phys. Chem.*, **70,** 1970 (1966).
99. J. W. Simons, C. J. Mazak, and G. W. Taylor, *J. Phys. Chem.*, **72,** 749 (1968).
100. D. W. Placzek, D. F. Ring, and B. S. Rabinovitch, *J. Phys. Chem.*, **69,** 1782 (1965).
101. J. P. Chesick and M. R. Willcott, III, *J. Phys. Chem.*, **67,** 2850 (1963).
102. J. W. Simons and B. S. Rabinovitch, *J. Am. Chem. Soc.*, **85,** 1023 (1963).
103. J. W. Simons and B. S. Rabinovitch, *J. Phys. Chem.*, **68,** 1322 (1964).
104. P. S. T. Lee, R. L. Russell, and F. S. Rowland, *Chem. Commun.*, **1970,** 18.
105. W. Braun, A. M. Bass, and M. Pilling, *J. Chem. Phys.*, **52,** 5131 (1970).
106. W. J. Dunning and C. C. McCain, *J. Chem. Soc.* (*B*), **1966,** 68.
107. J. W. Powell-Wiffen and R. P. Wayne, *Photochem. Photobiol.*, **8,** 131 (1968).
108. M. Jones, Jr., Ph.D. thesis, Yale University, New Haven, Conn., 1963.
109. W. von E. Doering and J. F. Coburn, *Tetrahedron Lett.*, **1965,** 991.
110. R. Srinivasan, *J. Am. Chem. Soc.*, **90,** 2752 (1968).
111. H. M. Frey and R. Walsh, *Chem. Commun.*, **1969,** 159.
112. R. Huisgen, *Angew. Chem.*, **67,** 439 (1955).
113. C. D. Gutsche and M. Hillman, *J. Am. Chem. Soc.*, **76,** 2236 (1954).
114. W. von E. Doering, L. H. Knox, and M. Jones Jr., *J. Org. Chem.*, **24,** 136 (1959).
115. V. Franzen and L. Fikentscher, *Ann.*, **617,** 1 (1958).
116. H. M. Frey, *Rec. Trav. Chim.*, **83,** 117 (1964).
117. H. M. Frey and M. A. Voisey, *Trans. Faraday Soc.*, **64,** 954 (1968).
118. M. A. Voisey, *Trans. Faraday Soc.*, **64,** 3058 (1968).
119. W. Kirmse and M. Buschhoff, *Chem. Ber.*, **102,** 1087 (1969).
120. H. M. Frey and M. A. Voisey, *Chem. Commun.*, **1966,** 454.
121. J. A. Kerr, B. V. O'Grady, and A. F. Trotman-Dickenson, *J. Chem. Soc.* (*A*), **1967,** 897.
122. G. G. Moore, Ph.D. thesis, Yale University, New Haven, Conn., 1961.
123. W. H. Urry and J. R. Eiszner, *J. Am. Chem. Soc.*, **73,** 2977 (1951).
124. W. H. Urry and J. R. Eiszner, *J. Am. Chem. Soc.*, **74,** 5822 (1952).
125. W. H. Urry, J. R. Eiszner, and J. W. Wilt, *J. Am. Chem. Soc.*, **79,** 918 (1957).
126. W. H. Urry and N. Bilow, *J. Am. Chem. Soc.*, **86,** 1815 (1964).

127. V. Franzen, *Ann.*, **627,** 22 (1959).
128. P. P. Gaspar, Ph.D. thesis, Yale University, New Haven, Conn., 1961.
129. W. von E. Doering and R. Sampson, unpublished work.
130. W. von E. Doering and H. Wiegandt, unpublished work.
131. H. D. Roth, *J. Am. Chem. Soc.*, **93,** 1527 (1971).
132. H. D. Roth, *J. Am. Chem. Soc.*, **93,** 4935 (1971).
133. K. Dees, D. W. Setser, and W. G. Clark, *J. Phys. Chem.*, **75,** 2231 (1971).
134. D. W. Setser, R. Littrell, and J. C. Hassler, *J. Am. Chem. Soc.*, **87,** 2062 (1965).
135. J. C. Hassler and D. W. Setser, *J. Chem. Phys.*, **45,** 3237 (1966).
136. J. C. Hassler, D. W. Setser, and R. L. Johnson, *J. Chem. Phys.*, **45,** 3231 (1966).
137. C. H. Bamford, J. E. Casson, and R. P. Wayne, *Proc. Roy. Soc.* (*London*), **A289,** 287 (1965).
138. R. S. B. Johnstone and R. P. Wayne, *Nature*, **211,** 1396 (1966).
139. C. H. Bamford, J. E. Casson, and A. N. Hughes, *Chem. Commun.*, **1967,** 1096.
140. R. S. B. Johnstone and R. P. Wayne, *Photochem. Photobiol.*, **6,** 531 (1967).
141. B. A. DeGraff and G. B. Kistiakowsky, *J. Phys. Chem.*, **71,** 1553 (1967).
142. R. A. Cox and R. J. Cvetanović, *J. Phys. Chem.*, **72,** 2236 (1968).
143. K. Dees and D. W. Setser, *J. Phys. Chem.*, **75,** 2240 (1971).
144. R. L. Russell and F. S. Rowland, *J. Am. Chem. Soc.*, **92,** 7508 (1970).
145. D. C. Montague and F. S. Rowland, *J. Am. Chem. Soc.*, **93,** 5381 (1971).
146. C. H. Bamford, J. E. Casson, and A. N. Hughes, *Proc. Roy. Soc.* (*A*), **306,** 135 (1968).
147. C. H. Bamford and J. E. Casson, *Proc. Roy. Soc.* (*A*), **312,** 141 (1969).
148. G. O. Pritchard, J. T. Bryant, and R. L. Thommarson, *J. Phys. Chem.*, **69,** 2804 (1965).
149. J. A. Kerr, B. V. O'Grady, and A. F. Trotman-Dickenson, *J. Chem. Soc.* (*A*), **1966,** 1621.
150. I. L. Knunyants and E. Ya. Pervava, *Izv. Akad. Nauk. SSSR*, *Ser. Khim.*, **1965,** 894.
151. V. Franzen and H. Kuntze, *Ann.*, **627**, 15 (1959).
152. K. A. W. Kramer and A. N. Wright, *J. Chem. Soc.*, **1963,** 3604.
153. C. J. Mazac and J. W. Simons, *J. Am. Chem. Soc.*, **90,** 2484 (1968).
154. J. W. Simons and C. J. Mazac, *Can. J. Chem.*, **45,** 1717 (1967).
155. C. B. Moore and G. C. Pimentel, *J. Chem. Phys.*, **41,** 3504 (1964).
156. A. E. Shilov, A. A. Shteinman, and M. B. Tjabin, *Tetrahedron Lett.*, **1968,** 4177.
157. Yu. G. Borod'ko, A. E. Shilov, and A. A. Shteinman, *Dokl. Akad. Nauk. SSSR*, **168,** 581 (1966).
157a. W. von E. Doering and T. Mole, *Tetrahedron*, **10,** 65 (1960).
158. H. Lind and A. J. Deutschman, Jr. *J. Org. Chem.*, **32,** 326 (1967).
159. H. M. Frey, *Chem. Ind.*, **1960,** 1266.
160. T. Terao, N. Sakai, and S. Shida, *J. Am. Chem. Soc.*, **85,** 3919 (1963).
161. M. E. Jacox and D. E. Milligan, *J. Am. Chem. Soc.*, **85,** 278 (1963).
162. W. von E. Doering and L. H. Knox, *J. Am. Chem. Soc.*, **72,** 2305 (1950).
163. W. von E. Doering, L. H. Knox, and F. Detert, *J. Am. Chem. Soc.*, **75,** 297 (1953).
164. R. M. Lemmon and W. Strohmeier, *J. Am. Chem. Soc.*, **81,** 106 (1959).
165. G. A. Russell and D. G. Hendry, *J. Org. Chem.*, **28,** 1933 (1963).
166. E. Müller and H. Kessler, *Ann.*, **692,** 58 (1966).
167. R. J. Bussey and R. C. Neuman, Jr., *J. Org. Chem.*, **34,** 1323 (1969).
167a. R. A. Moss and A. Mamantov, *J. Am. Chem. Soc.*, **92,** 6951 (1970).
168. H. M. Frey, *Chem. Ind.*, **1962,** 218; *J. Chem. Soc.*, **1962,** 2293.
169. V. Franzen, H.-J. Schmidt, and Ch. Mertz, *Chem. Ber.*, **94,** 2942 (1961).
170. K. A. W. Kramer and A. N. Wright, *Tetrahedron Lett.*, **1962,** 1095.
171. R. N. Haszeldine, A. E. Tipping, and R. O'B. Watts, *Chem. Commun.*, **1969,** 1364.

172. L. Friedman and H. Shechter, *J. Am. Chem. Soc.*, **81,** 5512 (1959).
173. W. Kirmse, H.-D. von Scholz, and H. Arold, *Ann.*, **711,** 22 (1968).
174. M. J. Goldstein and S. J. Baum, *J. Am. Chem. Soc.*, **85,** 1885 (1963).
175. W. Kirmse and W. von E. Doering, *Tetrahedron*, **11,** 266 (1960).
176. W. Kirmse and H. Arold, *Angew. Chem. Int. Ed.*, **7,** 539 (1968).
177. G. L. Closs, *J. Am. Chem. Soc.*, **84,** 809 (1962).
178. W. Kirmse and G. Wächtershäuser, *Tetrahedron*, **22,** 63 (1966).
179. A. M. Mansorr and I. D. R. Stevens, *Tetrahedron Lett.*, **1966,** 1733.
180. H. M. Frey and I. D. R. Stevens, *J. Chem. Soc.*, **1965,** 3101.
181. H. M. Frey and A. W. Scaplehorn, *J. Chem. Soc.* (*A*), **1966,** 968.
182. W. Kirmse and K. Horn, *Chem. Ber.*, **100,** 2698 (1967).
183. W. Kirmse and M. Buschhoff, *Angew. Chem. Int. Ed.*, **4,** 692 (1965).
184. W. Kirmse and M. Buschhoff, *Chem. Ber.*, **100,** 1491 (1967).
185. W. Kirmse, H. Schladetsch, and H.-W. Bücking, *Chem. Ber.*, **99,** 2579 (1966).
186. W. Kirmse and H.-W. Bücking, *Ann.*, **711,** 31 (1968).
187. E. Schmitz, C. Hörig, and C. Gründemann, *Chem. Ber.*, **100,** 2093 (1967).
188. D. M. Lemal, F. Menger, and G. W. Clark, *J. Am. Chem. Soc.*, **85,** 2529 (1963).
189. W. Kirmse and D. Grassmann, *Chem. Ber.*, **99,** 1746 (1966).
190. H. Tsuruta, K. Kurabayashi, and T. Mukai, *Tetrahedron Lett.*, **1967,** 3775.
191. W. Grimme, *Chem. Ber.*, **100,** 113 (1967).
192. T. Mukai, H. Tsuruta, T. Nakazawa, K. Isobe, and K. Kurabayashi, *Sci. Rep., Tohoko Univ. Ser. 1*, **51,** 113 (1968).
193. G. Büchi and J. D. White, *J. Am. Chem. Soc.*, **86,** 2884 (1964).
194. G. L. Closs, L. E. Closs, and W. Böll, *J. Am. Chem. Soc.*, **85,** 3796 (1963).
195. H. Dürr, *Chem. Ber.*, **103,** 369 (1970).
196. T. Severin, H. Krämer, and P. Adhikary, *Chem. Ber.*, **104,** 972 (1971).
197. C. D. Hund and S. C. Lui, *J. Am. Chem. Soc.*, **57,** 2656 (1935).
198. I. Tabushi, K. Takagi, M. Okano, and R. Oda, *Tetrahedron*, **23,** 2621 (1967).
199. M. Jones, Jr., and Y. H. Shen, unpublished work.
200. K. Geibel and H. Mäder, *Ber.*, **103,** 1645 (1970).
201. H. Meier and I. Menzer, *Synthesis*, **1971,** 215.
202. H.-K. Lee, *Daehan Hwahak Hwoejee*, **13,** 333 (1969); *Chem. Abstr.* **73,** 87353w.
203. P. S. Skell and J. Klebe, *J. Am. Chem. Soc.*, **82,** 247 (1960).
204. J. V. Gramas, Ph.D. thesis, Pennsylvania State University, University Park, Pa., 1965.
205. K. D. Keil, unpublished work quoted in reference 204.
206. R. A. Bernheim, R. J. Kempf, J. V. Gramas, and P. S. Skell, *J. Chem. Phys.*, **43,** 196 (1965).
207. R. A. Bernheim, R. J. Kempf, P. W. Humer, and P. S. Skell, *J. Chem. Phys.*, **41,** 1156 (1964).
208. E. Wassermann, L. Barash, and W. A. Yager, *J. Am. Chem. Soc.*, **87,** 2075 (1965).
209. R. Selvarajan and J. H. Boyer, *J. Org. Chem.*, **36,** 1679 (1971).
210. R. Fields and R. N. Haszeldine, *Proc. Chem. Soc.*, 22 (1960).
211. J. H. Atherton, R. Fields, and R. N. Haszeldine, *J. Chem. Soc.*, (*A*) **1971,** 366.
212. R. Fields and R. N. Haszeldine, *J. Chem. Soc.*, **1964,** 1881.
213. J. H. Atherton and R. Fields, *J. Chem. Soc.* (*C*), **1968,** 2276.
214. E. Wassermann, L. Barash, and W. A. Yager, *J. Am. Chem. Soc.*, **87,** 4974 (1965).
215. J. H. Atherton and R. Fields, *J. Chem. Soc.* (*C*), **1967,** 1450.
216. D. M. Gale, W. J. Middleton, and C. G. Krespan, *J. Am. Chem. Soc.*, **88,** 3617 (1966).
217. D. M. Gale, W. J. Middleton, and C. G. Krespan, *J. Am. Chem. Soc.*, **87,** 657 (1965).

218. D. M. Gale, *J. Org. Chem.*, **33,** 2536 (1968).
219. W. Mahler, *J. Am. Chem. Soc.*, **90,** 523 (1968).
220. L. Friedman and H. Shechter, *J. Am. Chem. Soc.*, **82,** 1002 (1960).
221. P. B. Shevlin and A. P. Wolf, *J. Am. Chem. Soc.*, **88,** 4735 (1966).
222. J. W. Wilt, J. M. Kosturik, and R. C. Orlowski, *J. Org. Chem.*, **30,** 1052 (1965).
223. J. W. Wilt, J. F. Zawadzki, and D. G. Schultenover, *J. Org. Chem.*, **31,** 876 (1966).
224. A. Guarino and A. P. Wolf, *Tetrahedron Lett.*, **1969,** 655.
225. J. A. Berson, W. Bauer, and M. M. Campbell, *J. Am. Chem. Soc.*, **92,** 7515 (1970).
226. Professor R. G. Bergman, California Institute of Technology and M. B. Sohn, Princeton, private communications.
227. C. L. Bird, H. M. Frey, and I. D. R. Stevens, *Chem. Commun.*, **1967,** 707.
228. I. D. R. Stevens, H. M. Frey, and C. L. Bird, *Angew. Chem. Int. Ed.*, **7,** 646 (1968).
229. M. Jones, Jr., and M. B. Sohn, unpublished work.
230. W. Kirmse, B.-G. von Bülow, and H. Schepp, *Ann.*, **691,** 41 (1966).
231. R. H. Shapiro and M. J. Heath, *J. Am. Chem. Soc.*, **89,** 5734 (1967).
232. R. H. Shapiro, *Tetrahedron Lett.*, **1968,** 345.
233. H. M. Ensslin and M. Hanack, *Angew. Chem. Int. Ed.*, **6,** 702 (1967).
234. S. Nishida, I. Moritani, E. Tsuda, and T. Teraji, *Chem. Commun.*, **1969,** 781.
235. W. Kirmse and K.-H. Pook, *Chem. Ber.*, **98,** 4022 (1965).
236. W. Kirmse and K.-H. Pook, *Angew. Chem. Int. Ed.*, **5,** 594 (1966).
237. K. B. Wiberg, G. J. Burgmaier, and P. Warner, *J. Am. Chem. Soc.*, **93,** 246 (1971).
238. M. Jones, Jr., S. D. Reich, and L. T. Scott, *J. Am. Chem. Soc.*, **92,** 3118 (1970).
239. S. Masamune, C. G. Chin, K. Hojo, and R. T. Seidner, *J. Am. Chem. Soc.*, **89,** 4804 (1967).
240. M. Jones, Jr., and L. E. Sullivan, unpublished work. See also *Angew. Chem. Int. Ed.*, **7,** 644 (1968).
241. L. R. Sousa, *Diss. Abstr. B*, **1971,** 3269, H. E. Zimmerman and L. R. Sousa, *J. Am. Chem. Soc.*, **94,** 834 (1972).
242. E. H. White, G. E. Meier, R. Graeve, U. Zirngibl, and E. W. Friend, *J. Am. Chem. Soc.*, **88,** 611 (1966).
243. S. Masamune and M. Kato, *J. Am. Chem. Soc.*, **88,** 610 (1966).
244. R. F. Peterson, Jr., R. T. K. Baker, and R. L. Wolfgang, *Tetrahedron Lett.*, **1969,** 4749.
245. P. B. Shevlin and A. P. Wolf, *J. Am. Chem. Soc.*, **92,** 406 (1970).
246. G. L. Closs and V. N. M. Rao, *J. Am. Chem. Soc.*, **88,** 4116 (1966).
247. D. H. Paskovich and P. W. N. Kwok, *Tetrahedron Lett.*, **1967,** 2227.
248. C. H. DePuy and D. H. Froemsdorf, *J. Am. Chem. Soc.*, **82,** 634 (1960).
249. W. Kirmse and K. Pöhlmann, *Chem. Ber.*, **100,** 3564 (1967).
250. D. M. Lemal and K. S. Shim, *Tetrahedron Lett.*, **1964,** 3231.
251. G. L. Closs and R. B. Larrabee, *Tetrahedron Lett.*, **1965,**287.
252. H. Babad, W. Flemon, and J. B. Wood, III, *J. Org. Chem.*, **32,** 2871 (1967).
253. M. Rey, R. Begrich, W. Kirmse, and A. S. Dreiding, *Helv. Chim. Acta*, **51,** 1001 (1968).
254. W. Kirmse and C. Hase, *Angew. Chem. Int. Ed.*, **7,** 891 (1968).
255. J. Casanova and B. Waegell, *Bull. Chim. Soc. France*, **1971,** 1295.
256. M. Schwarz, A. Besold, and E. R. Nelson, *J. Org. Chem.*, **30,** 2425 (1965).
257. J. W. Wilt, C. A. Schneider, H. F. Dabek, Jr., J. F. Kraemer, and W. J. Wagner, *J. Org. Chem.*, **31,** 1543 (1966)
258. J. R. Wiseman and J. A. Chong, *J. Am. Chem. Soc.*, **91,** 7775 (1969).
259. J. A. Marshall and H. Faubl, *J. Am. Chem. Soc.*, **92,** 948 (1970).
260. K. Geibel, *Chem. Ber.*, **103,** 1637 (1970).

261. M. Jones, Jr., and R. T. Conlin, unpublished results.
262. W. von E. Doering and P. M. LaFlamme, *Tetrahedron*, **2,** 75 (1958).
263. W. M. Jones, M. H. Grasley, and D. G. Baarda, *J. Am. Chem. Soc.*, **86,** 912 (1964).
264. W. M. Jones and J. M. Walbrick, *J. Org. Chem.*, **34,** 2217 (1969).
265. W. M. Jones and J. W. Wilson, Jr., *Tetrahedron Lett.*, **1965,** 1587.
266. W. M. Jones and J. M. Walbrick, *Tetrahedron Lett.*, **1968,** 5229.
267. J. M. Walbrick, J. W. Wilson, Jr., and W. M. Jones, *J. Am. Chem. Soc.*, **90,** 2895 (1968).
268. M. Rey, U. A. Huber, and A. S. Dreiding, *Tetrahedron Lett.*, **1968,** 3583.
269. J. Meinwald, J. W. Wheeler, A. A. Nimetz, and J. S. Liu, *J. Org. Chem.*, **30,** 1038 (1965).
270. K. B. Wiberg, J. E. Hiatt, and G. Burgmaier, *Tetrahedron Lett.*, **1968,** 5855.
271. J. R. Chapman, *Tetrahedron Lett.*, **1966,** 113.
272. B. Singh, *J. Org. Chem.*, **31,** 181 (1966).
273. R. Kalish and W. H. Pirkle, *J. Am. Chem. Soc.*, **89,** 2781 (1967).
274. G. Maier, *Tetrahedron Lett.*, **1965,** 3603.
275. F. T. Bond and D. E. Bradway, *J. Am. Chem. Soc.*, **87,** 4977 (1965).
276. P. K. Freeman and R. C. Johnson, *J. Org. Chem.*, **34,** 1751 (1969).
277. W. Kirmse and L. Ruetz, *Ann.*, **726,** 36 (1969).
278. L. Friedman and H. Shechter, *J. Am. Chem. Soc.*, **83,** 3159 (1961).
279. C. Swithenbank and M. C. Whiting, *J. Chem. Soc.*, **1963,** 4573.
280. R. W. Alder and M. C. Whiting, *J. Chem. Soc.*, **1963,** 4595.
281. A. P. Krapcho and R. Donn, *J. Org. Chem.*, **30,** 641 (1965).
282. P. K. Freeman and D. G. Kruper, *J. Org. Chem.*, **30,** 1047 (1965).
283. J. W. Wheeler, R. H. Chung, Y. N. Vaishnav, and C. C. Shroff, *J. Org. Chem.*, **34,** 545 (1969).
284. J. E. Baldwin and H. C. Krauss, Jr., *J. Org. Chem.*, **35,** 2426 (1970).
285. W. Kirmse and G. Münscher, *Ann.* **726,** 42 (1969).
286. A. C. Cope and S. S. Hecht, *J. Am. Chem. Soc.*, **89,** 6920 (1967).
287. J. Casanova and B. Waegell, *Bull. Chim. Soc. France*, **1971,** 1289.
288. J. Meinwald and F. Uno, *J. Am. Chem. Soc.*, **90,** 800 (1968).
289. S. Masamune, K. Fukumoto, Y. Yasunari, and D. Darwish, *Tetrahedron Lett.*, **1966,** 193.
290. J. W. Powell and M. C. Whiting, *Tetrahedron*, **7,** 305 (1959).
291. W. Reusch, M. W. DiCarlo, and L. Traynor, *J. Org. Chem.*, **26,** 1711 (1961).
292. H. Krieger, S.-E. Masar, and H. Ruotsalainen, *Suom. Kemistil.*, **39B,** 237 (1966).
293. R. H. Shapiro, *Tetrahedron Lett.*, **1966,** 3401.
294. R. A. Moss and J. R. Whittle, *Chem. Commun.*, **1969,** 341.
295. P. K. Freeman and D. M. Balls, *J. Org. Chem.*, **32,** 2354 (1967).
296. P. K. Freeman, V. N. M. Rao, and G. E. Bigam, *Chem. Commun.*, **1965,** 511.
297. D. M. Lemal and A. J. Fry, *J. Org. Chem.*, **29,** 1673 (1964).
298. A. C. Udding, J. Strating, H. Wynberg, and J. L. M. A. Schlatmann *Chem. Commun.*, **1966,** 657.
299. Z. Majerski, S. H. Liggero, and P. von R. Schleyer, *Chem. Commun.*, **1970,** 949.
299a. H. W. Geluk and Th. J. de Boer, *Chem. Commun.*, **1972,** 3.
300. S. Oida and E. Ohki, *Chem. Pharm. Bull. (Tokyo)*, **15,** 545 (1967).
301. T. Sasaki, S. Eguchi, and T. Kiriyama, *J. Am. Chem. Soc.*, **91,** 212 (1969).
302. M. R. Vegar and R. J. Wells, *Tetrahedron Lett.*, **1969,** 2565.
303. M. H. Fisch and H. D. Pierce, Jr., *Appl. Spectroscopy*, **23** 637 (1969).
304. M. H. Fisch and H. D. Pierce, Jr., *Chem. Commun.*, **1970,** 503.
305. R. D. Allan and R. J. Wells, *Austr. J. Chem.*, **23,** 1625 (1970).

306. R. Gleiter and R. Hoffmann, *J. Am. Chem. Soc.*, **90,** 5457 (1968).
307. R. A. Moss, U.-H. Dolling, and J. R. Whittle, *Tetrahedron Lett.*, **1971,** 931.
308. P. K. Freeman, R. S. Raghavan, and D. G. Kuper, *J. Am. Chem. Soc.*, **93,** 5288 (1971).
309. P. B. Shevlin and A. P. Wolf, *Tetrahedron lett.*, **1970** 3987.
310. W. M. Jones, M. E. Stowe, E. E. Wells, Jr., and E. W. Lester, *J. Am. Chem. Soc.*, **90,** 1849 (1968).
311. R. Breslow, J. Brown, and J. J. Gajewski, *J. Am. Chem. Soc.*, **89,** 4383 (1967).
312. R. Breslow, L. J. Altman, A. Krebs, E. Mohacsi, I. Murata, R. A. Peterson, and J. Posner, *J. Am. Chem. Soc.*, **87,** 1326 (1965).
313. W. M. Jones and J. M. Denham, *J. Am. Chem. Soc.*, **86,** 944 (1964).
314. W. M. Jones and M. E. Stowe, *Tetrahedron Lett.*, **1964,** 3459.
315. W. M. Jones and C. L. Ennis, *J. Am. Chem. Soc.*, **89,** 3069 (1967); **91,** 6391 (1969).
316. T. Mukai, T. Nakazawa, and K. Isobe, *Tetrahedron Lett.*, **1968,** 565.
317. W. M. Jones, B. N. Hamon, R. C. Joines, and C. L. Ennis, *Tetrahedron Lett.*, **1969,** 3909.
318. A. T. Blomquist and C. F. Heins, *J. Org. Chem.*, **34,** 2906 (1969).
319. W. Kirmse, L. Horner, and H. Hoffmann, *Ann.*, **614,** 19 (1958).
320. J. E. Basinski, Ph.D. thesis, Yale University, New Haven, Conn., 1961.
321. R. A. Moss, *J. Org. Chem.*, **31,** 3296 (1966).
322. R. A. Moss and J. R. Przybyla, *J. Org. Chem.*, **33,** 3816 (1968).
323. M. Jones, Jr., unpublished work.
324. E. Wasserman, L. Barash, A. M. Trozzolo, R. W. Murray, and W. A. Yager, *J. Am. Chem. Soc.*, **86,** 2304 (1964).
325. D. Schönleber, *Chem. Ber.*, **102,** 1789 (1969).
326. M. Jones, Jr., R. N. Hochman, and J. D. Walton, *Tetrahedron Lett.*, **1970,** 2617.
327. D. Schönleber, *Angew. Chem.*, *Int. Ed.* **8,** 76 (1969).
328. M. Jones, Jr., *J. Org. Chem.*, **33,** 2538 (1969).
329. See, for instance, J. F. M. Oth, R. Merényi, H. Röttele, and G. Schröder, *Tetrahedron Lett.*, **1968,** 3941.
330. H. Dürr and G. Scheppers, *Chem. Ber.*, **100,** 3236 (1967).
331. H. Dürr and L. Schrader, *Chem. Ber.*, **102,** 2026 (1969).
332. H. Dürr, G. Scheppers, and L. Schrader, *Chem. Commun.*, **1969,** 257.
333. H. Dürr and L. Schrader, *Angew. Chem. Int. Ed.*, **8,** 446 (1969).
334. H. Dürr, L. Schrader, and H. Seidl, *Chem. Ber.*, **104,** 391 (1971).
335. H. Dürr and G. Scheppers, *Angew. Chem. Int. Ed.*, **7,** 371 (1968).
336. H. Dürr and G. Scheppers, *Ann.*, **734,** 141 (1970).
337. H. Dürr and G. Scheppers, *Chem. Ber.*, **103,** 380 (1970).
338. E. T. McBee, J. A. Bosoms, and C. J. Morton, *J. Org. Chem.*, **31,** 768 (1966).
339. E. T. McBee, G. W. Calundann, and T. Hodgins, *J. Org. Chem.*, **31,** 4260 (1966).
340. D. Lloyd and M. I. C. Singer, *Chem. Ind.*, **1967,** 510.
341. D. Lloyd and M. I. C. Singer, *Chem. Ind.*, **1967,** 118.
342. D. Lloyd and M. I. C. Singer, *Chem. Ind.*, **1967,** 787.
343. D. Lloyd, M. I. C. Singer, M. Regitz, and A. Liedhegener, *Chem. Ind.*, **1967,** 324.
344. I. B. M. Band, D. Lloyd, M. I. C. Singer, and F. T. Wasson, *Chem. Commun.*, **1966,** 544.
345. D. Lloyd and M. I. C. Singer, *J. Chem. Soc.* (*C*), **1971,** 2939.
346. D. Lloyd and M. I. C. Singer, *Chem. Commun.*, **1967,** 390.
347. B. H. Freeman and D. Lloyd, *Chem. Commun.*, **1970,** 924.
348. D. Lloyd and N. W. Preston, *Chem. Ind.*, **1966,** 1039.
349. M. Jones, Jr., A. M. Harrison, and K. R. Rettig, *J. Am. Chem. Soc.*, **91,** 7462 (1969).
350. A. J. Fry, *J. Am. Chem. Soc.*, **87,** 1816 (1965).

351. M. Jones, Jr., *Angew. Chem. Int. Ed.*, **8,** 76 (1969).
352. M. Jones, Jr., and A. M. Harrison, unpublished work.
353. M. Jones, Jr., T. Berdick, and R. H. Levin, unpublished work.
354. H. Dannenberg and H. J. Gross, *Tetrahedron*, **21**, 1611 (1965).
355. W. G. Dauben, M. E. Lorber, N. D. Vietmeyer, R. A. Shapiro, J. H. Duncan, and K. Tomer, *J. Am. Chem. Soc.*, **90,** 4762 (1968).
356. W. Kirmse and L. Ruetz, *Ann.*, **726,** 30 (1969).
357. M. Jones, Jr., and K. R. Rettig, unpublished work.
358. G. F. Koser and W. H. Pirkle, *J. Org. Chem.*, **32,** 1992 (1967).
359. M. J. S. Dewar and K. Narayanaswami, *J. Am. Chem. Soc.*, **86,** 2422 (1964).
360. W. H. Pirkle and G. F. Koser, *J. Am. Chem. Soc.*, **90,** 3598 (1968).
361. H. D. Hartzler, *J. Am. Chem. Soc.*, **86,** 2174 (1964).
362. D. Y. Curtin, J. A. Kampmeier, and B. R. O'Connor, *J. Am. Chem. Soc.*, **87,** 863 (1965).
363. M. S. Newman and A. O. M. Okorodudu, *J. Am. Chem. Soc.*, **90,** 4189 (1968).
364. M. S. Newman and T. B. Patrick, *J. Am. Chem. Soc.*, **91,** 6461 (1969).
365. M. S. Newman and T. B. Patrick, *J. Am. Chem. Soc.*, **92,** 4312 (1970).
366. M. S. Newman and T. B. Patrick, *J. Am. Chem. Soc.*, **92,** 1793 (1970).
367. M. S. Newman and C. D. Beard, *J. Org. Chem.*, **35,** 2412 (1970).
368. D. J. Northington and W. M. Jones, *Tetrahedron Lett.*, **1971,** 317.
369. C. D. Gutsche, G. L. Bachman, and R. S. Coffey, *Tetrahedron*, **18,** 617 (1962).
370. G. L. Bachman, *Diss. Abstr.*, **25,** 1556 (1964).
371. G. L. Closs and R. A. Moss, *J. Am. Chem. Soc.*, **86,** 4042 (1964).
372. H. Dietrich, G. W. Griffin, and R. C. Petterson, *Tetrahedron Lett.*, **1968,** 153.
373. C. D. Gutsche, G. L. Bachman, W. Odell, and S. Baüerlein, *J. Am. Chem. Soc.*, **93,** 5172 (1971).
374. T. A. Baer and C. D. Gutsche, *J. Am. Chem. Soc.*, **93,** 5180 (1971).
375. G. W. Griffin, private communication with permission to cite.
376. H. Kristinsson and G. W. Griffin, *J. Am. Chem. Soc.*, **88,** 1579 (1966).
377. P. Scheiner, *J. Org. Chem.*, **34,** 199 (1969); *Tetrahedron Lett.*, **1971,** 4489.
378. A. M. Trozzolo, R. W. Murray, and E. Wasserman, *J. Am. Chem. Soc.*, **84,** 4990 (1962).
379. E. Wasserman, A. M. Trozzolo, W. A. Yager, and R. W. Murray, *J. Chem. Phys.*, **40,** 2408 (1964).
380. R. A. Moss and U.-H. Dolling, *J. Am. Chem. Soc.*, **93,** 954 (1971).
380a. J. F. Ogilvie, *Photochem. and Photobiol.*, **9,** 65 (1969).
381. C. G. Overberger and J.-P. Anselme, *J. Org. Chem.*, **29,** 1188 (1964).
382. I. Moritani, Y. Yamamoto, and S. I. Murahashi, *Tetrahedron Lett.*, **1968,** 5697.
383. I. Moritani, Y. Yamamoto, and S.-I. Murahashi, *Tetrahedron Lett.*, **1968,** 5755.
384. P. B. Sargeant and H. Shechter, *Tetrahedron Lett.*, **1971,** 3957.
385. H. M. R. Hoffmann, *Angew. Chem. Int. Ed.*, **8,** 556 (1969).
386. M. E. Hendrick and M. Jones, Jr., unpublished observation.
387. W. Kirmse and H. Dietrich, *Chem. Ber.*, **100,** 2710 (1967).
388. R. Garner, Tetrahedron Letters, 221 (1968).
389. H. Dürr, *Angew. Chem. Int. Ed.*, **6,** 1084 (1967).
390. J. A. Landgrebe and A. G. Kirk, *J. Org. Chem.*, **32,** 3499 (1967).
391. H. E. Zimmerman and J. H. Munch, *J. Am. Chem. Soc.*, **90,** 187 (1968).
392. J. H. Robson and H. Shechter, *J. Am. Chem. Soc.*, **89,** 7112 (1967).
393. R. M. Etter, H. S. Skovronek, and P. S. Skell, *J. Am. Chem. Soc.*, **81,** 1008 (1959).
394. G. L. Closs and L. E. Closs, *Angew. Chem., Int. Ed.* **1,** 334 (1962).
395. P. W. Humer, Ph.D. thesis, Pennsylvania State University, College Park, 1964.

395a. G. L. Closs and L. E. Closs, unpublished work reported in footnote 42 of reference 380.
396. M. Jones, Jr., W. J. Baron, and Y. H. Shen, *J. Am. Chem. Soc.*, **92,** 4745 (1970).
397. J. E. Hodgkins and M. P. Hughes, *J. Org. Chem.*, **27,** 4187 (1962).
398. I. A. D'yakanov, G. U. Golodnikov, and I. B. Repinskaya, *Zh. Org. Khim.*, **1,** 220. (1965); *J. Org. Chem. USSR*, **1,** 210 (1965).
399. I. A. D'yakonov, I. B. Repinskaya, and T. D. Marinina, *Zh. Obsch. Khim.*, **39,** 717, (1969); *J. Gen. Chem. USSR*, **39,** 684 (1969).
400. I. A. D'yakanov, I. M. Stroiman, and A. G. Vitenberg, *Zh. Org. Khim.*, **6,** 42 (1970); *J. Org. Chem. USSR*, **6,** 41 (1970).
401. I. A. D'yakonov and T. A. Kornilova, *Zh. Org. Khim.*, **2,** 1317 (1966), *J. Org. Chem. USSR*, **2,** 1314 (1966).
402. R. W. Murray, A. M. Trozzolo, E. Wasserman, and W. A. Yager, *J. Am. Chem. Soc.*, **84,** 3213 (1962).
403. R. W. Brandon, G. L. Closs, and C. A. Hutchinson, Jr., *J. Chem. Phys.*, **37,** 1878 (1962).
404. A. M. Trozzolo, R. W. Murray, and E. Wasserman, *J. Am. Chem. Soc.*, **84,** 4990 (1962).
405. E. Wasserman, A. M. Trozzolo, W. A. Yager, and R. W. Murray, *J. Chem. Phys.*, **40,** 2408 (1964).
406. A. M. Trozzolo, E. Wasserman, and W. A. Yager, *J. Chim. Phys.*, **61,** 1663 (1964).
407. R. W. Brandon, G. L. Closs, C. E. Davoust, C. A. Hutchinson, Jr., B. E. Kohler, and R. Silbey, *J. Chem. Phys.*, **43,** 2006 (1965).
408. A. M. Trozzolo, W. A. Yager, G. W. Griffin, H. Kirstinsson, and I. Sarkar, *J. Am. Chem. Soc.*, **89,** 3357 (1967).
409. R. E. Moser, J. M. Fritsch, and C. N. Matthews, *Chem. Commun.* **1967,** 770.
410. L. Barasch, E. Wasserman, and W. A. Yager, *J. Am. Chem. Soc.*, **89,** 3931 (1967).
411. F. Bölsing, *Tetrahedron Lett.*, **1968,** 4299.
412. C. A. Hutchinson, Jr., and B. E. Kohler, *J. Chem. Phys.*, **51,** 3327 (1969).
413. For a review see A. M. Trozzolo, *Acc. Chem. Res.*, **1,** 329 (1968).
414. N. D. Epiotis, Princeton University, private communication with permission to cite.
415. H. E. Zimmerman and D. H. Paskovich, *J. Am. Chem. Soc.*, **86,** 2149 (1964).
416. H. Reimlinger and R. Paulisson, *Tetrahedron Lett.*, **1970,** 3143, and references therein.
417. R. Hoffmann, R. Gleiter, and F. B. Mallory, *J. Am. Chem. Soc.*, **92,** 1460 (1970). See also 418.
418. H. Basch, *J. Chem. Phys.*, **55,** 1700 (1971).
419. M. Jones, Jr., and K. R. Rettig, *J. Am. Chem. Soc.*, **87,** 4013, 4015 (1965).
420. W. J. Baron, unpublished work.
421. D. Bethell, A. R. Newall, and D. Whittaker, *J. Chem. Soc. (B)*, 23 (1971).
422. D. Bethel, G. Stevens, and P. Tickle, *Chem. Commun.*, **1970,** 792.
423. G. A. Hamilton and J. R. Giacin, *J. Am. Chem. Soc.*, **8,** 1584 (1966).
424. P. D. Bartlett and T. G. Traylor, *J. Am. Chem. Soc.*, **84,** 3408 (1962).
425. R. W. Murray and A. Suzui, *J. Am. Chem. Soc.*, **93,** 4963 (1971).
426. M. E. Hendrick, W. J. Baron, and M. Jones, Jr., *J. Am. Chem. Soc.*, **93,** 1554 (1971).
427. J. O. Stoffer and H. R. Musser, *Chem. Commun.*, **1970,** 481.
428. J. C. Sheehan and I. Lengyel, *J. Org. Chem.*, **28,** 3252 (1963).
429. W. J. Middleton and W. H. Sharkey, *J. Org. Chem.*, **30,** 1384 (1965).
430. A. Schönberg, W. Knöfel, E. Frese, and K. Praefche, *Chem. Ber.*, **103,** 949 (1970).
431. A. N. Pudovik, R. D. Gareev, and L. A. Stabrovskaya, *Zh. Obsch. Khim.*, **40,** 698 (1970). *J. Gen. Chem. USSR*, **40,** 668 (1970).

432. R. Wheland and P. D. Bartlett, *J. Am. Chem. Soc.*, **92,** 6057 (1970).
433. See, however, references 94 and 95.
434. T. D. Walsh and D. R. Powers, *Tetrahedron Lett.*, **1970,** 3855.
435. W. J. Baron, unpublished observation.
436. G. Frater and O. P. Strausz, *J. Am. Chem. Soc.*, **92,** 6654 (1970).
437. A. Schönberg, E. Frese, W. Knöfel, and K. Praefcke, *Chem. Ber.*, **103,** 938 (1970).
438. M. M. Bagga, G. Ferguson, J. A. D. Jeffreys, C. M. Mansell, P. L. Pauson, I. C. Robertson, and J. G. Sime, *Chem. Commun.*, **1970,** 672.
439. A. A. Lamola, B.S. thesis, Massachusetts Institute of Technology, Cambridge, Mass, 1961.
440. See, for example, I. A. D'yakonov, V. P. Dushina, and G. V. Golodnikov, *Zh. Obsch. Khim.*, **39,** 923 (1969); *J. Gen. Chem. USSR*, **39,** 887 (1969).
441. E. Funakubo, I. Moritani, T. Nagai, S. Nishida, and S.-I. Murahashi, *Tetrahedron Lett.*, **1963,** 1069.
442. S.-I. Murahashi, I. Moritani, and T. Nagai, *Bull. Chem. Soc. Japan*, **40,** 1655 (1967).
443. C. A. Hutchinson, Jr., and G. A. Pearson, *J. Chem. Phys.*, **43,** 2545 (1965).
444. C. A. Hutchinson, Jr., and G. A. Pearson, *J. Chem. Phys.*, **47,** 520 (1967).
445. J. D. Walton, A.B. thesis, Princeton University, Princeton, N.J., 1968.
446. M. Jones, Jr., W. Ando, M. E. Hendrick, A. Kulczycki, Jr., P. M. Howley, K. M. Hummel, and D. S. Malament, *J. Am. Chem. Soc.*, **94,** 7469 (1972).
447. D. J. Atkinson, M. J. Perkins, and P. Ward, *J. Chem. Soc. (C)*, **1971,** 3247.
448. H. E. Zimmerman, D. S. Crumrine, D. Döpp, and P. S. Huyffer, *J. Am. Chem. Soc.*, **91,** 434 (1969).
449. H. Dürr and H. Kober, *Angew. Chem. Int. Ed.*, **10,** 342 (1971).
450. M. Jones, Jr., Ph.D. thesis, Yale University, New Haven, Conn., 1963.
451. J. E. Baldwin and A. H. Andrist, *Chem. Commun.*, **1971,** 1512.
452. I. Moritani, S.-I. Murahashi, K. Yoshinaga, and H. Ashitaka, *Bull. Chem. Soc. Japan*, **40,** 1506 (1967).
453. A. E. Greene, A.B. thesis, Princeton University, Princeton, N.J., 1966.
454. I. Moritani, S.-I. Murahashi, H. Ashitaka, K. Kimura, and H. Tsubomura, *J. Amer. Chem. Soc.*, **90,** 5918 (1968).
455. I. Moritani, S.-I. Murahashi, M. Nishino, K. Kimura, and H. Tsubomura, *Tetrahedron Lett.*, **1966,** 1373.
456. S.-I. Murahashi, I. Moritani, and M. Nishino, *Tetrahedron*, **27,** 5131 (1971).
457. S.-I. Murahashi, I. Moritani, and M. Nishino, *J. Am. Chem. Soc.*, **89,** 1257 (1967).
458. I. Moritani, S.-I. Murahashi, M. Nishino, Y. Yamamoto, K. Itoh, and N. Mataga, *J. Am. Chem. Soc.*, **89,** 1259 (1967).
459. Y. Yamamoto, I. Moritani, Y. Magda, and S.-I. Murahashi, *Tetrahedron*, **26,** 251 (1970).
460. J. C. Fleming and H. Shechter, *J. Org. Chem.*, **34,** 3962 (1969).
461. N. Filipescu and J. W. Pavlik, *J. Chem. Soc. (C)*, **1970,** 1851.
462. G. Cauquis and G. Reverdy, *Tetrahedron Lett.*, **1967,** 1493.
463. G. Cauquis and G. Reverdy, *Tetrahedron Lett.*, **1968,** 1085.
464. G. Cauquis and G. Reverdy, *Tetrahedron Lett.*, **1971,** 3771.
465. G. Cauquis and G. Reverdy, *Tetrahedron Lett.*, **1971,** 4289.
466. G. Cauquis, B. Divisia, M. Rastoldo, and G. Reverdy, *Bull. Soc. Chim. France*, **1971,** 3022.
467. G. Cauquis, B. Divisia, and G. Reverdy, *Bull. Soc. Chim. France*, **1971,** 3027.
468. G. Cauquis, B. Divisia, and G. Reverdy, *Bull. Soc. Chim. France*, **1971,** 3031.
469. M. E. Hendrick, unpublished work.
470. R. A. Moss and J. D. Funk, *J. Chem. Soc. (C)*, **1967,** 2026.

471. A. Sonoda, I. Moritani, T. Saraie, and T. Wada, *Tetrahedron Lett.*, **1969,** 2943.
472. A. Sonoda and I. Moritani, *Bull. Chem. Soc. Japan*, **43,** 3522 (1970).
473. A. Sonoda and I. Moritani, *J. Organometal. Chem.*, **26,** 133 (1971).
474. A. Sonoda, I. Moritani, S. Yasuda, and T. Wada, *Tetrahedron*, **26,** 3075 (1970).
475. A. Sonoda and I. Moritani, *J. Organomet. Chem.*, **26,** 133 (1971).
476. M. Imoto, T. Nakaya, T. Tomomoto, and O. Ohashi, *J. Polymer Sci. (B)*, **4,** 955 (1966).
477. T. Nakaya, K. Ohashi, and M. Imoto, *Makromol. Chem.*, **111,** 115 (1968).
478. T. Nakaya, T. Tomomoto, and M. Imoto, *Bull. Chem. Soc. Japan*, **40,** 691 (1967).
479. A. M. Trozzolo, E. Wasserman, and W. A. Yager, *J. Am. Chem. Soc.*, **87,** 129 (1965).
480. E. Wasserman, V. J. Kuck, W. A. Yager, R. S. Hutton, F. D. Greene, V. P. Abegg, and N. M. Weinshenker, *J. Am. Chem. Soc.*, **93,** 6335 (1971).
481. R. V. Hoffman and H. Shechter, *J. Am. Chem. Soc.*, **93,** 5940 (1971).
482. A. M. Trozzolo, R. W. Murray, G. Smolinsky, W. A. Yager, and E. Wasserman, *J. Am. Chem. Soc.*, **85,** 2526 (1963).
483. E. Wasserman, R. W. Murray, W. A. Yager, A. M. Trozzolo, and G. Smolinsky, *J. Am. Chem. Soc.*, **89,** 5076 (1967).
484. K. Itoh, *Chem. Phys. Lett.*, **1,** 235 (1967).
485. K. Itoh, H. Konishi, and N. Mataga, *J. Chem. Phys.*, **48,** 4789 (1968).
486. R. W. Murray and M. L. Kaplan, *J. Am. Chem. Soc.*, **88,** 3527 (1966).
487. A. Schönberg and K. Junghans, *Chem. Ber.*, **95,** 2137 (1962).
488. H. K. Reimlinger, *Chem. Ber.*, **97,** 3493 (1964).
489. A. Schönberg and K. Junghans, *Chem. Ber.*, **98,** 820 (1965).
490. D. J. Abraham, T. G. Cochran, and R. D. Rosenstein, *J. Am. Chem. Soc.*, **93,** 6279 (1971), and references therein.
491. M. Regitz, H. Scherer, and W. Anschütz, *Tetrahedron Lett.*, **1970,** 753.
491a. H. Gunther, B. D. Tunggal, M. Regitz, H. Scherer, and T. Keller, *Angew. Chem. Int. Ed.*, **10,** 563 (1971).
492. H. Staudinger and R. Endle, *Ber.*, **46,** 1437 (1913).
493. H. Staudinger, E. Anthes, and F. Pfenninger, *Ber.*, **49,** 1928 (1916).
494. R. M. Magid, P. P. Gaspar, M. R. Willcott, III, and M. Jones, Jr., private communications with permission to cite.
495. A. G. Harrison and F. P. Lossing, *J. Am. Chem. Soc.*, **82,** 1052 (1960).
496. F. O. Rice and J. D. Michaelsen, *J. Phys. Chem.*, **66,** 1535 (1962).
497. V. Franzen and H.-I. Joschek, *Ann.*, **633,** 7 (1960).
498. W. D. Crow and C. Wentrup, *Tetrahedron Lett.*, **1965,** 6149, and references therein.
499. C. Wentrup, *Tetrahedron*, **27,** 367 (1971), and references therein.
500. K. E. Krajac, T. Mitsuhashi, and W. M. Jones, *J. Am. Chem. Soc.*, **94,** 3661 (1972), and references therein.
501. R. C. Joines, A. B. Turner, and W. M. Jones, *J. Am. Chem. Soc.*, **91,** 7754 (1969).
502. C. Wentrup and K. Wilczek, *Helv. Chim. Acta*, **53,** 1459 (1970).
503. P. Ashkenazi, S. Lupan, A. Schwarz, and M. Cais, *Tetrahedron Lett.*, **1969,** 817.
504. W. J. Baron, M. Jones, Jr., and P. P. Gaspar, *J. Am. Chem. Soc.*, **92,** 4739 (1970).
505. G. G. Vander Stouw, A. R. Kraska, and H. Shechter, *J. Am. Chem. Soc.*, **94,** 1655 (1972).
506. See also G. G. Vander Stouw, Ph.D. thesis, Ohio State University, Columbus, Ohio, 1964.
507. P. Schissel, M. E. Kent, D. J. McAdoo, and E. Hedaya, *J. Am. Chem. Soc.*, **92,** 2147 (1970).
508. E. Hedaya and M. E. Kent, *J. Am. Chem. Soc.*, **93,** 3283 (1971).
509. J. A. Myers, R. C. Joines, and W. M. Jones, *J. Am. Chem. Soc.*, **92,** 4740 (1970).

510. F. Weygand and H. J. Bestmann, *Angew. Chem.*, **72,** 535 (1960).
511. W. von E. Doering and L. H. Knox, *J. Am. Chem. Soc.*, **83,** 1989 (1961).
512. R. R. Sauers and R. J. Kiesel, *J. Am. Chem. Soc.*, **89,** 4695 (1967).
513. L. E. Helgen, Ph.D. thesis, Yale University, New Haven, Conn., 1965.
514. A. Ritter and L. H. Sommer, *Sci. Commun. Prague*, **1965,** 279.
515. A. G. Brook, J. M. Duff, and D. G. Anderson, *J. Am. Chem. Soc.*, **92,** 7567 (1970).
516. W. Kirmse, H. Dietrich, and H. W. Bücking, *Tetrahedron Lett.*, **1967,** 1833.
517. J. A. Kaufman and S. J. Weininger, *Chem. Commun.*, **1969,** 593.
518. H. Ledon, G. Cannic, G. Linstrumelle, and S. Julia, *Tetrahedron Lett.*, **1970,** 3971.
519. G. Lowe and J. Parker, *Chem. Commun.*, **1971,** 577.
520. Y. Yamamoto and I. Moritani, *Tetrahedron Lett.*, **1969,** 3087.
521. H. Lind and A. J. Deutschman, Jr., *J. Org. Chem.*, **32,** 326 (1967).
522. M. Cocivera and H. D. Roth, *J. Am. Chem. Soc.*, **92,** 2573 (1970).
523. R. R. Rando, *J. Am. Chem. Soc.*, **92,** 6706 (1970).
524. E. Buchner and J. Geronimus, *Ber.*, **36,** 3782 (1903).
525. I. A. D'yakanov, M. I. Komendantov, F. Gui-siya, and L. G. Korichev, *Zh. Obsch. Khim.*, **32,** 928 (1962). *J. Gen. Chem. USSR.*, **32,** 917 (1962).
526. J. Warkentin, E. Singleton, and J. F. Edgar, *Can. J. Chem.*, **43,** 3456 (1965).
527. P. S. Skell and R. M. Etter, *Proc. Chem. Soc.*, **1961,** 443.
528. H. Nozaki, S. Moriuti, H. Takaya, and K. Noyori, *Tetrahedron Lett.*, **1966,** 5239.
529. H. Nozaki, H. Takaya, S. Moriuti, and K. Noyori, *Tetrahedron*, **24,** 3655 (1968).
530. W. R. Moser, *J. Am. Chem. Soc.*, **91,** 1135 (1969).
531. W. Kirmse and H. Dietrich, *Chem. Ber.*, **98,** 4027 (1965).
532. M. Jones, Jr., A. Kulczycki, Jr., and K. F. Hummel, *Tetrahedron Lett.*, **1967,** 183.
533. M. Jones, Jr., W. Ando, and A. Kulczycki, Jr., *Tetrahedron Lett.*, **1967,** 1391.
534. The ground states of such species are not known, since ESR signals have not been obtained.
535. J. A. Berson, D. R. Hartter, H. Klinger, and P. W. Grubb, *J. Org. Chem.*, **33,** 1669 (1968).
536. W. von E. Doering, G. Laber, R. Vonderwahl, N. F. Chamberlain, and R. B. Williams, *J. Am. Chem. Soc.*, **78,** 5448 (1956).
537. G. Linsrumelle, *Tetrahedron Lett.*, **1970,** 85.
538. J. E. Baldwin and R. A. Smith, *J. Org. Chem.*, **32,** 3511 (1967).
539. J. E. Baldwin and R. A. Smith, *J. Am. Chem. Soc.*, **89,** 1886 (1967).
540. E. Büchner and S. Hediger, *Ber.*, **36,** 3502 (1903).
541. R. Huisgen and G. Juppe, *Chem. Ber.*, **94,** 2332 (1961).
542. W. von E. Doering and M. J. Goldstein, *Tetrahedron*, **5,** 53 (1959).
543. S. H. Graham, D. M. Pugh, and A. J. S. Williams, *J. Chem. Soc. (C)*, **1969,** 68.
544. T. V. Domareva-Mandel'shtam and I. A. D'yakanov, *Zh. Obsh. Khim.*, **34,** 3844 (1964). *J. Gen. Chem. USSR.*, **34,** 3896 (1964).
545. T. V. Domareva-Mandel'shram, I. A. D'yakanov, and L. D. Kristol, *Zh. Org. Khim.*, **2,** 2263 (1966). *J. Org. Chem. USSR.*, **2,** 2223 (1966).
546. J. Meinwald and E. G. Miller, *Tetrahedron Lett.*, **1961,** 253.
547. G. E. Hall and J. P. Ward, *Tetrahedron Lett.*, **1965,** 437.
548. A. S. Kende and P. T. MacGregor, *J. Am. Chem. Soc.*, **86,** 2088 (1964).
549. E. Ciganek, *J. Org. Chem.*, **30,** 4366 (1965).
550. M. Vidal, F. Massot, and P. Arnoud, *Comp. Rend. Ser. C.*, **268,** 423 (1969).
551. I. A. D'yakonov, M. I. Komendantov, and S. P. Korshunov, *J. Gen. Chem. USSR*, **32,** 912 (1962).
552. I. Moritani, T. Hosokawa, and N. Obata, *J. Org. Chem.*, **34,** 670 (1969).
553. See reference 1, pages 100–101, for a discussion of furan formation.

554. M. E. Hendrick, *J. Am. Chem. Soc.*, **93**, 6337 (1971).
555. E. Buchner, *Ber.*, **28**, 215 (1895).
556. U. Schöllkopf and N. Rieber, *Angew. Chem.*, **79**, 906 (1967).
557. U. Schöllkopf, D. Hoppe, N. Rieber, and V. Jacobi, *Ann.*, **730**, 1 (1969).
558. U. Schöllkopf and N. Rieber, *Angew. Chem. Int. Ed.*, **6**, 261 (1967).
559. U. Schöllkopf and N. Rieber, *Chem. Ber.*, **102**, 488 (1969).
560. F. Gerhart, U. Schöllkopf, and H. Schumacher, *Angew. Chem. Int. Ed.*, **6**, 74 (1967).
561. U. Schöllkopf, F. Gerhart, M. Reetz, H. Frasnelli, and H. Schumacher, *Ann.*, **716**, 204 (1968).
562. U. Schöllkopf and M. Reetz, *Tetrahedron Lett.*, **1969**, 1541.
563. U. Schöllkopf and H. Frasnelli, *Angew. Chem., Int. Ed.*, **9**, 301 (1970).
564. T. DoMinh, H. E. Gunning, and O. P. Strausz, *J. Am. Chem. Soc.*, **89**, 6785 (1967).
565. O. P. Strausz, T. DoMinh, and J. Font, *J. Am. Chem. Soc.*, **90**, 1930 (1968).
566. J. Kučera and Z. Arnold, *Tetrahedron Lett.*, **1966**, 1109.
567. Z. Arnold, *Chem. Commun.*, **1967**, 299.
568. J. Kučera, Z. Janoušek, and Z. Arnold, *Coll. Czech. Chem. Commun.*, **35**, 3618 (1970).
569. S. Masamune, *J. Am. Chem. Soc.*, **86**, 735 (1964).
570. W. von E. Doering and M. Pomerantz, *Tetrahedron Lett.*, **1964**, 961.
571. P. K. Freeman and D. G. Kuper, *Chem. Ind.*, **1965**, 424.
572. J. Meinwald and G. H. Wahl, Jr., *Chem Ind.*, **1965**, 425.
573. S. Masamune and N. T. Castellucci, *Proc. Chem. Soc.*, **1964**, 298.
574. S. Masamune and K. Fukumoto, *Tetrahedron Lett.*, **1965**, 4647.
575. E. J. Moriconi and J. J. Murray, *J. Org. Chem.*, **29**, 3577 (1964).
576. U. Simon, O. Süs, and L. Horner, *Ann.*, **697**, 17 (1966).
577. O. Tsuge, I. Shinkai, and M. Koga, *J. Org. Chem.*, **36**, 745 (1971).
578. W. von E. Doering, B. M. Ferrier, E. T. Fossel, J. H. Hartenstein, M. Jones, Jr., G. Klumpp, R. M. Rubin, and M. Saunders, *Tetrahedron*, **23**, 3943 (1967).
579. W. von E. Doering, E. T. Fossel, and R. L. Kaye, *Tetrahedron* **21**, 25 (1965).
580. M. M. Fawzi and C. D. Gutsche, *J. Org. Chem.*, **31**, 1390 (1961).
581. K. Mori and M. Matsui, *Tetrahedron Lett.*, **1969**, 2729.
582. A. S. Monahan, *J. Org. Chem.*, **33**, 1441 (1968).
583. H. O. House and C. J. Blankley, *J. Org. Chem.*, **33**, 53 (1968).
584. M. Takebayashi, T. Ibata, H. Kohara, and K. Ueda, *Bull. Chem. Soc. Japan*, **42**, 2938 (1969).
585. L. T. Scott, Ph.D. thesis, Harvard University, Cambridge, Mass., 1970.
586. A. Costantino, G. Linstrumelle, and S. Julia, *Bull. Soc. Chim. France*, 907, 912 (1970).
587. E. Vogel, A. Vogel, H.-K. Kübbeler, and W. Sturm, *Angew. Chem. Int. Ed.*, **9**, 514 (1970).
588. F. Weygand, W. Schwenke, and H. J. Bestmann, *Angew. Chem.*, **70**, 506 (1958).
589. F. Weygand, H. Dworschak, K. Koch, and S. Konstas, *Angew. Chem.*, **73**, 409 (1961).
590. D. W. Kurtz and H. Shechter, *Chem. Commun.*, **1966**, 689.
591. J. K. Chakrabarti, S. S. Szinai, and A. Todd, *J. Chem. Soc. (C)*, 1303 (1970).
592. R. Huisgen, in *Aromaticity*, special publication No. 21 of the Chemical Society (1967).
593. W. Ried and H. Mengler, *Ann.*, **678**, 113 (1964).
594. R. Huisgen, G. Binsch, and L. Ghosez, *Chem. Ber.*, **97**, 2628 (1964),
595. R. Huisgen, H. J. Sturm, and G. Binsch, *Chem. Ber.*, **97**, 2864 (1964).
596. H. Dworschak and F. Weygand, *Chem. Ber.*, **101**, 302 (1968).

597. H. Dworschak and F. Weygand, *Chem. Ber.*, **101,** 289 (1968).
598. G. Binsch, R. Huisgen, and H. König, *Chem. Ber.*, **97,** 2893 (1964).
599. R. Huisgen, G. Binsch, and H. König, *Chem. Ber.*, **97,** 2884 (1964).
600. R. Huisgen, G. Binsch, and H. König, *Chem. Ber.*, **97,** 2868 (1964).
601. W. Ried and W. Radt, *Ann.*, **688,** 170 (1965).
602. W. Ried and E.-A. Baumbach, *Ann.*, **713,** 139 (1968).
603. W. Ando, T. Yagihara, S. Tozune, and T. Migita, *J. Am. Chem. Soc.*, **91,** 2786 (1969).
604. I. Zugravescu, E. Rucinschi, and G. Surpateanu, *Tetrahedron Lett.*, **1970,** 941.
605. Reference 1, pages 106–111.
606. H. Nozaki, H. Takaya, and R. Noyori, *Tetrahedron Lett.*, **1965,** 2563.
607. H. Nozaki, H. Takaya, and R. Noyori, *Tetrahedron*, **22,** 3393 (1966).
608. A. Schönberg and K. Praefcke, *Chem Ber.*, **99,** 196 (1966).
609. W. Ando, K. Nakayama, K. Ichibori, and T. Migita, *J. Am. Chem. Soc.*, **91,** 5164 (1969).
610. W. Ando, T. Yagihara, S. Kondo, K. Nakayama, H. Yamato, S. Nakaido, and T. Migita, *J. Org. Chem.*, **36,** 1732 (1971).
611. W. Ando, S. Konda, and T. Migita, *J. Am. Chem. Soc.*, **91,** 6516 (1969).
612. W. Ando, S. Kondo, and T. Migita, *Bull. Chem. Soc. Japan*, **44,** 571 (1971).
613. E. Ciganek, *J. Org. Chem.*, **35,** 862 (1970).
614. W. E. Bachmann and W. S. Struve, *Organic Reactions*, Vol. I, 1942, page 38.
615. F. Weygand and H. S. Bestmann, *Syntheses Using Diazoketones. Newer Methods of Preparative Organic Chemistry*, Vol. III, Academic Press, New York, 1964, pp. 451–508.
616. J. L. Mateos, A. Dosal, and C. Carbajal, *J. Org. Chem.*, **30,** 3578 (1965).
617. T. Gibson and W. F. Erman, *J. Org. Chem.*, **31,** 3028 (1966).
618. K. B. Wiberg and A. de Meijere, *Tetrahedron Lett.*, **1969,** 519.
619. A. J. Ashe, *Tetrahedron Lett.*, **1969,** 523.
620. M. P. Cava and R. J. Spangler, *J. Am. Chem. Soc.*, **89,** 4550 (1967).
621. W. Lwowski, *Angew. Chem. Int. Ed.*, **6,** 897 (1967).
622. F. Kaplan and G. K. Meloy, *J. Am. Chem. Soc.*, **88,** 950 (1966).
623. M. S. Newman and A. Arkell, *J. Org. Chem.*, **24,** 385 (1959).
624. P. Kinson and B. M. Trost, *Tetrahedron Lett.*, **1969,** 1075.
625. W. Jugelt and D. Schmidt, *Tetrahedron*, **25,** 969 (1969).
626. W. Bartz and M. Regitz, *Chem. Ber.*, **103,** 1463 (1970).
627. A. Padwa and R. Layton, *Tetrahedron Lett.*, **1965,** 2167.
628. A. L. Wilds, N. F. Woolsey, J. Van Den Berghe, and C. H. Winestock, *Tetrahedron Lett.*, **1965,** 4841.
629. E. Fahr, K. Keil, H. Lind, and F. Scheckenbach, *Z. Natur.*, **20b,** 526 (1965).
630. A. Melzer and E. F. Jenny, *Tetrahedron Lett.*, **1968,** 4503.
631. D. O. Cowan, M. M. Couch, K. R. Kopecky, and G. S. Hammond, *J. Org. Chem.*, **29,** 1922 (1964).
632. M. Jones, Jr., and W. Ando, *J. Am. Chem. Soc.*, **90,** 2200 (1968).
633. A. M. Trozzolo and S. R. Fahrenholtz, Abstracts of the 151st Meeting of the American Chemical Society, March, 1966, K23.
634. U. Franzen, *Ann.*, **614,** 31 (1958).
635. G. Frater and O. P. Strausz, *J. Am. Chem. Soc.*, **92,** 6654 (1970).
636. I. G. Csizmadia, J. Font, and O. P. Strausz, *J. Am. Chem. Soc.*, **90,** 7360 (1968).
637. D. G. Thornton, R. K. Gosavi, and O. P Strausz, *J. Am. Chem. Soc.*, **92,** 1768 (1970).
638. S. A. Matlin and P. G. Sammes, *Chem. Commun.*, **1972,** 11.

639. J. Shafer, P. Baronowsky, R. Laursen, F. Finn, and F. H. Westheimer. *J. Biol. Chem.*, **241,** 421 (1965).
640. H. Chaimovich, R. J. Vaughan, and F. H. Westheimer, *J. Am. Chem. Soc.*, **90**, 4088 (1968).
641. O. P. Strausz, T. DoMinh, and H. E. Gunning, *J. Am. Chem. Soc.*, **90,** 1660 (1968).
642. T. DoMinh and O. P. Strausz, *J. Am. Chem. Soc.*, **92,** 1766 (1970).
643. G. O. Schenck and A. Ritter, *Tetrahedron Lett.*, **1968,** 3189.
644. T. DoMinh, O. P. Strausz, and H. E. Gunning, *J. Am. Chem. Soc.*, **91,** 1261 (1969).
645. J. Ciabattoni, R. A. Campbell, C. A. Renner, and P. W. Concannon, *J. Am. Chem. Soc.*, **92,** 3826 (1970).
646. D. C. Richardson, M. E. Hendrick, and M. Jones, Jr., *J. Am. Chem. Soc.*, **93,** 3790 (1971).
647. J. Gehlhaus and R. W. Hoffmann, *Tetrahedron*, **26,** 5901 (1970).
648. G. L. Closs and J. J. Coyle, *J. Am. Chem. Soc.*, **87,** 4270 (1965).
649. G. L. Closs and G. M. Schwartz, *J. Am. Chem. Soc.*, **82,** 5729 (1960).
650. R. A. Mitsch, *J. Am. Chem. Soc.*, **87,** 758 (1965).
651. A. M. van Leusen, and J. Strating *Rec. Trav. Chim.*, **84,** 151 (1965).
652. M. Regitz, *Angew. Chem. Int. Ed.*, **6,** 733 (1967).
653. M. Regitz, *Transfer of Diazo Groups. Newer Methods of Preparative Organic Chemistry*, Vol. VI, Academic Press, New York, 1970.
654. A. M. van Leusen, R. J. Mulder, and J. Strating, *Rec. Trav. Chim.*, **86,** 225 (1967).
655. R. A. Abramovitch and J. Roy. *Chem. Commun.*, **1965,** 542.
656. R. J. Mulder, Thesis, Groningen University, The Netherlands, 1968.
657. R. J. Mulder, A. M. van Leusen, and J. Strating, *Tetrahedron Lett.*, **1967,** 3057.
658. J. Diekmann, *J. Org. Chem.*, **30,** 2272 (1965).
659. A. M. van Leusen, P. M. Smid, and J. Strating, *Tetrahedron Lett.*, **1967,** 1165.
660. D. M. Lemal and E. H. Banitt, *Tetrahedron Lett.*, **1964,** 245.
661. U. Schöllkopf and E. Wiscott, *Ann.*, **694,** 44 (1966).
662. R. M. McDonald and R. A. Krueger, *J. Org. Chem.*, **31,** 488 (1966).
663. R. J. Crawford and R. Raap, *Can. J. Chem.*, **43,** 356 (1965).
664. R. J. Crawford and R. Raap, *Proc. Chem. Soc.*, **1963,** 370.
665. D. M. Lemal, E. P. Gosselink, and S. D. McGregor, *J. Am. Chem. Soc.*, **88,** 582 (1966).
666. W. C. Agosta and A. M. Foster, *Chem. Commun.*, **1971,** 433.
668. E. Ciganek, *J. Am. Chem. Soc.*, **88,** 1979 (1966).
669. E. Ciganek, *J. Am. Chem. Soc.*, **87,** 652 (1965).
670. E. Ciganek, *J. Am. Chem. Soc.*, **87,** 1149 (1965).
671. A. G. Anastassiou, R. P. Cellura, and E. Ciganek, *Tetrahedron Lett.*, **1970,** 5267.
672. R. E. Moser, J. M. Fritsch, T. L. Westman, R. M. Kliss, and C. N. Matthews, *J. Am. Chem. Soc.*, **89,** 5673 (1967).
673. M. Franck-Neumann and C. Buchecker, *Angew. Chem. Int. Ed.*, **9,** 526 (1970).
674. U. Schöllkopf, P. Tonne, H. Schäfer, and P. Markusch, *Ann.*, **722,** 45 (1969).
675. U. Schöllkopf and P. Markusch, *Tetrahedron Lett.*, **1966,** 6199.
676. U. Schöllkopf and P. Markusch, *Angew. Chem., Int. Ed.*, **8,** 612 (1969).
677. L. Horner, H. Hoffmann, H. Ertel, and G. Klahre, *Tetrahedron Lett.*, **1961,** 9.
678. N. Kreutzkamp, E. Schmidt-Samoa, and K. Herberg, *Angew. Chem. Int. Ed.*, **4,** 1078 (1965).
679. D. Seyferth, P. Hilbert, and R. S. Marmor, *J. Am. Chem. Soc.*, **89,** 4811 (1967).
680. D. Seyferth and R. S. Marmor, *Tetrahedron Lett.*, **1970,** 2493.
681. M. Regitz and W. Anschütz, *Ann.*, **730,** 194 (1969).
682. M. Regitz, W. Anschütz, and Liedhegenger, *Chem. Ber.*, **101, 3734** (1968).

683. T. DoMinh, O. P. Strausz, and H. E. Gunning, *Tetrahedron Lett.*, **1968,** 5237.
684. P. Yates and F. X. Garneau, *Tetrahedron Lett.*, **1967,** 71.
685. H. D. Hartzler, Abstracts of the 155th Meeting of the American Chemical Society, P 205.
686. M. F. Lappert and J. Lorberth, *Chem. Commun.*, **1907,** 836.
687. See, however, O. J. Scherer, and M. Schmidt, *Z. Naturforsch.* **20b,** 1009 (1965).
688. K. D. Kaufman, B. Auräth, P. Träger, and K. Rühlmann, *Tetrahedron Lett.*, **1968,** 4973.
689. D. Seyferth, A. W. Dow, H. Menzel, and T. C. Flood, *J. Am. Chem. Soc.*, **90,** 1080 (1968).
690. D. Seyferth and T. C. Flood, *J. Organomet. Chem.*, **29,** C25 (1971).
691. J. Lorberth, *J. Organomet. Chem.*, **27,** 303 (1971).
692. E. Müller and W. Rundel, *Chem. Ber.*, **90,** 1299 (1957).
693. R. Beutler, B. Zeeh, and E. Müller, *Chem. Ber.*, **102,** 2636 (1969).

CHAPTER 2

The Application of Relative Reactivity Studies to the Carbene Olefin Addition Reaction

ROBERT A. MOSS*
Wright Laboratory, School of Chemistry, Rutgers, The State University of New Jersey, New Brunswick, New Jersey

I. INTRODUCTION

Kinetic studies of the generation and fate of dihalocarbenes were reported by Hine's group in the 1950s.[1] Doering and Hoffmann[2] focused on the interception of such carbenes by olefins, and in 1956 Skell and Garner[3] determined relative rate constants for competitive reactions of dibromocarbene with a variety of olefinic substrates. Soon thereafter, Doering and Henderson offered analogous data for dichlorocarbene and compared the abilities of the two carbenes to discriminate between olefins.[4]

* I Thank the National Institutes of Health and the National Science Foundation for financial support.

Since 1958, many relative reactivity studies of the carbene-olefin addition reaction have appeared. It is often claimed that these studies inform us about the relative stabilization which carbenes derive from interaction with their substituents, the polarization of the transition state during a carbene's addition to an olefin, and even the multiplicity of the reactant carbene.

Absolute rate measurements would be preferable to the relative reactivity data, but examples of the former are rare. CF_2, which is quite selective toward olefins, adds to C_2F_4 (gas phase, 0.2–2.5 Torr of C_2F_4, mixed with N_2; total pressure, 50 Torr) with a rate constant estimated to be 35 liters/(mole)(sec) (300° K) and an activation energy on the order of 6 kcal/mole.[5] Addition reactions of less selective (presumably less stabilized) carbenes, in the condensed phase, will surely be very rapid, and absolute rates will not be *simply* obtained, even with flash photolytic techniques.[5a]

We have had to be satisfied with relative reactivity studies; it seems likely that we will so continue, at least for a time. Since much of these data have accumulated since 1956, it seemed valuable to collect it in one place and review the claims and interpretations which have been drawn from it. This chapter comprises an introduction, tables of relative reactivity data, commentary on the tables, and an appendix in which peripheral material is gathered. One virtue of this arrangement is that a reader dissatisfied with the commentary can ignore it and still have a reasonably complete collection of "facts" on which to build his own interpretation.*

In Skell and Garner's original study,[3] relative rate constants were calculated by Eq. 1. Here k is the rate constant for reactions of the carbene with olefins

$$\frac{k_y}{k_x} = \frac{\log (y/y_0)}{\log (x/x_0)} \tag{1}$$

X and Y, x_0 and y_0 are the initial concentrations of these olefins, and x and y are the corresponding final concentrations. This analysis was previously used by Ingold and Smith in their study of aromatic nitration.[6] In view of the doubt which often surrounds the nature of a carbenic reactant, it seems worth quoting some remarks of Ingold and Smith on the nature of Eq. 1. Application to carbene addition reactions is apparent.

"The one essential assumption is that the *attack of the active nitrating entity* (whatever that may be) *on the aromatic compound is unimolecular with respect to the aromatic compound*. Nothing else matters: the active agent may

* We have somewhat arbitrarily excluded the reactions of olefins with elemental carbon and C_3 from this review.

be unknown and its attack may be of any total order; it may be produced in some preliminary change—possibly slow, possibly reversible; it, or some material essential to it, may be simultaneously concerned in a dozen independent, or mutually interdependent, side reactions; all circumstances which would render hopeless any direct kinetic attack on the problem of relative rates of nitration: yet the competition method will give the right result."

A requirement for application of Eq. 1 is that the reactions be homogeneous. The frequent use of partially soluble bases in carbene generation therefore gives cause for concern. Evidence available for dichlorocarbene (Section IV.C.5) indicates essentially identical selectivities for the species produced from chloroform with either soluble or partially soluble bases. In these cases, at least, the heterogeneity seems unimportant.

Doering and Henderson[4] used a simplified competition expression, Eq. 2. Here P_i is the mole fraction of product cyclopropane, and O_i is the initial

$$\frac{k_y}{k_x} = \frac{P_y O_x}{P_x O_y} \tag{2}$$

mole fraction of the precursor olefin. Successful use of Eq. 2 depends on a large excess of olefins over carbene progenitor, so that the concentrations of the olefins remain essentially constant. Less than a 10% error is estimated in use of Eq. 2, provided that O_i is six to seven times greater than the concentration of carbene precursor.[4] If the proper excess is provided, the reaction can be carried to any analytically convenient extent of conversion. In contrast, for use of Eq. 1, it is necessary to carry the reaction to high degrees of conversion (~60%) and to estimate the olefin concentrations by direct measurement in order to obtain the most accurate results.[7]

Most experimenters have used Eq. 2 and analyzed products. However, when competitions are carried out with difficult-to-obtain substrates, the excess substrate necessary for use of Eq. 2 is a serious drawback, and employment of Eq. 1 is indicated. Use of Eq. 1 can be made experimentally simple,[7] but in both cases one must be certain that only reaction with the carbene removes the olefins.

A List of Tables appears directly after this introduction. Occasionally, for ease of discussion, data within the tables have been renormalized. Such changes are noted. When separately published tables have been combined, the references so indicate. In general, however, the many data which are often available for a simple carbene (e.g., CCl_2) were not combined into a single large table. It seemed of little advantage to propagate individual variations of experimental and analytical conditions into greater uncertainty by manipulating the products of chains of relative reactivities from one table to another. Large tables are often more confusing than several smaller ones. More cogently, many of the original tables highlight a particular point. For ease of discussion, it seemed best to leave well enough alone.

The quality of the data is variable, of course. In Skell and Garner's original work,[3] fractional distillation was used to determine product yields. The final olefin concentrations, necessary for the use of Eq. 1, were then obtained by difference from the initial values. The advent of gas chromatography, however, allowed almost all subsequent investigations to make use of this convenient and accurate tool. Where product cyclopropanes were thermally labile, Eq. 2 coupled with NMR product analysis was used. With these exceptions, the reproducibility of the tabulated data was generally within 5%. Poorer reported reproducibility is noted. It was not always possible to ascertain the reproducibility from the publication; 5% is a probable value if gas chromatography was employed. Reproducibility for the exceptions noted above is considered to be within 15%.

Although most investigators used Eq. 2 (product analysis), they seldom showed that product yield and final olefin concentrations quantitatively account for initial olefin concentrations. Frequently, however, the internal consistency of the data has been checked by allowing sets of three olefins (X, Y, and Z) to compete for a carbene in all binary combinations. An observed k_x/k_y is then compared with one calculated from the observed k_x/k_z and k_z/k_y. Such "crosschecks" support the data, and it is to be hoped that they will be included in all future studies. Crosscheck experiments do not, however, generate unique data and are therefore not included in the tables. Only data determined directly against the standard are tabulated, when a choice exists.

Use of Eq. 1 in a table is noted by the heading "Relative Rate Constants"; use of Eq. 2 is noted by the heading "Relative Reactivities." The data are treated without distinction. An effort has been made to include generative methods and conditions within each table. In some cases, it is noted that the species is suspected to be reacting as a triplet (e.g., on the basis of physical evidence or the stereochemical course of its addition reactions). Otherwise, the tables include data for carbenes which appear to be reacting largely as singlets. The tables also note if a species is known or strongly suspected to be reacting as a "carbenoid" or "carbene-complex."

II. LIST OF TABLES

III. TABLES OF RELATIVE REACTIVITY DATA

TABLE 1[8]

Methylene

H_2C: from (1) $CH_2{=}C{=}O + h\nu$, 2600 Å, gas phase (350–700 Torr), at 23°–24° [*singlet methylene*]; and from (2) $CH_2{=}C{=}O + h\nu$, Hg photosensitization, 2537 Å, gas phase (350–700 Torr), at 23°–24° [*triplet methylene*].

	Relative reactivities[a]	
Olefin structures	Reaction (1)[b]	Reaction (2)[c]
$(CH_3)_2C{=}C(CH_3)_2$	0.91	0.96
$(CH_3)_2C{=}C(CH_3)H$	1.07	0.64
$(CH_3)_2C{=}CH_2$	1.00	1.00
cis-$(CH_3)HC{=}CH(CH_3)$ (CH₃ groups on the same side)	0.58	0.33
trans-$(CH_3)HC{=}CH(CH_3)$	0.65	0.31

TABLE 1[8] (*Continued*).

Olefin structures	Relative reactivities[a]	
	Reaction (1)[b]	Reaction (2)[c]
$CH_2{=}CH{-}CH{=}CH_2$	3.20	6.64
$CH_3CH_2CH{=}CH_2$	0.82	0.56
$CH_3CH{=}CH_2$	0.78	0.35
$CH_2{=}CH_2$	0.68	0.35

[a] Derivation of the data is considerably more complicated than usual. See the original for complete discussion of the analysis.

[b] Data are from total relative rate of reaction of $^1CH_2$ with the indicated olefin, corrected for the fraction of reaction which represents addition to the double bond. The tetramethylethylene and trimethylethylene points are corrected with *estimated* factors only. The data are assumed to represent pure $^1CH_2$. If 10% $^3CH_2$ reaction is assumed, a new set of data, quite similar to the original set, can be generated (see original). However, more than 10% $^3CH_2$ could be present.[8a] Data are reproducible to within 10%.

[c] The overall relative rates of $^3CH_2$ reaction with the olefins are believed to be a "good approximation" of the relative rates of addition to the double bonds. $^3CH_2$ is believed to be essentially pure triplet; see, however, references 8b and 8c; some $^1CH_2$ could be present (an estimate of 13% has been given[8d]) and would react rapidly with the alkenes. Data are reproducible to within 10%.

TABLE 2[9]

Methylene[a]

H_2C: from CH_2I_2 + Zn(Cu) (Simmons-Smith reagent), in ether, at reflux.

Olefin structures	Relative rate constants
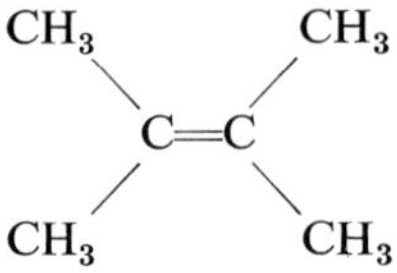 $(CH_3)_2C{=}C(CH_3)_2$	1.29
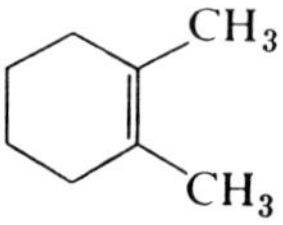 1,2-dimethylcyclohexene (CH_3, CH_3)	0.58
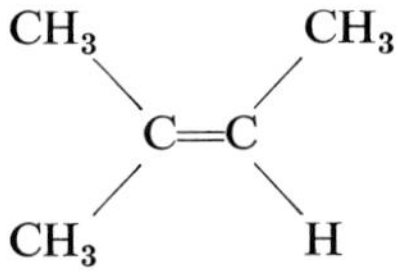 $(CH_3)_2C{=}C(CH_3)H$	2.18
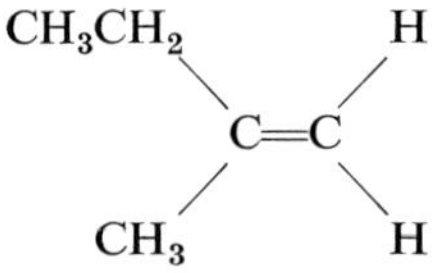 $(CH_3CH_2)(CH_3)C{=}CH_2$	2.53
 cyclohexene	1.00

TABLE 2[9] (*Continued*)

Olefin structures	Relative rate constants
CH_3CH_2 CH_2CH_3 C=C H H	0.83
CH_3CH_2 H C=C H CH_2CH_3	0.42
n-C_4H_9 H C=C H H	0.36
n-C_5H_{11} H C=C H H	0.39
$(CH_3)_3C$ H C=C H H	0.14

[a] Not a free carbene.

TABLE 3[10]

Methylene[a]

H_2C: from CH_2I_2 + Zn(Cu) (Simmons-Smith reagent), in ether, at reflux.

Olefin structures	Relative rate constants
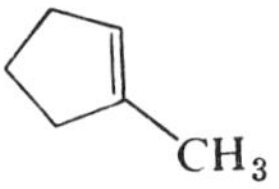 CH_3	5.14
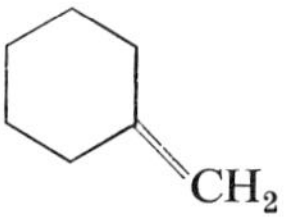 CH_2	3.84
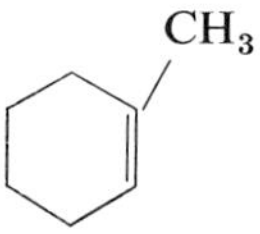 CH_3	2.14
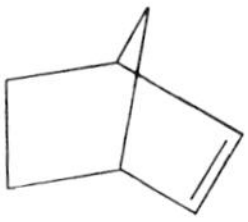	1.70
	1.60
	1.18
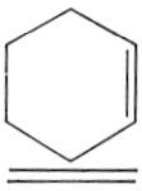	1.00
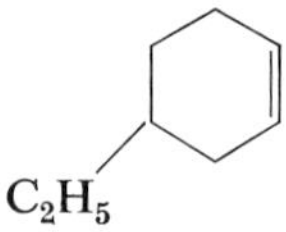 C_2H_5	0.95

TABLE 3[10] (*Continued*)

Olefin structures	Relative rate constants
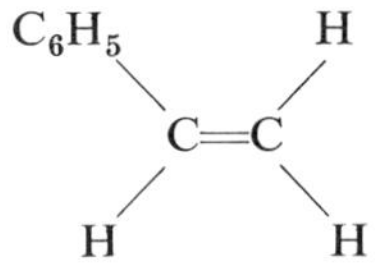 C_6H_5, H, C=C, H, H	0.95
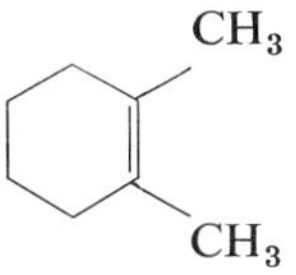 CH_3, CH_3	0.94[b]
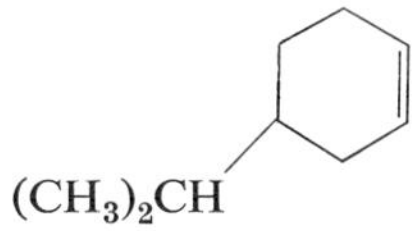 $(CH_3)_2CH$	0.94
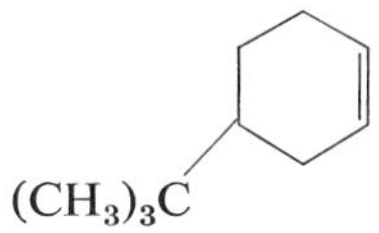 $(CH_3)_3C$	0.92
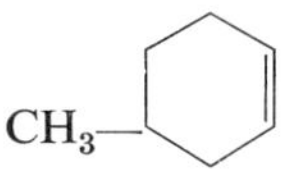 CH_3—	0.91
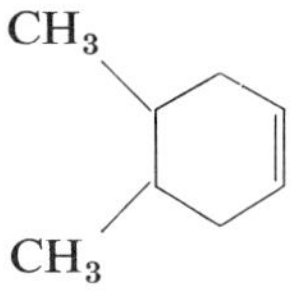 CH_3, CH_3	0.84
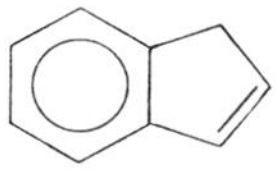	0.68
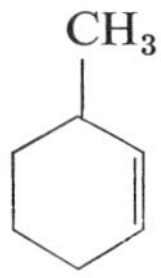 CH_3	0.58
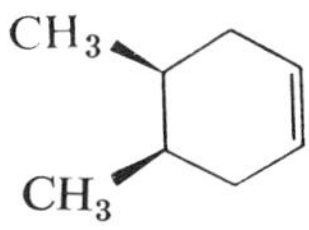 CH_3, CH_3	0.55

TABLE 3[10] (*Continued*)

Olefin structures	Relative rate constants
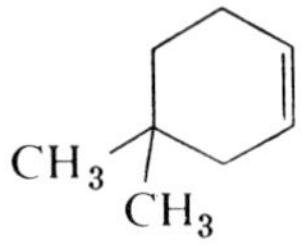 CH_3, CH_3	0.48
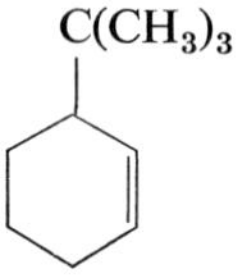 $C(CH_3)_3$	0.44
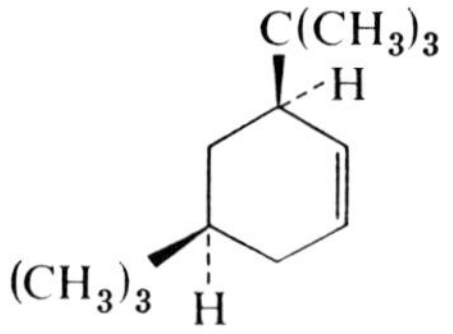 $C(CH_3)_3$, H, $(CH_3)_3$, H	0.40
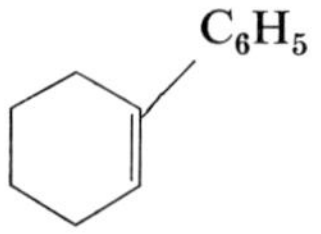 C_6H_5	0.30

[a] Not a free carbene.

[b] There is a discrepancy with the corresponding value in Table 2.

TABLE 4[11]

Methylene[a]

H_2C: from CH_2I_2 + Zn(Cu) (Simmons-Smith reagent), in ether, at reflux.

Olefin structures	Relative rate constants
OH	1.00[b]
OCH_3	0.50[b]
CH_2OH	0.46
OH, CH_3	1.54[b]
OH, CH_3	0.46[b]
OH	0.091[b,c]

TABLE 4[11] (*Continued*)

Olefin structures	Relative rate constants
OCH_3 (on cyclohexene ring)	0.059
CH_2OH (on cyclohexene ring)	"Very slow"[d]

[a] Not a free carbene. Data reproducible to within 10%.

[b] Only one product isomer, in which the cyclopropyl group is thought to be *cis* to the oxygen function, was observed. Stereochemical assignments for products of the fourth to sixth entries are presumptive.

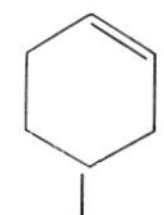

[c] This olefin reacts "immeasurably faster" than $C(CH_3)_3$, which itself is about 10% less reactive than cyclohexene in the Simmons-Smith reaction (cf. Table 3). Thus most olefins in the present table react at least one and probably two orders of magnitude more rapidly with the Simmons-Smith reagent than do the various alkylcyclohexenes.

[d] Stereochemistry of methylenation was 55% *trans* and 45%

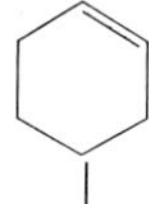

cis. The rate of reaction, relative to $C(CH_3)_3$, was 1.6 ± 0.3.

TABLE 4A[11a]

Methylene[a]

H_2C: from (1), $BrClCH_2 + n\text{-}C_4H_9Li$ in pentane, at −50°; from (2), $C_2H_5AlCl_2 + CH_2N_2$ in pentane, at −50°.

Olefin structures	Relative rate constants (1)	Relative rate constants (2)
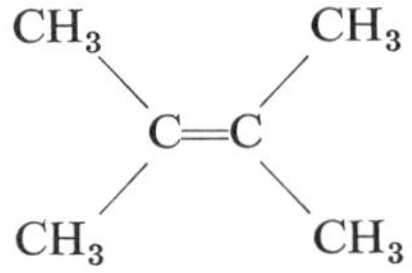 $(CH_3)_2C{=}C(CH_3)_2$	0.29	0.01
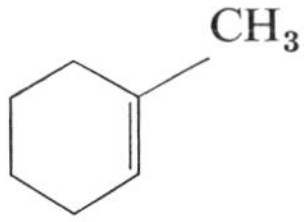 1-methylcyclohexene (CH_3 on ring double bond)	0.68[b]	0.47
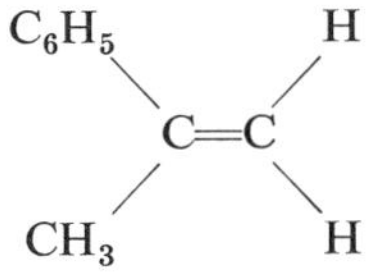 $C_6H_5(CH_3)C{=}CH_2$	3.5	—
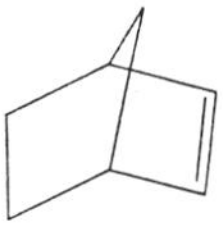 norbornene	1.10	0.65
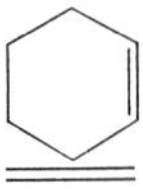 cyclohexene	1.00	1.00
 cyclopentene	—	0.80
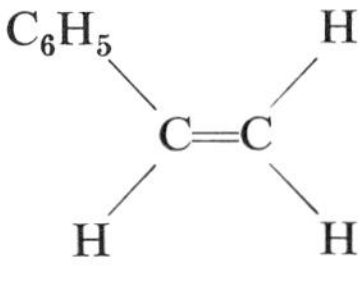 $C_6H_5(H)C{=}CH_2$	3.7[c]	—
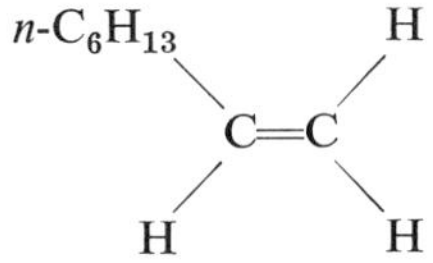 $n\text{-}C_6H_{13}(H)C{=}CH_2$	1.23	0.13

[a] Not free carbenes.
[b] The identical result was obtained using $t\text{-}C_4H_9Li$.
[c] With "$LiCH_2Br$" the rate constant was 5.4.

TABLE 4B[11b]

Methylene[a]

H_2C: from (1), $CH_3Cl + C_6H_5Na$ (decane), at 5–10°; from (2), $BrClCH_2$ + n-C_4H_9Li (pentane), at −20°; from (3), $BrClCH_2$ + *neo*-$C_5H_{11}Li$ (pentane), at +20°; from (4), Br_2CH_2 + n-C_4H_9Li (pentane), at −20°; and from (5), Br_2CH_2 + *neo*-$C_5H_{11}Li$ (pentane), at −20°.

Olefin structures	Relative reactivities[b] (1)	(2)	(3)	(4)	(5)
$(CH_3)_2C{=}C(CH_3)_2$	0.86[c]	0.44	0.07	1.28	0.09
$(CH_3)_2C{=}CH_2$	1.32	1.81[d]	0.13	3.48[e]	0.16
cyclohexene	1.00	1.00	1.00	1.00	1.00

[a] Not free carbenes.

[b] Only double to triple excesses of olefins over carbene precursors were maintained.

[c] 0.85 at 25–30°.

[d] 0.71 with halide added to olefins + RLi (inverse addition); otherwise all results are for addition of RLi to halide + olefins.

[e] 0.37 by inverse addition.

TABLE 5[12]

Methylene[a]

H_2C: from $(BrCH_2)_2Hg$, in benzene, at 80°.

Olefin structures	Relative reactivities
$(CH_3)_2C=C(CH_3)CH_2CH_3$	26.1
$(CH_3CH_2)_2C=C(CH_3)H$	4.16
cyclohexene	1.00
$n\text{-}C_5H_{11}HC=CH_2$	0.223

[a] Not a free carbene.

TABLE 5A[12a]

Methylene[a]

H_2C: from $(C_2H_5)_2Zn + CH_2I_2$, in (1) ether, (2) pentane, and (3) benzene, at 24.4°.

Olefin structures	Relative reactivities (1)	(2)	(3)
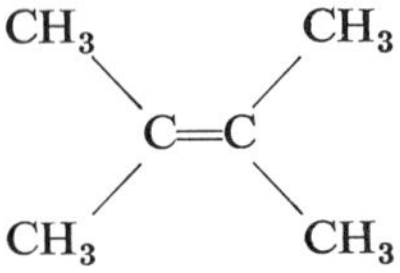 $(CH_3)_2C{=}C(CH_3)_2$	8.82	16.2	12.5
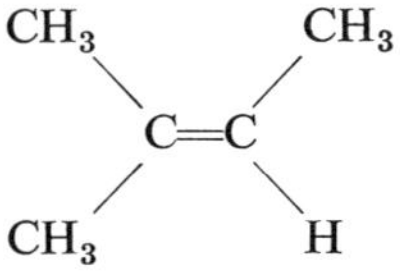 $(CH_3)_2C{=}CH(CH_3)$	—	—	9.65
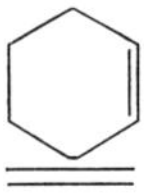 cyclohexene	1.00	1.00	1.00
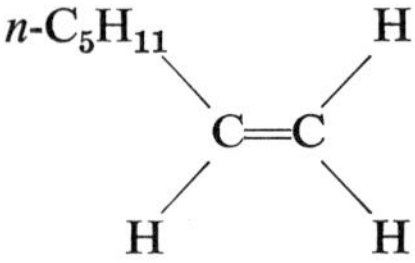 n-$C_5H_{11}CH{=}CH_2$	0.15	0.27	0.15

[a] Not a free carbene.

TABLE 5B[12a]

Methylene[a]

H_2C: from $(C_2H_5)_2Zn + CH_2I_2$, in ether, at 24.4°.

Olefin structures	Relative reactivities
$(CH_3)_3CO(H)C=C(H)H$ (cis: $(CH_3)_3CO$ and H upper; H, H lower)	3.09
n-$C_3H_7O(H)C=C(H)CH_3$ (n-C_3H_7O and CH_3 cis; H, H)	2.98
$CH_3CH_2O(H)C=C(H)CH_2CH_3$ (CH_3CH_2O and CH_2CH_3 cis; H, H)	2.87
n-$C_3H_7O(H)C=C(H)CH_3$ (n-C_3H_7O and H upper; H, CH_3 lower)	1.98
$CH_3CH_2O(H)C=C(H)CH(CH_3)_2$ (CH_3CH_2O and $CH(CH_3)_2$ cis; H, H)	1.58
$(CH_3)_2CHO(H)C=C(H)H$ ($(CH_3)_2CHO$ and H upper; H, H lower)	1.52
3,4-dihydro-2H-pyran (ring with O)	1.43
$CH_3CH_2O(H)C=C(H)CH_2CH_3$ (CH_3CH_2O and H upper; H, CH_2CH_3 lower)	1.07

TABLE 5B[12a] (*Continued*)

Olefin structures	Relative reactivities
$(CH_3)_2CHCH_2O(H)C{=}CH_2$ (cis-H, H)	1.00[b]
$CH_3CH_2O(H)C{=}CH_2$	0.36
$(CH_3)_2CHCH_2OCH_2(H)C{=}CH_2$	0.28
$ClCH_2CH_2O(H)C{=}CH_2$	0.28
Cyclohexene	0.18

[a] Not a free carbene.

[b] To adjust these data to a cyclohexene standard, multiply by 5.55.

TABLE 5C[12a]

Methylene[a]

H_2C: from $(C_2H_5)_2Zn + CH_2I_2$, in benzene, at 78.6°.

Olefin structures[b]	Relative reactivities
p-CH_3	1.64
H	1.00
p-F	0.681
p-Cl	0.464
m-CF_3	0.175

[a] Not a free carbene.

[b] Entries are substituents in X–C_6H_4–$CH{=}CH_2$

TABLE 5D[12b]

Methylene[a]

H_2C: from CH_2N_2 + Cu^{II}[b], in *n*-hexane, at −15° to −20°

Olefin structures	Relative reactivities
$(CH_3)_2C{=}C(CH_3)_2$	3.46
$(CH_3)_2C{=}CH(CH_3)$	3.32
$C_6H_5(CH_3)C{=}CH_2$	4.52
$C_6H_5CH{=}CHCH_3$	2.79
cyclohexene	1.00

TABLE 5D[12b] (*Continued*)

Olefin structures	Relative reactivities
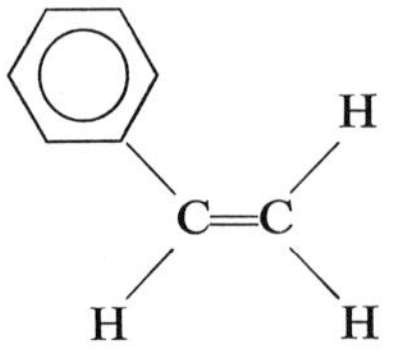	3.32
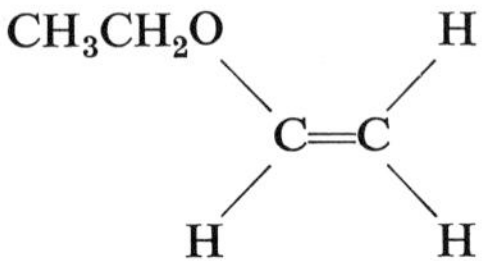	3.06
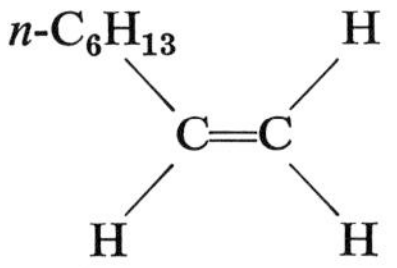	0.60

[a] Not a free carbene.
[b] Cu is supplied in the form of a chelate:

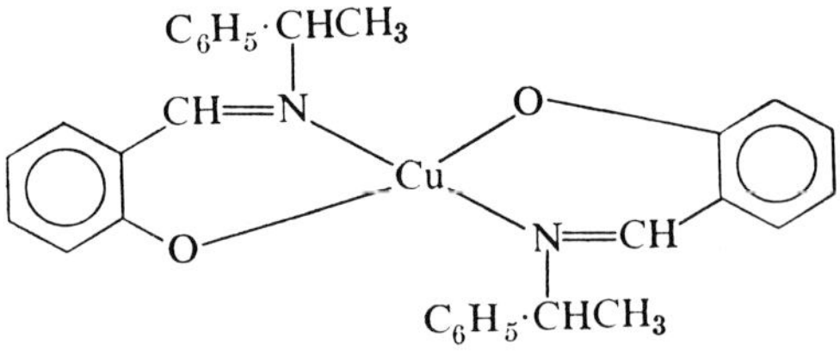

TABLE 6[13a]

trans-Dimethylcyclopropylidene

^{3}(trans-dimethylcyclopropylidene, CH_3, CH_3) from $C_1(^3P)$ + trans-$CH_3CH{=}CHCH_3$, neopentane matrix, —100° to −150°.

Olefin structures	Relative reactivities[a,b]
$(CH_3)_2C{=}CH_2$	0.2
cis-$CH_3CH{=}CHCH_3$	0.7
trans-$CH_3CH{=}CHCH_3$	1.0 (standard)
$CH_3CH{=}CH_2$	0.4
$CH_2{=}CH{-}CH{=}CH_2$	25

[a] See original for a discussion of the complexities of the analyses.

[b] There are large discrepancies with previous data[13b] for this system, and a reinvestigation has cast serious doubt on the entire original studies.[13c]

TABLE 7[14]

2,2-Diphenylcyclopropylidene

C_6H_5, C_6H_5 (2,2-diphenylcyclopropylidene) : from $(C_6H_5)_2$-cyclopropyl-$N(NO)CONH_2$ + $Li^+\ ^-OC_2H_5 \cdot C_2H_5OH$, at 0° (i.e., from $(C_6H_5)_2$-cyclopropylidene$=N_2$).

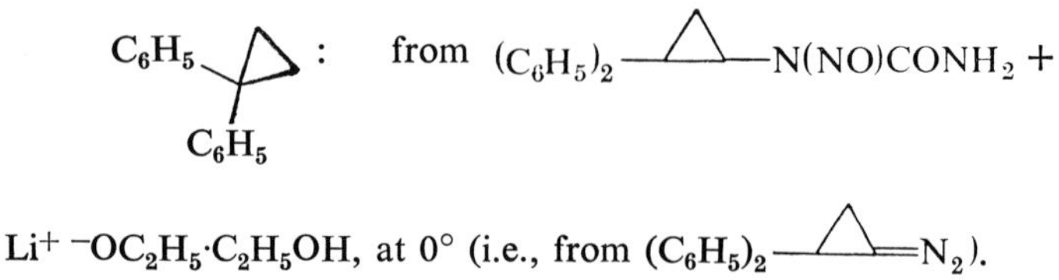

Olefin structures	Relative reactivities
$(CH_3)_2C=C(CH_3)_2$	0.41
$(CH_3)_2C=CH_2$	1.00
Cyclohexene	1.23
$CH_3(H)C=C(CH_3)_2$	1.15
$CH_3(H)C=CH_2$	0.42
$CH_3CH_2(H)C=CH_2$	0.22

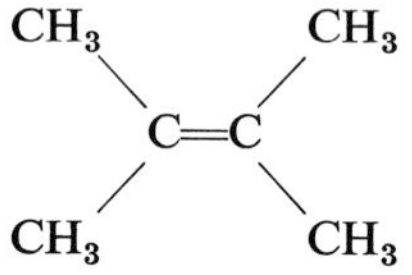

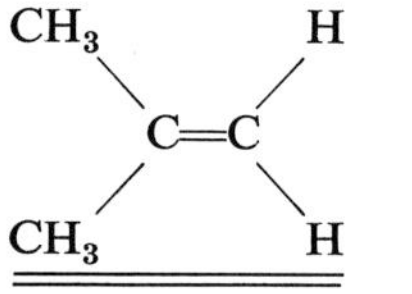

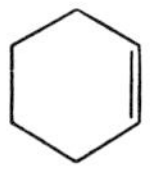

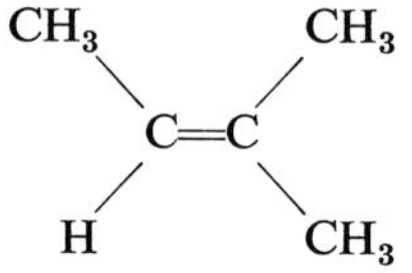

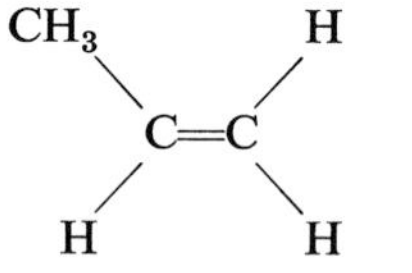

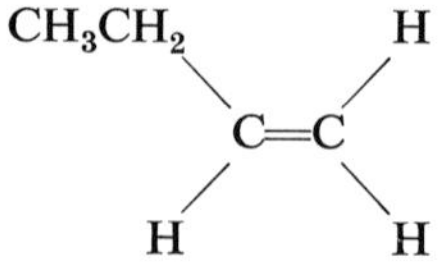

TABLE 8[15]

Dimethylethylidene Carbene

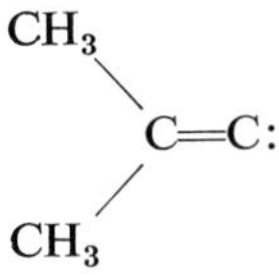

$(CH_3)_2C=C\colon$ from

$(CH_3)_2C$—O

C=O

CH_2—N

NO

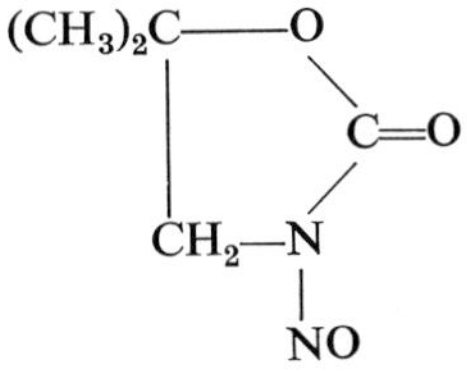

+ $LiOCH_2CH_2OC_2H_5$, at 40°

[i.e., from $(CH_3)_2C=CN_2$].

Olefin structures	Relative reactivities
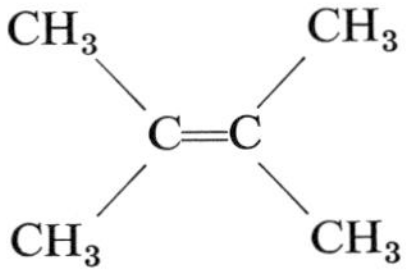 CH_3, CH_3 / C=C / CH_3, CH_3	0.02
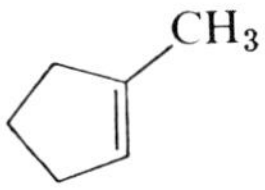 CH_3 (1-methylcyclopentene)	0.3
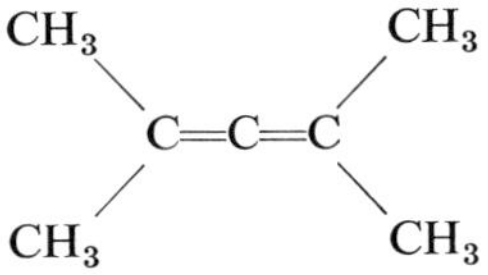 CH_3, CH_3 / C=C=C / CH_3, CH_3	0.9
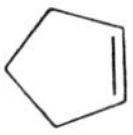 (cyclopentene)	1.1
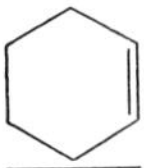 (cyclohexene)	1.0
 (cycloheptene)	0.6
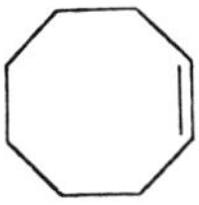 (cyclooctene)	0.7
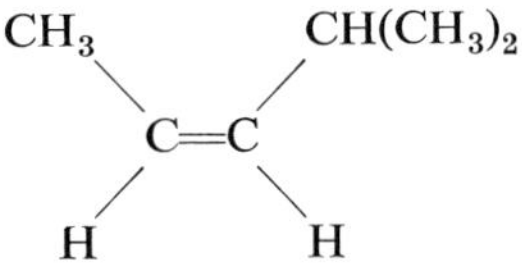 CH_3, $CH(CH_3)_2$ / C=C / H, H	0.5

TABLE 8[15] (*Continued*)

Olefin structures	Relative reactivities
$CH_3CH{=}CHCH(CH_3)_2$	0.07
$CH_3C_6H_4CH{=}CH_2$	71.
$C_6H_5CH{=}CH_2$	6.2
$ClC_6H_4CH{=}CH_2$	2.6
$(CH_3)_3C{-}O{-}CH{=}CH_2$	0.4
$n\text{-}C_6H_{13}CH{=}CH_2$	0.2
$Cl_2C{=}CCl_2$	0.0

TABLE 9[16a,16b]

Dimethylvinylidene Carbene

$(CH_3)_2C{=}C{=}C$: from (1), $(CH_3)_2C{=}C{=}CHCl$ + $K^{+}{}^{-}O\text{-}t\text{-}C_4H_9$, at $-10°$;
from (2), $(CH_3)_2CCl{-}C{\equiv}CH$ + $K^{+}{}^{-}O\text{-}t\text{-}C_4H_9$, at $-10°$.

Olefin structures	Relative reactivities[a] (1)	(2)
$(CH_3)_2C{=}C(CH_3)_2$	16.0	20.0
$(CH_3)_2C{=}CHCH_3$	4.9	4.7
$(CH_3CH_2)(CH_3)C{=}CH_2$	4.2	5.0
Cyclohexene	1.0	1.0
$n\text{-}C_4H_9CH{=}CH_2$	0.23	0.25

[a] Data converted from log (data) given in the original. Reproducibility within 10%.

TABLE 9A[16d]

Diphenylvinylidene Carbene

$(C_6H_5)_2C{=}C{=}C{:}$ from (1), $(C_6H_5)_2C{=}C{=}C{=}N_2$, at 0°;[a] from (2), $(C_6H_5)_2C(OOCCH_3){-}C{\equiv}CH$ + base.[b]

Olefin structures	Relative reactivities (1)	Relative reactivities (2)
$(CH_3)_2C{=}C(CH_3)_2$	8.0[c]	9.0
cyclohexene	1.0	1.0
$n\text{-}C_4H_9CH{=}CH_2$	0.70[c]	0.61

[a] Generated from 3,3-diphenyl-1,2-*trans*-bis(*N*-nitrosourethano)cyclopropane by the action of sodium methoxide. See the original for mechanistic details.

[b] Base and temperature are not specified, but are probably potassium *t*-butoxide at 0°.

[c] This reaction goes in very low yield, and the reproducibility is poor, deviations in the relative reactivities of ±1.5 and 0.18, respectively, are noted.

TABLE 9B[16c]

di-*t*-Butylvinylidene Carbene[a]

$$\begin{matrix} (CH_3)_3C \\ \quad\diagdown \\ \quad\quad C{=}C{=}C\text{:} \\ \quad\diagup \\ (CH_3)_3C \end{matrix}$$ from $[(CH_3)_3C]_2C(OOCCH_3)C{\equiv}CH + K^{+-}O\text{-}t\text{-}C_4H_9$, at 25–35°

Olefin structures	Relative reactivities
$(CH_3)_2C{=}C(CH_3)_2$	16.
$(CH_3)_2C{=}CH(CH_3)$	4.9
$(CH_3CH_2)(CH_3)C{=}CH_2$	2.0
Cyclohexene	<u>1.0</u>
$n\text{-}C_4H_9CH{=}CH_2$	0.20

[a] Converted from the log (data) given in the original. Reproducibility within 10%.

[b] In tetrahydrofuran solution.

TABLE 10[17a]

Carbonylcarbene

O=C=C: from (1), O=C=C=C=O + $h\nu$ (3000 Å), gas phase, at various temperatures (triplet); from (2), O=C=C=C=O + $h\nu$ (2500 Å), gas phase, at 304° K (singlet).

Olefin structures	Relative rate constants[a,b]					
	Reaction (1)			Activation parameters, reaction (1)[c]		Reaction (2)
	273° K	302° K	326° K	$A_i/A_{ethylene}$	$E^{a}_{ethylene}$-E_i^{a}	304° K
$(CH_3)_2C=C(CH_3)_2$	477	245	127	0.2	4.4	2.17
$(CH_3)_2C=CH(CH_3)$	186	95.8	58.1	0.1	4.0	—
$(CH_3)_2C=CH_2$	89.5	56.7	37.0	0.5	2.8	—

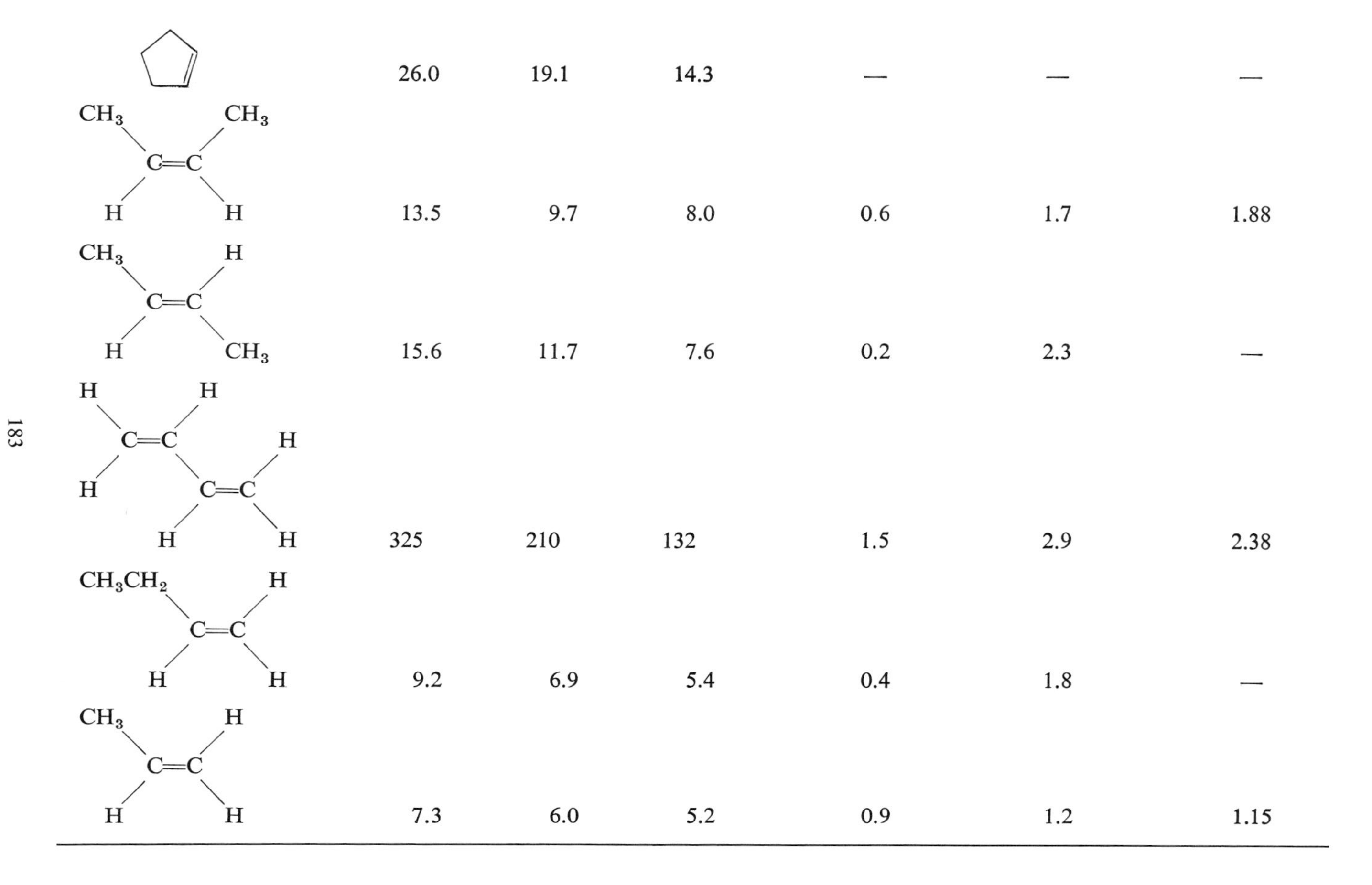

	26.0	19.1	14.3	—	—	—
	13.5	9.7	8.0	0.6	1.7	1.88
	15.6	11.7	7.6	0.2	2.3	—
	325	210	132	1.5	2.9	2.38
	9.2	6.9	5.4	0.4	1.8	—
	7.3	6.0	5.2	0.9	1.2	1.15

TABLE 10[17a] (*Continued*)

	Relative rate constants[a] [b]					
	Reaction (1)			Activation parameters, reaction (1)[c]		Reaction (2)
Olefin structures	273° K	302° K	326° K	$A_i/A_{ethylene}$	$E^a_{ethylene}$-E_i^a	304° K
$(CH_3)_2C{=}C{=}C(CH_3)_2$	—	110.	—	—	—	—
$CH_3—C{\equiv}C—CH_3$	—	8.9	—	—	—	—
$H—C{\equiv}C—H$	—	0.34	—	—	—	—
$H_2C{=}CH_2$	1.00	1.00	1.00	1.0	0	1.00

[a] There is a complete disagreement between the data shown here for :C═C═O and the data determined by Trotman-Dickenson and co-workers [*J. Chem. Soc.* (*A*), **1966,** 975]. The data quoted here are direct comparisons of olefin reactivities; the other data involve indirect and more complicated comparisons. The data must still be regarded as tentative and the discrepancy as unresolved. For a more complete discussion, see the original. See also references 17b and 17c.

[b] Relative rate constants were determined by comparison of product allene yields from photolyses of ethylene and C_2O_3 in the presence and absence of olefin$_i$; see original for details of analysis.

[c] ΔE^a in kcal/mole.

TABLE 11[18]

Cyclopentadienylidene

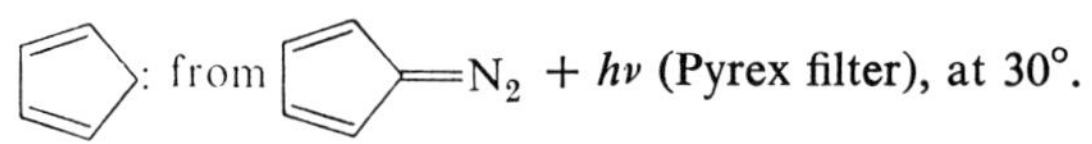

Olefin structures	Relative reactivities
$(CH_3)_2C{=}C(CH_3)_2$	0.74
$(CH_3)_2C{=}CH(CH_3)$	0.75
C_6H_{10}	<u>1.00</u>
$C_6H_5CH{=}CH_2$	0.94
$n\text{-}C_4H_9CH{=}CH_2$	0.94
$(CH_3)_3CCH{=}CH_2$	0.70

TABLE 12[19,19a]

4,4-Dimethylcyclohexadienylidene

(4,4-dimethylcyclohexadienylidene structure): from (1), 4,4-dimethylcyclohexadienylidene$=N_2$ + $h\nu$ (Pyrex filter), at ~25°;

from (2), (spiro[bicyclo[4.1.0]heptadiene-7,1'-(4',4'-dimethylcyclohexadiene)] structure) + $h\nu$ (quartz), at ~25°.

Olefin structures	Relative reactivities (1)	Relative reactivities (2)
$(CH_3)_2C{=}C(CH_3)_2$	1.23	1.07
$(CH_3)_2C{=}CH(CH_3)$	1.00	1.00
$CH_3CH{=}CHCH(CH_3)_2$ (cis: CH_3 and $CH(CH_3)_2$ on same side)	0.19	—
$CH_3CH{=}CHCH(CH_3)_2$ (trans)	0.21	—
$n\text{-}C_3H_7CH{=}CH_2$	0.24	—
$(CH_3)_3CCH{=}CH_2$	0.21	0.23
$CH_2{=}C(CH_3)C(CH_3){=}CH_2$	3.0	3.07

TABLE 13[20]

Arylcarbene

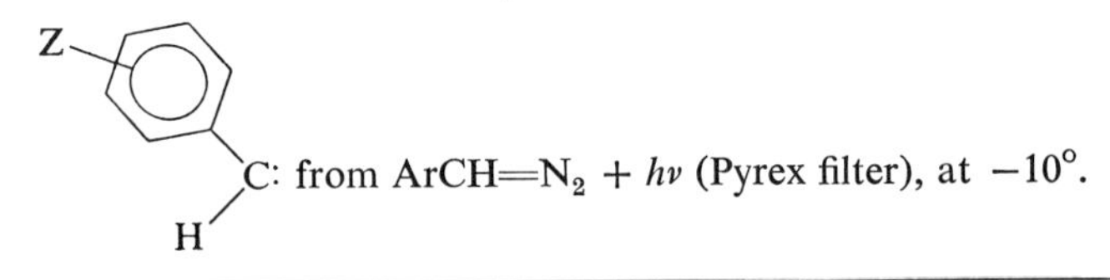

Olefin structures	Relative reactivities, function of Z[a]				
	m-Cl	p-Cl	H	p-CH$_3$	p-CH$_3$O
$(CH_3)_2C{=}C(CH_3)H$	—	—	1.8 (1.1)	1.4 (1.2)	—
$(CH_3)_2C{=}CH_2$	1.00	1.00	1.00	1.00	1.00
$CH_3(H)C{=}C(H)CH_3$ (*cis*: CH$_3$, CH$_3$ same side)	0.70 (1.2)	0.93 (1.1)	0.99 (1.1)	1.1 (1.7)	1.1 (2.8)
$CH_3(H)C{=}C(H)CH_3$ (*trans*: CH$_3$ opposite)	0.80	0.90	1.1	1.2	1.3
$CH_3CH_2(H)C{=}CH_2$	0.50 (1.2)	0.51 (1.1)	0.56 (1.0)	0.46 (1.2)	0.48 (1.4)

[a] Data in parentheses are stereoselectivities, *syn*-aryl/*syn*-hydrogen.

TABLE 14[20]

Arylcarbene[a]

Z-substituted phenyl-C(H): from $ArCHBr_2 + n\text{-}C_4H_9Li$, in pentane, at $-10°$.

Olefin structures	Relative reactivities, function of Z[b]				
	m-Cl	*p*-Cl	H	*p*-CH_3	*p*-CH_3O
$(CH_3)_2C{=}C(CH_3)H$	—	—	0.97 (1.3)	1.1 (1.3)	—
$(CH_3)_2C{=}CH_2$	1.00	1.00	1.00	1.00	1.00
cis-$CH_3CH{=}CHCH_3$	0.93 (3.7)	0.93 (2.9)	0.90 (2.4)	0.85 (4.5)	0.75 (8.5)
trans-$CH_3CH{=}CHCH_3$	0.49	0.55	0.59	0.55	0.51
$CH_3CH_2CH{=}CH_2$	0.32 (2.5)	0.39 (2.1)	0.46 (2.1)	0.46 (2.6)	0.28 (3.0)

[a] Not free carbenes.

[b] Data in parentheses are stereoselectivities, *syn*-aryl/*syn*-hydrogen.

TABLE 15[21]

p-Tolylcarbene[a]

$p\text{-}CH_3C_6H_4\ddot{C}H$ (structure: CH_3–C_6H_4–$\ddot{C}$–H)

from (1), p-$CH_3C_6H_4CH{=}N_2 + h\nu$ (Pyrex filter), at 0° or −10°; (2), p-$CH_3C_6H_4CHBr_2 + CH_3Li$, in ether, at 0°; (3), p-$CH_3C_6H_4CH{=}N_2 + LiBr$, in ether, at 0°; (4), p-$CH_3C_6H_4CH{=}N_2 + ZnCl_2$, in ether, at −10°; (5), p-$CH_3C_6H_4CH{=}N_2 + ZnBr_2$, in ether, at −10°; (6), p-$CH_3C_6H_4CH{=}N_2 + ZnI_2$, in ether, at −10°; (7), p-$CH_3C_6H_4CH{=}N_2 + CoBr_2$, in ether, at −10°.

Olefin structures	Relative reactivities[b]						
	(1)	(2)	(3)	(4)	(5)	(6)	(7)
$(CH_3)_2C{=}C(CH_3)H$	2.40 (1.2)	1.50 (1.3)	9.6 (2.4)	2.0 (1.3)	2.7 (1.5)	5.0 (2.6)	2.20 (2.5)
$(CH_3)_2C{=}CH_2$	0.83	1.35	3.62	1.82	2.54	3.90	2.55
cis-$CH_3CH{=}CHCH_3$	1.76 (1.7)	1.35 (2.3)	1.65 (7.5)	1.44 (6.6)	2.11 (8.5)	4.96 (21.0)	2.93 (18.0)

TABLE 15[21] (*Continued*)

Olefin							
trans-$CH_3CH=CHCH_3$	1.00	1.00	1.00	1.00	1.00	1.00	1.00
$CH_3CH_2CH=CH_2$	0.89 (1.2)	0.72 (1.3)	0.28 (2.0)	0.43 (3.5)	0.44 (4.9)	0.51 (5.4)	0.37 (3.4)
$CH_2=CH-CH=CH_2$	1.7 (1.2)	1.9 (1.6)	16.0 (0.8)	1.2 (4.2)	1.4 (4.9)	1.5 (5.9)	—

[a] Except for reaction (1), the intermediates are not free carbenes.
[b] Data in parentheses are stereoselectivities, *syn*-*p*-tolyl/*syn*-hydrogen.

TABLE 16[22]

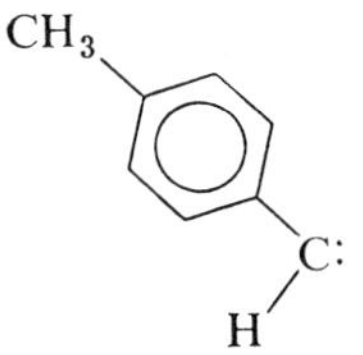

p-Tolylcarbene[a]

from (1), p-$CH_3C_6H_4CHI_2$ + n-C_4H_9Li, in pentane, at $-10°$;
(2), p-$CH_3C_6H_4CHBr_2$ + n-C_4H_9Li, in pentane, at $-10°$;
(3), p-$CH_3C_6H_4CHBr_2$ + $CH_3Li \cdot LiBr$, in ether, at $-10°$

Olefin structures	Relative reactivities[b] (1)	(2)	(3)
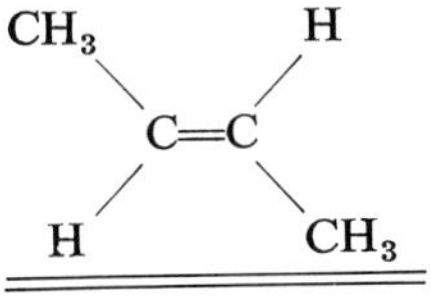 $CH_3(H)C{=}C(H)CH_3$ (trans-2-butene)	1.00	1.00	1.00
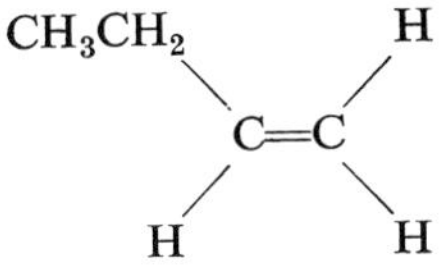 $CH_3CH_2(H)C{=}CH_2$	0.68 (2.7)	0.90 (2.6)	0.55 (2.1)
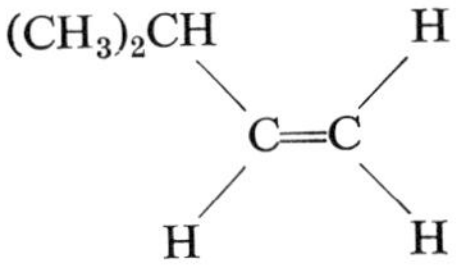 $(CH_3)_2CH(H)C{=}CH_2$	0.35 (1.9)	0.39 (1.9)	0.28 (1.4)
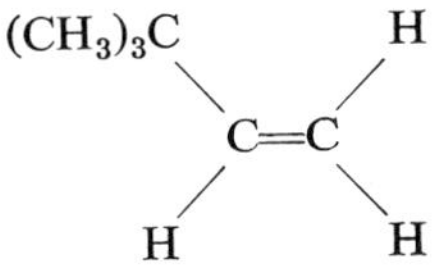 $(CH_3)_3C(H)C{=}CH_2$	0.12 (0.42)	0.15 (0.72)	0.14 (0.45)

[a] Not a free carbene.
[b] Data in parentheses are stereoselectivities, *syn-p*-tolyl/*syn*-hydrogen.

TABLE 17[23]

Diphenylcarbene

$(C_6H_5)_2C$: from $(C_6H_5)_2C{=}N_2$, in pyridine, at 80° (triplet).

Olefin structures	Relative reactivities
$p\text{-}CH_3OC_6H_4CH{=}CH_2$	83.7
Vinylferrocene (Fe, C_5H_5, $C_5H_4CH{=}CH_2$)	6.8
$C_6H_5CH{=}CH_2$	1.0
Cyclohexene	0.0

TABLE 17A[24,26a]

Diphenylcarbene

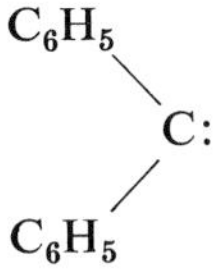

$(C_6H_5)_2C$: from $(C_6H_5)_2C{=}N_2 + h\nu$ (Pyrex filter), temperature not given (triplet).

Olefin structures	Relative reactivities[a]
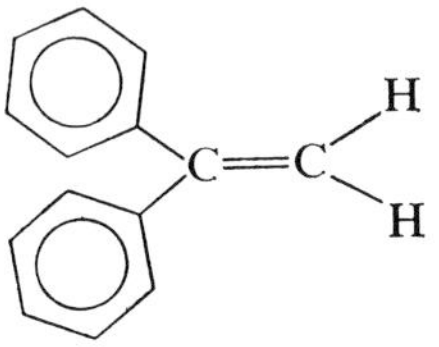 $(C_6H_5)_2C{=}CH_2$	>300[b]
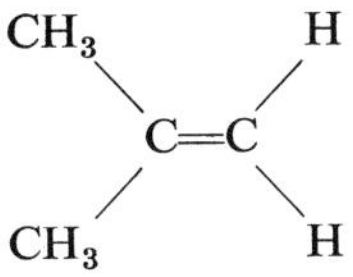 $(CH_3)_2C{=}CH_2$	1.03
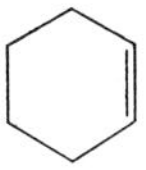 cyclohexene	4.03
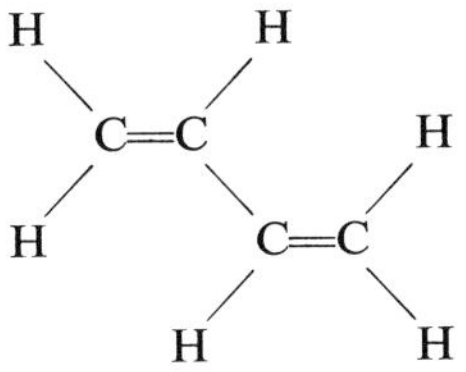 $CH_2{=}CH{-}CH{=}CH_2$	>100
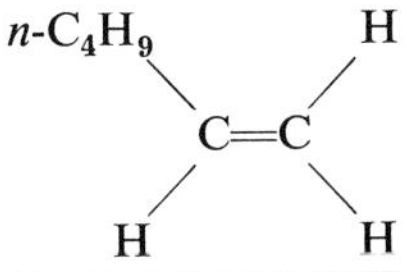 $n\text{-}C_4H_9CH{=}CH_2$	1.00

[a] The numerical data are from reference 26a, and are summarized in reference 24. Reproducibility is not indicated.[26a] Analysis was by gas chromatography except for one case.[b]

[b] Analysis was carried out by distillation and refractive index measurements.

TABLE 18[25,26]

Carboethoxycarbene

$C_2H_5OC(=O)\text{–}\ddot{C}\text{–}H$: from (1), $C_2H_5OOCCH{=}N_2 + h\nu$ (Pyrex filter), at $-35°$; from (2), $C_2H_5OOCCH{=}N_2 + CuSO_4$[a], at reflux temperature of a mixture of olefin and cyclohexene.

Olefin structures	Relative reactivities[b]	
	Reaction (1)[c]	Reaction (2)
$(CH_3)_2C{=}C(CH_3)_2$	1.0	1.0
$(CH_3)_2C{=}CHCH_3$	1.0	1.0[d]
$(CH_3)_2C{=}CH_2$	0.91	—

TABLE 18[25,26] (*Continued*)

Olefin structures	Relative reactivities[b] Reaction (1)[c]	Reaction (2)
$CH_3(H)C{=}C(H)CH_3$ (cis)	0.82	—
cyclohexene	—	0.61 (16.0)[e]
$CH_3(H)C{=}C(H)CH_3$ (trans)	0.82	—
$n\text{-}C_4H_9(H)C{=}CH_2$	—	0.56[c]

[a] Not a free carbene.

[b] Both sets are adjusted to a trimethylethylene standard from data in the original. The data for reaction (2) are slightly different in the Ph.D. thesis[26a] related to this publication: 1.00, 1.00, 0.68, 0.60, in order, top to bottom.

[c] Stereoselectivity not reported.

[d] Only one cyclopropane, presumably the *exo* isomer was reported.

[e] Stereoselectivity, *exo*/*endo*-7-carboethoxynorcarane.

TABLE 19[27]

Carboethoxycarbene

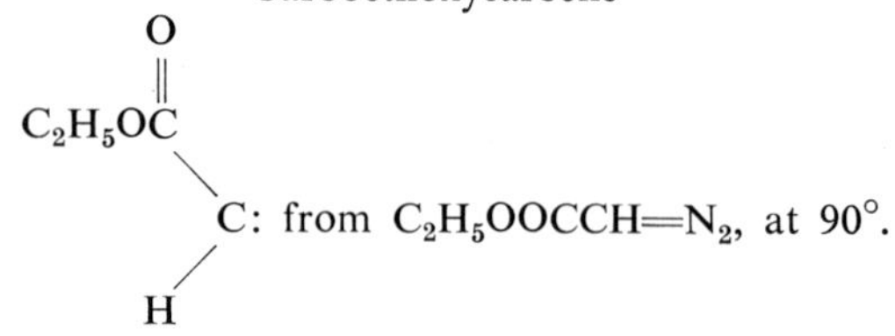

Olefin structures	Relative reactivities[a,b]
$(CH_3)(CH_3)C{=}C(H)(H)$	3.47
$(CH_3)(H)C{=}C(CH_3)(H)$	1.87
	1.00
$(CH_3)(H)C{=}C(H)(CH_3)$	0.67
$(CH_3CH_2)(H)C{=}C(H)(H)$	1.33

[a] Adjusted to a cyclohexene standard from data in the original.

[b] Data are not corrected for relative detector response in the gas chromatographic analysis. The cyclohexene value is thus probably "too large." Stereoselectivity not given for these conditions.

TABLE 19A[27a]

Carboethoxycarbene[a]

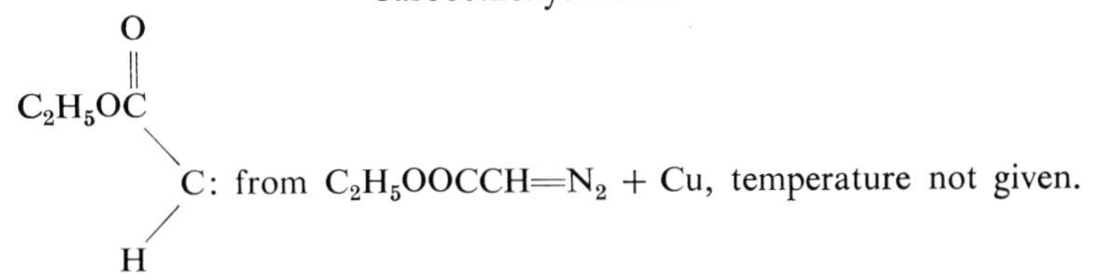

Olefin structures	Relative reactivities
$CH_3C{\equiv}C{-}CH_2CH_2CH_3$	1.31
cyclohexene	1.00
cis-$CH_3CH{=}CHCH_2CH_2CH_3$	0.68
cis-$CH_3CH{=}CHC{\equiv}CCH_3$	0.57[b] 0.37[c]

[a] Not a free carbene.
[b] Reactivity of triple bond.
[c] Reactivity of double bond.

TABLE 20[28,29]

Dicarbomethoxycarbene

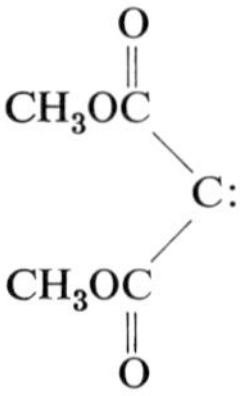

$(CH_3OC(=O))_2C$: from (1), $(CH_3OOC)_2C{=}N_2 + h\nu$ (Prexy filter), at 25° (singlet); from (2), $(CH_3OOC)_2C{=}N_2 + h\nu$ (benzophenone photosensitization), at 25° (triplet).

Olefin structures	Relative reactivities (1)	Relative reactivities (2)
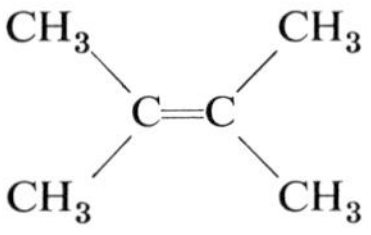 $(CH_3)_2C{=}C(CH_3)_2$	0.88	0.33
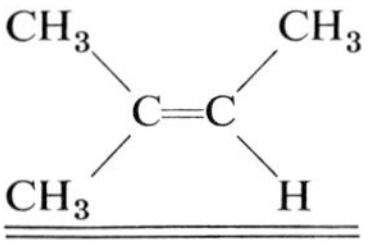 $(CH_3)_2C{=}CH(CH_3)$	1.00	1.00
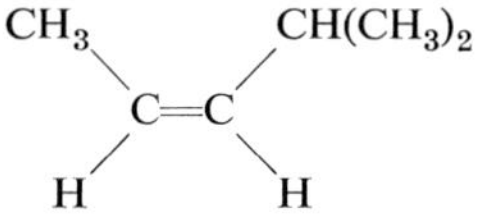 $CH_3CH{=}CHCH(CH_3)_2$ (cis)	0.55	0.15

TABLE 20[28,29] (*Continued*)

Olefin structures	Relative reactivities (1)	Relative reactivities (2)
CH_3 H C=C H $CH(CH_3)_2$	0.23	0.13
n-C_3H_7 H C=C H H	0.47	0.46
$(CH_3)_3C$ H C=C H H	0.48	0 48
H H C=C H H C=C H H	—	4.5
H CH_3 C=C H H C=C CH_3 H	1.3	4.4

TABLE 21[27]

Dicarbomethoxycarbene

$$(CH_3OOC)_2C{:}$$ from $(CH_3OOC)_2C{=}N_2 + h\nu$ (2537 Å), at 25°.

Olefin structures	Relative reactivities[a]
$(CH_3)_2C{=}CH_2$	5.61
cis-$CH_3CH{=}CHCH_3$	2.11
Cyclohexene	1.00
trans-$CH_3CH{=}CHCH_3$	1.20
$CH_2{=}CH{-}CH{=}CH_2$	4.34
$CH_3CH_2CH{=}CH_2$	0.41

[a] Adjusted to a cyclohexene standard from data in the original.

TABLE 22[30]

Phenoxycarbene[a]

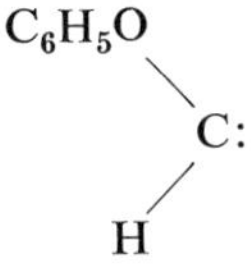

$C_6H_5O(H)C$: from $C_6H_5OCH_2Cl$ + n-C_4H_9Li, in ether, at −15°.

Olefin structures	Relative reactivities
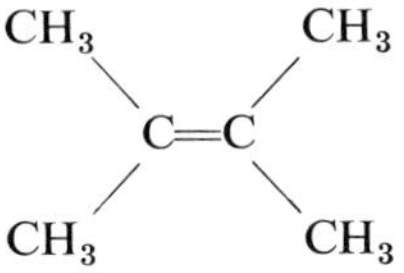 $(CH_3)_2C{=}C(CH_3)_2$	0.43
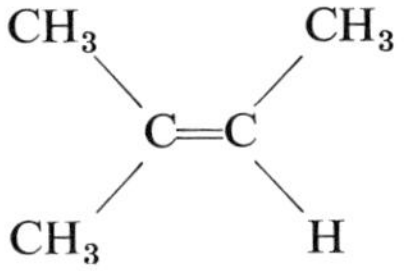 $(CH_3)_2C{=}C(CH_3)H$	1.05 (3.15)[b]
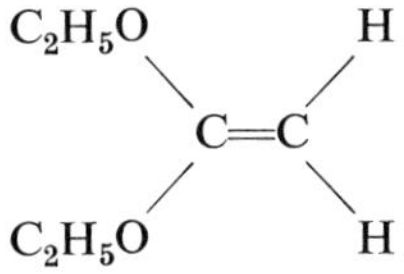 $(C_2H_5O)_2C{=}CH_2$	16.
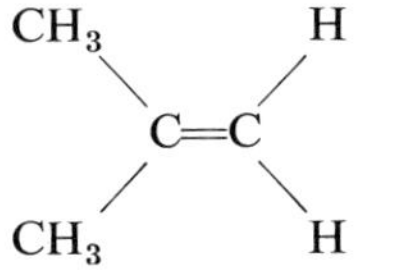 $(CH_3)_2C{=}CH_2$	1.00
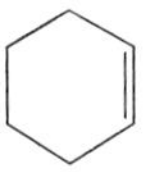	0.2 (0.4)[b]

[a] The reactive intermediate is most likely a carbenoid.
[b] Stereoselectivity, *syn*-phenoxy/*syn*-hydrogen.

TABLE 22A[30a]

Phenylthiocarbene

$C_6H_5S\diagdown C{:}\diagup H$ from $C_6H_5SCH_2Cl + K^{+-}O\text{-}t\text{-}C_4H_9$, at $-15°$.

Olefin structures	Relative reactivities
$(CH_3)_2C{=}C(CH_3)_2$	4.51
$(CH_3)_2C{=}C(CH_3)H$	2.17 (1.02)[a]
$(CH_3)_2C{=}CH_2$	1.00
$CH_3((CH_3)_2CH)C{=}CH_2$	0.372[b]
cis-$CH_3HC{=}CHCH_3$	0.743 (9.5)[a]
cyclohexene	0.326 (1.02)[a]
trans-$CH_3HC{=}CHCH_3$	1.41
$CH_3HC{=}CH_2$	0.602 (19.5)

[a] Stereoselectivity, *syn*-phenylthio/*syn*-hydrogen.

[b] Isomer ratio of product cyclopropanes is 1:5, but product stereochemistry is not yet assigned.

TABLE 23[31]

Fluoro(tritio)carbene

FTC: from CH_2F_2 and recoil T, gas phase, at 22°.

Olefin structures	Relative reactivities[a]
$(CH_3)_2C=C(CH_3)_2$	2.10
$(CH_3)_2C=CHCH_3$	1.48[b]
$(CH_3)_2C=CH_2$	1.00
$CH_3CH=CHCH_3$ (cis)	1.08[b]

TABLE 23 (*Continued*)

Olefin structures	Relative reactivities[a]
$CH_3(H)C{=}C(H)CH_3$ (*trans*: CH_3 and H on first C; H and CH_3 on second C)	1.40
$CH_3CH_2(H)C{=}CH_2$	0.73[b]
$CH_3(H)C{=}CH_2$	0.74[b]
$H_2C{=}CH_2$	0.36[c]
$D_2C{=}CD_2$	0.305

[a] Adjusted to an isobutene standard from data in the original.

[b] Stereoselectivity is unity, within experimental error.

[c] An activation energy difference of 770 cal/mole is reported for the CFT addition to *trans*-butene relative to ethylene.

TABLE 24[32,33]

Chlorocarbene

$$\mathrm{ClHC\!:}$$ from (1), CH_2Cl_2 + n-C_4H_9Li (ether), at −35°[a];

from (2), $ClCH{=}N_2$, at −30°.[b]

Olefin structures	Relative reactivities (1)	Relative reactivities (2)
$(CH_3)_2C{=}C(CH_3)_2$	2.81	1.20
$(CH_3)_2C{=}CH(CH_3)$	1.78 (1.6)[c]	1.18 (1.0)[c]
$(CH_3)_2C{=}CH_2$	1.00	1.00
$(CH_3)HC{=}CH(CH_3)$ (cis)	0.91 (5.5)[c]	0.99 (1.0)[c]

TABLE 24 (*Continued*)

Olefin structures	Relative reactivities	
	(1)	(2)
Cyclohexene	0.60 (3.2)[c]	—
$CH_3(H)C{=}C(H)CH_3$ (trans)	0.45	1.09
$CH_3CH_2(H)C{=}C(H)H$	—	0.74 (1.0)[c]
$n\text{-}C_3H_7(H)C{=}C(H)H$	0.23 (3.4)[c]	—

[a] Not a free carbene.

[b] Side products from the preparation of $ClCHN_2$ are present; see original.

[c] Stereoselectivity, *syn*-chloro/*syn*-hydrogen; see reference 32a.

TABLE 25[33]

Bromocarbene*

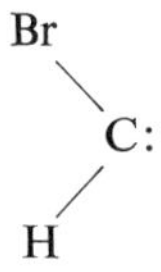

$Br(H)C{:}$ from $BrCH{=}N_2$, at $-30°$.[a]

Olefin structures	Relative reactivities
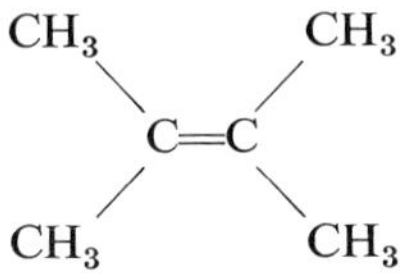 $(CH_3)_2C{=}C(CH_3)_2$	1.18
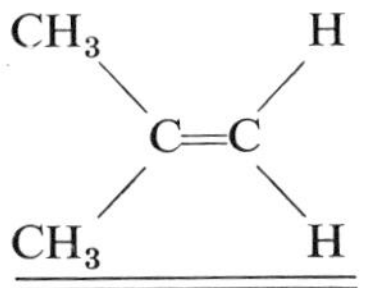 $(CH_3)_2C{=}CH_2$	1.00
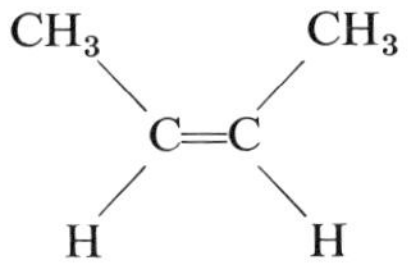 cis-$CH_3CH{=}CHCH_3$	1.02[b]
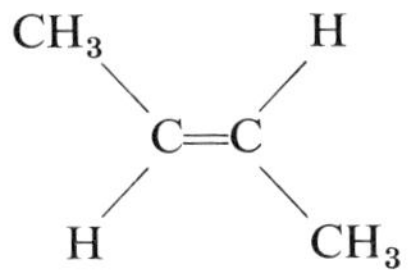 trans-$CH_3CH{=}CHCH_3$	1.10
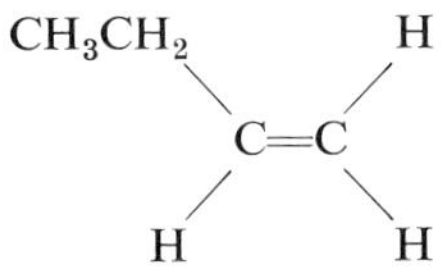 $CH_3CH_2CH{=}CH_2$	0.75[b]

[a] Side products from the preparation of $BrCHN_2$ are present; see original.

[b] Stereoselectivity is unity.

* See Addenda, number 1.

TABLE 26[34]

Fluorophenylcarbene

$F(C_6H_5)C{:}$ from $C_6H_5CHBrF + K^{+-}O\text{-}t\text{-}C_4H_9$, at 25°.

Olefin structures	Relative reactivities
$(CH_3)_2C{=}C(CH_3)_2$	2.7
$(CH_3)_2C{=}CH(CH_3)$	1.2 (0.76)[a]
$(CH_3)_2C{=}CH_2$	1.00
cis-$CH_3CH{=}CHCH_3$	0.12 (1.23)[a]
trans-$CH_3CH{=}CHCH_3$	0.10

[a] Stereoselectivity, *syn*-fluoro/*syn*-phenyl.

TABLE 27[34a]

Chloromethylcarbene

$$\mathrm{Cl(CH_3)C\colon}\ \text{from}\ \mathrm{Cl(CH_3)C\langle\overset{N}{\underset{N}{\|}}} + h\nu\ \text{(Pyrex), at } 25^\circ.$$

Olefin structures	Relative reactivities
$(CH_3)_2C{=}C(CH_3)_2$	3.87
$(CH_3)_2C{=}C(CH_3)H$	2.44 (1.45)[a]
$(CH_3)_2C{=}CH_2$	1.00
$CH_3(H)C{=}C(H)CH_3$ (cis)	0.74 (2.84)[a]
$CH_3(H)C{=}C(H)CH_3$ (trans)	0.52

[a] Stereoselectivity, *syn*-chloro/*syn*-methyl. Similar stereoselectivities were observed for the additions to these olefins of the "methylchlorocarbene" generated by the action of *n*-butyllithium on ethylidene chloride.[34b] The stereoselectivity of addition of this species to trimethylethylene was invariant in hexane, ether, tetrahydrofuran, and diglyme solvents. The species also afforded competitive results of ~5.2 and 3.6 for tetramethylethylene versus *cis*-butene and trimethylethylene versus *trans*-butene, respectively. (Compare with 5.2 and 4.7 for the diazirine-generated carbene, as calculated from the data in the table.)

TABLE 28[35]

Chlorophenylcarbene

$Cl(C_6H_5)C$: from (1), $C_6H_5CHCl_2 + K^{+-}O\text{-}t\text{-}C_4H_9$, at 25°;

from (2), $Cl(C_6H_5)C\langle N{=}N \rangle$ (3-chloro-3-phenyldiazirine) + $h\nu$ (Pyrex filter), at 25°.

Olefin structures	Relative reactivities (1)	Relative reactivities (2)
$(CH_3)_2C{=}C(CH_3)_2$	2.6	5.1
$(CH_3)_2C{=}CH(CH_3)$	1.6 (1.18)[a]	3.2 (1.29)[a]
$(CH_3)_2C{=}CH_2$	1.00	1.00
$CH_3CH{=}CHCH_3$ (cis)	0.31 (2.10)[a]	0.37 (1.97)[a]
$CH_3CH{=}CHCH_3$ (trans)	0.11	0.20

[a] Stereoselectivity, *syn*-chloro/*syn*-phenyl.

TABLE 29[25]

Bromocarboethoxycarbene

$$\mathrm{(Br)(C_2H_5OC(=O))C:}$$ from $C_2H_5OOCCBr{=}N_2 + h\nu$ (Pyrex filter), at −35°.

Olefin structures	Relative reactivities[a]
$(CH_3)_2C{=}C(CH_3)_2$	1.5
$(CH_3)_2C{=}C(CH_3)H$	1.0[b]
$(CH_3)_2C{=}CH_2$	0.45
cis-$CH_3CH{=}CHCH_3$	0.50[b]
$trans$-$CH_3CH{=}CHCH_3$	0.36

[a] Adjusted to a trimethylethylene standard from data in the original.

[b] Stereoselectivity not reported.

TABLE 30[36,37]

Bromophenylcarbene

$Br(C_6H_5)C$: from (1), $C_6H_5CHBr_2 + K^{+-}O\text{-}t\text{-}C_4H_9$, at 25°;

from (2), $Br(C_6H_5)C{<}(N{=}N)$ (diazirine) + $h\nu$ (Pyrex filter), at 23°.

Olefin structures	Relative reactivities (1)	Relative reactivities (2)
$(CH_3)_2C{=}C(CH_3)_2$	1.6	4.4
$(CH_3)_2C{=}CH(CH_3)$	1.3 (1.28)[a]	2.5 (1.31)[a]
$(CH_3)_2C{=}CH_2$	1.00	1.00
cis-$CH_3CH{=}CHCH_3$	0.29 (1.35)[a]	0.53 (1.55)[a]
trans-$CH_3CH{=}CHCH_3$	0.15	0.25

[a] Stereoselectivity, *syn*-bromo/*syn*-phenyl.

TABLE 31[38]

Difluorocarbene

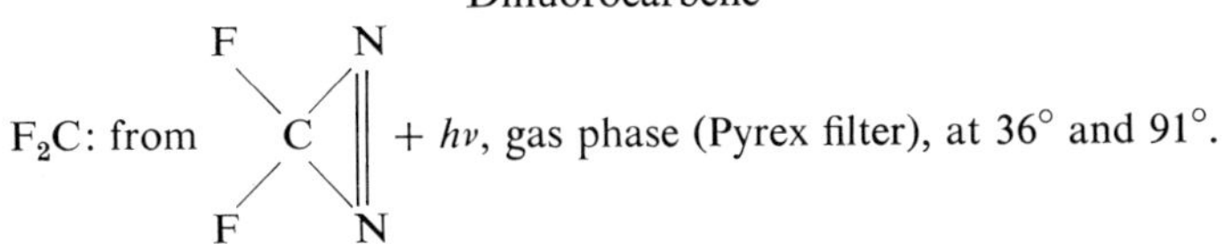

F_2C: from F_2CN_2 (diazirine) + $h\nu$, gas phase (Pyrex filter), at 36° and 91°.

Olefin structures	Relative reactivities 36°	Relative reactivities 91°	Activation parameters ΔE_a[a]	Activation parameters A_i/A_s[b]
$(CH_3)_2C=C(CH_3)_2$	13.1	6.46	0.00	1.00
$(CH_3)_2C=C(CH_3)H$	3.53	2.16	0.87	1.11
$(CH_3)_2C=CH_2$	1.00	1.00	2.88	8.41
$CH_3(H)C=C(H)CH_3$ (cis)	0.065	0.082	3.81	2.47

TABLE 31[38] (*Continued*)

Difluorocarbene

Olefin structures	Relative reactivities		Activation parameters	
	36°	91°	ΔE_a[a]	A_i/A_s[b]
CH_3 H C=C H CH_3	0.074	0.097	3.96	3.6
H H C=C H H C=C H H	0.274	0.404	4.47	15.1[c]
CH_3 H C=C H H	0.0181	0.0325	5.20	7.64
CH_3CH_2 H C=C H H	0.0105	0.0096	2.53	0.047

[a] E_a for olefin$_i$—E_a for $(CH_3)_2C{=}C(CH_3)_2$ in kcal/mole. Errors, about ±0.3 kcal/mole.

[b] A for olefin$_i$ divided by A for $(CH_3)_2C{=}C(CH_3)_2$. For errors, see original.

[c] Corrected by a factor of 0.5 for two double bonds.

TABLE 32[39]

Difluorocarbene

F_2C: from $(CH_3)_3SnCF_3$ + Na^+I^-, in dimethoxyethane, at 80°

Olefin structures	Relative reactivities
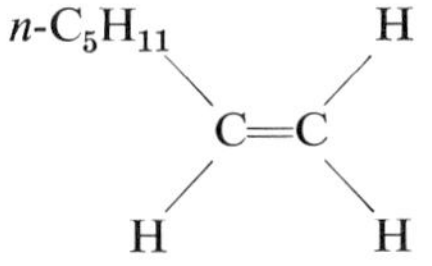 n-$C_5H_{11}CH{=}CH_2$	1.00
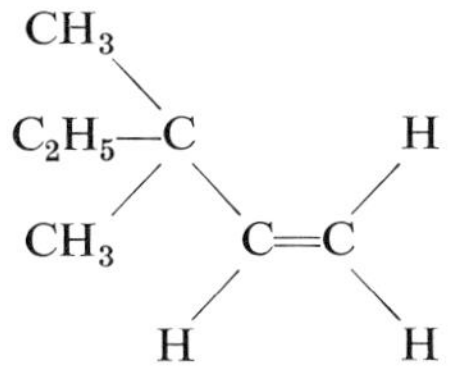 $C_2H_5C(CH_3)_2CH{=}CH_2$	0.12
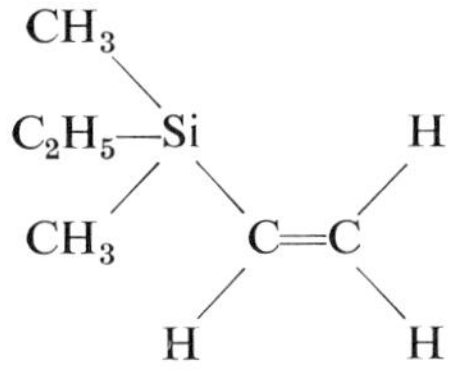 $C_2H_5Si(CH_3)_2CH{=}CH_2$	0.26
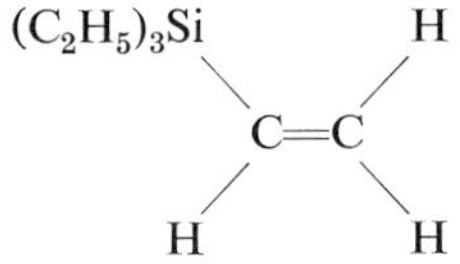 $(C_2H_5)_3SiCH{=}CH_2$	0.15

TABLE 33[4]

Dichlorocarbene

Cl_2C: from $CHCl_3 + K^{+-}O\text{-}t\text{-}C_4H_9$, $-10°$ to $-20°$.

Olefin structures	Relative reactivities
$(CH_3)_2C{=}C(CH_3)_2$	53.7[a]
$(CH_3)_2C{=}CH(CH_3)$	23.5[a]
$(CH_3)_2C{=}CH_2$	8.32
$(CH_3CH_2)(CH_3)C{=}CH_2$	5.50
$CH_3CH_2CH{=}CHCH_3$ (trans)	2.14[b]
$CH_3CH_2OCH{=}CH_2$	1.86[c]

TABLE 33[4] (*Continued*)

Olefin structures	Relative reactivities
$CH_3CH_2(H)C{=}C(H)CH_3$ (trans)	1.62[b]
cyclohexene	1.00
$n\text{-}C_4H_9(H)C{=}CH_2$	0.19[d]
$n\text{-}C_3H_7(H)C{=}CH_2$	0.14

[a] These rates, independently determined, are also given in Table 36, normalized to isobutene.

[b] The *trans*-pentene-2 rate, at least, must be in error, since with simple alkenes, CCl_2 invariably reacts more rapidly with a *cis* compound than its *trans* isomer. Redetermination of these rates gave *cis*-pentene-2, 1.52; and *trans*-pentene-2, 0.86 on the cyclohexane = 1.00 scale. See Table 34.

[c] There is a small discrepancy here from the value 1.28, Table 45, which may be due to a change in temperature and solvent conditions.

[d] The rate isobutene/1-hexene, 44, differs substantially from the value 73, which can be calculated from the data in Tables 36 and 37. The cause of this discrepancy is not known.

TABLE 34[40]

Dichlorocarbene
Cl_2C: from (1), $C_6H_5HgCCl_2Br$ + NaI, −15°; from (2), $CHCl_3$ + $K^{+}\,^{-}O\text{-}t\text{-}C_4H_9$, −15°.

Olefin structures	Relative reactivities (1)	Relative reactivities (2)
$(CH_3)_2C{=}C(CH_3)_2$	43.6	53.7[a]
$CH_3CH_2(CH_3)C{=}C(CH_3)_2$	22.7	—
$(CH_3CH_2)_2C{=}C(CH_3)H$	4.73	—
$CH_3CH_2(H)C{=}C(CH_3)H$	1.44	1.52
cyclohexene	1.00	1.00

TABLE 34[40] (*Continued*)

Olefin structures	Relative reactivities (1)	(2)
$CH_3CH_2(H)C{=}C(H)CH_3$	0.834	0.86
$n\text{-}C_3H_7(H)C{=}C(CH_2CH_3)H$	0.81	0.89
$n\text{-}C_3H_7(H)C{=}C(H)CH_2CH_3$	0.435	0.435
$n\text{-}C_5H_{11}(H)C{=}C(H)H$	0.11	—

[a] From Table 33.

TABLE 35[41]

Dichlorocarbene

Cl_2C: from $CHCl_3$ + K^{+} ^{-}O-t-C_4H_9, $-10°$ to $-15°$.

Olefin structures	Relative rate constants[a]
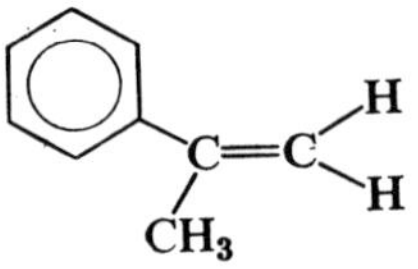 $C_6H_5(CH_3)C{=}CH_2$	2.2
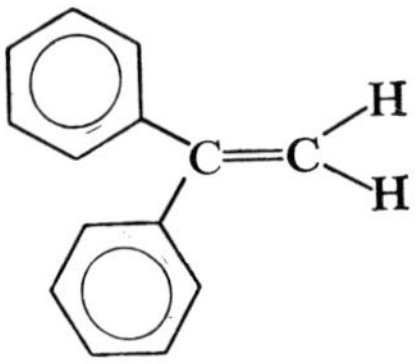 $(C_6H_5)_2C{=}CH_2$	1.8
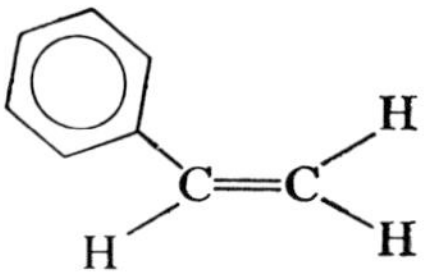 $C_6H_5CH{=}CH_2$	1.4
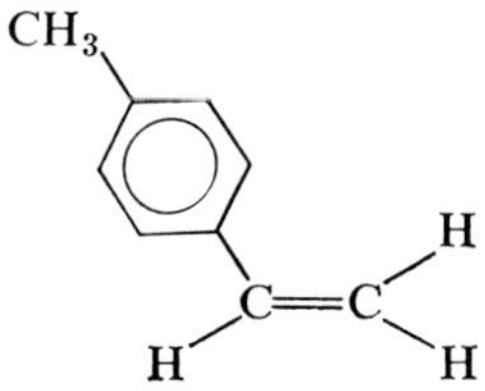 p-$CH_3C_6H_4CH{=}CH_2$	1.5
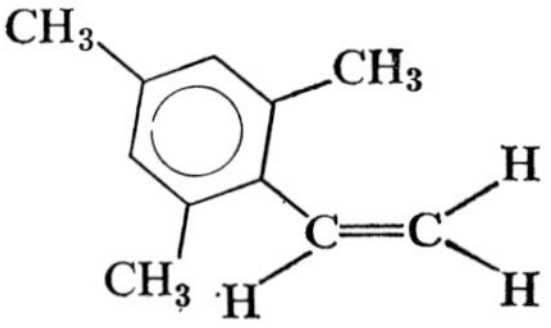 $2,4,6$-$(CH_3)_3C_6H_2CH{=}CH_2$	1.4
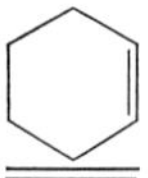 cyclohexene	1.00

[a] Data as calculated by the Ingold expression[6] are given here.

TABLE 36[42]

Dichlorocarbene

Cl_2C: from $CHCl_3$ + K^+ $^-O\text{-}t\text{-}C_4H_9$, at $-10°$ and $25°$.

	Relative reactivities		Activation parameters	
Olefin structures	$-10°$[a]	$25°$[b]	ΔE_a[c]	A_i/A_s[d]
$(CH_3)(CH_3)C{=}C(CH_3)(CH_3)$	7.40	8.98	861 ± 125	38.4 ± 8.97
$(CH_3)(CH_3)C{=}C(H)(CH_3)$	3.05	3.12	101 ± 271	3.70 ± 2
$(CH_3)(CH_3)C{=}C(H)(H)$	1.00	1.00	0	1.00
$(CH_3)(H)C{=}C(CH_3)(H)$	0.23	0.27	713 ± 111	0.90 ± 0.16
$(CH_3)(H)C{=}C(H)(CH_3)$	0.15	0.18	811 ± 72	0.71 ± 0.13
$(CH_3CH_2)(H)C{=}C(H)(H)$	0.011	—	—	—

[a] Data for *cis*- and *trans*-butene, and 1-butene, reference 42a; other data, reference 42b.

[b] Reference 42b.

[c] E_a for olefin$_i$ $-E_a$ for isobutene, in cal/mole. Errors are derived from extremes of the rate measurements at the two temperatures.

[d] A for olefin$_i$/A for isobutene, errors derived as in note c.

TABLE 37[43]

Dichlorocarbene

Cl_2C: from $CHCl_3$ + $K^{+}\,^{-}O\text{-}t\text{-}C_4H_9$, $-10°$.

Olefin structures	Relative reactivities
$CH_3CH_2(H)C{=}CH_2$	1.00[a]
$(CH_3)_2CH(H)C{=}CH_2$	0.43
$(CH_3)_3C(H)C{=}CH_2$	0.029
$n\text{-}C_4H_9(H)C{=}CH_2$	1.24
$CH_3{-}CH(CH_3){-}CH_2(H)C{=}CH_2$	0.99
$(CH_3)_3C{-}CH_2(H)C{=}CH_2$	0.97

[a] To adjust these data to an isobutene standard (for comparison with data of Table 36) multiply by 0.011.

TABLE 37A[43a]

Dichlorocarbene

Cl_2C: from $CHCl_3$ + NaOH (H_2O) in the presence of tricaprylymethylammonium chloride, 30°–45°.

Olefin structures	Relative reactivities
$(CH_3)_2C{=}C(CH_3)_2$	16.0
$CH_3CH_2CH_2(CH_3)C{=}CH_2$	7.2
$(CH_3)_2CH(H)C{=}C(H)CH_3$ (cis)	1.4
cyclohexene	1.0[a]
$(CH_3)_2CH(H)C{=}C(H)CH_3$ (trans)	1.0
$CH_3(H)C{=}C(H)CH_2CH_2CH_3$ (trans)	0.8
n-$C_4H_9(H)C{=}CH_2$	0.2
$CH_3(C_2H_5)CH(H)C{=}CH_2$	0.12

[a] Adjusted to a cyclohexene standard from data in the original.

TABLE 38[44,44a]

Dichlorocarbene

Cl_2C: from $CHCl_3$ + n-C_4H_9Li, in tetrahydrofuran, at 0° and −78°.[a]

Olefin structures	Relative reactivities[b] (0°)	Relative reactivities[b] (−78°)	Activation parameters[b] $\Delta\Delta F^{\ddagger}$ (0°) (kcal/mole)	$\Delta\Delta F^{\ddagger}$ (−78°) (kcal/mole)	$\Delta\Delta H^{\ddagger}$ (kcal/mole)	$\Delta\Delta S^{\ddagger}$ (eu)
$(CH_3O)(CH_3)C{=}C(CH_3)_2$	10.00	7.77	−1.26	−0.80	0.30	5.72
$(CH_3)_2C{=}C(CH_3)_2$[c]	7.59	5.25	−1.10	−0.65	0.53	5.97
$(CH_3O)(H)C{=}C(CH_3)_2$	7.77	6.31	—	—	—	—
$(CH_3)(CH_3O)C{=}C(CH_3)(H)$	5.63	4.17	−0.94	−0.57	0.43	5.02

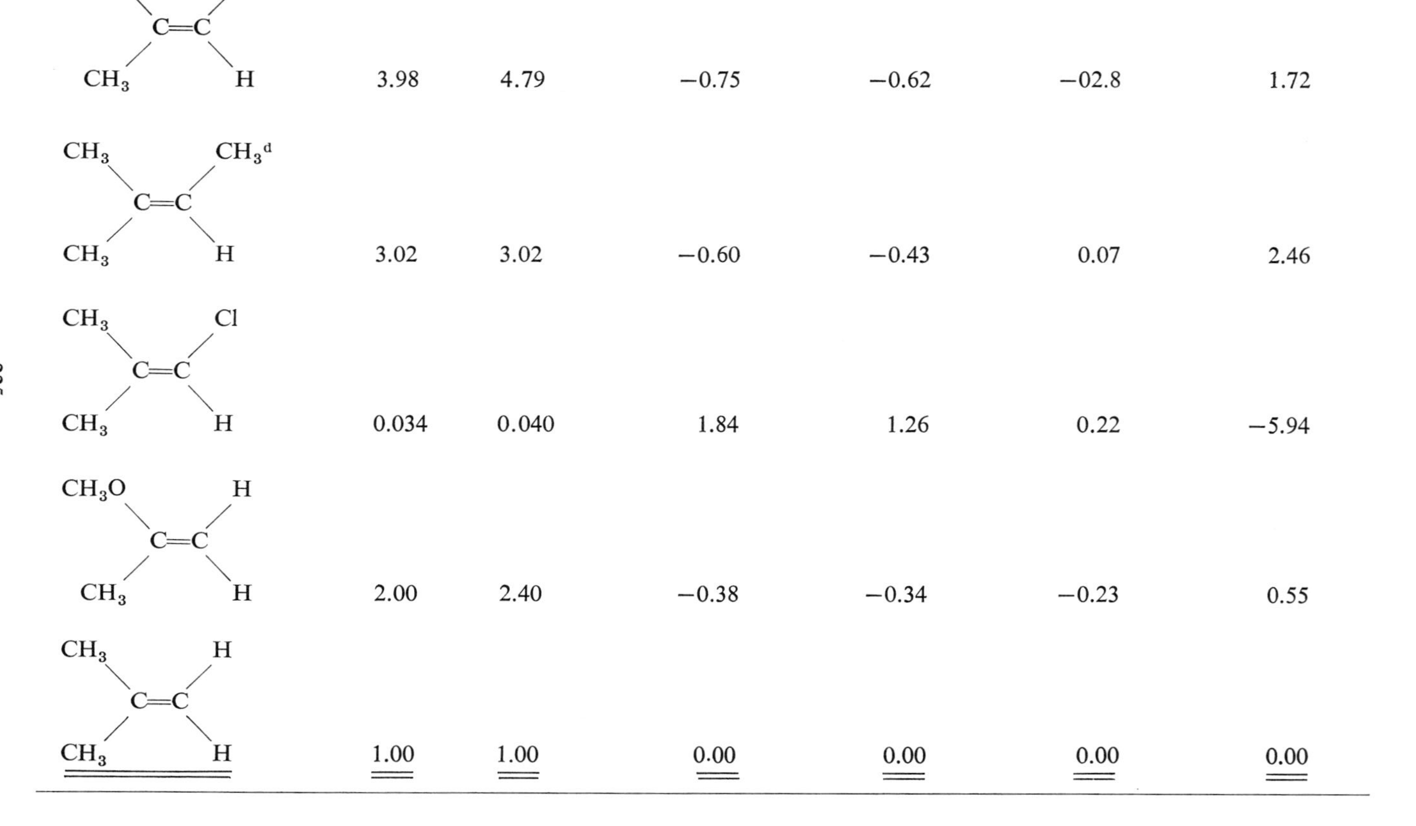

$(CH_3O)(CH_3)C{=}C(CH_3)(H)$	3.98	4.79	−0.75	−0.62	−02.8	1.72
$(CH_3)(CH_3)C{=}C(CH_3)(H)$[d]	3.02	3.02	−0.60	−0.43	0.07	2.46
$(CH_3)(CH_3)C{=}C(Cl)(H)$	0.034	0.040	1.84	1.26	0.22	−5.94
$(CH_3O)(CH_3)C{=}C(H)(H)$	2.00	2.40	−0.38	−0.34	−0.23	0.55
$(CH_3)(CH_3)C{=}C(H)(H)$	1.00	1.00	0.00	0.00	0.00	0.00

TABLE 38 (*Continued*)

Olefin structures	Relative reactivities[b] (0°)	(−78°)	Activation parameters[b] $\Delta\Delta F^{\ddagger}$ (0°) (kcal/mole)	$\Delta\Delta F^{\ddagger}$ (−78°) (kcal/mole)	$\Delta\Delta H^{\ddagger}$ (kcal/mole)	$\Delta\Delta S^{\ddagger}$ (eu)
$(CH_3OCH_2)(CH_3)C{=}CH_2$	0.28	0.25	0.69	0.54	0.14	−2.02
$CH_3CH{=}CHCH_3$ (cis)[e]	0.32	0.26	0.63	0.57	0.29	−1.25
Cyclohexene	0.19	0.14	0.90	0.77	0.14	−2.78
$CH_3CH{=}CH_2$	0.038	0.016	1.78	1.62	1.19	−2.16

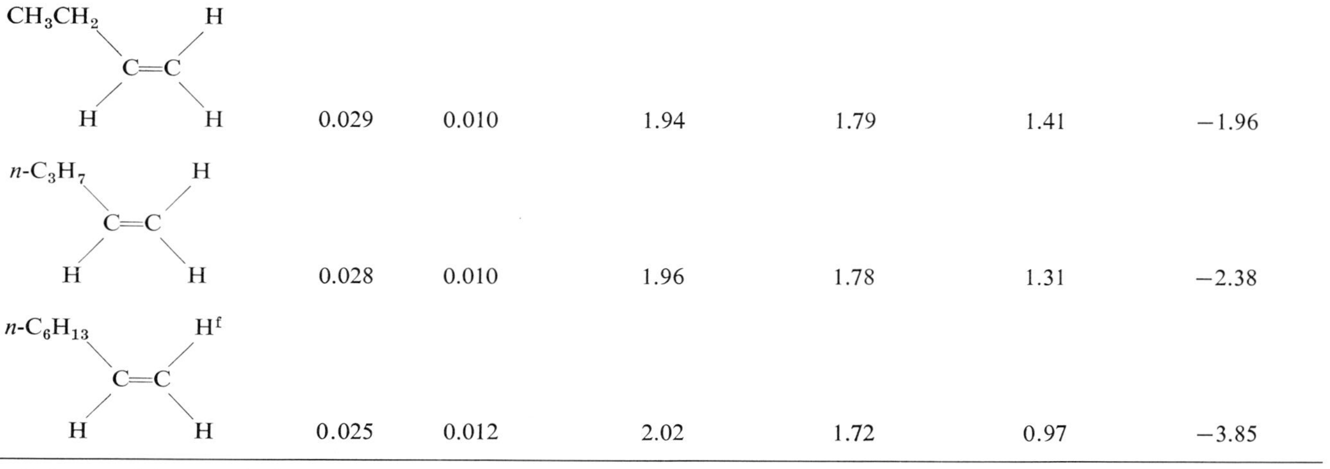

CH_3CH_2(H)C=C(H)H	0.029	0.010	1.94	1.79	1.41	−1.96
n-C_3H_7(H)C=C(H)H	0.028	0.010	1.96	1.78	1.31	−2.38
n-C_6H_{13}(H)C=C(H[f])H	0.025	0.012	2.02	1.72	0.97	−3.85

[a] Other conditions are indicated by notes.

[b] Relative reactivities are reconverted from the log values which appear in the original. Errors in relative reactivities are ±10%. Errors in $\Delta\Delta H^{\ddagger}$ are ±0.25 kcal/mole. Errors in $\Delta\Delta S^{\ddagger}$ are ±1.0 eu.

[c] Relative reactivities (0°, −78°) in 1,2-dimethoxyethane (DME) solvent, 8.51, 5.63; in ether, 7.95, 4.68; with n-C_4H_9Li·tetramethylethylenediamine (TMEDA) in isopentane as base, 10.48, 5.50.

[d] Relative reactivities with n-C_4H_9Li·TMEDA in isopentane as base, 4.17, 3.31.

[e] Relative reactivities with n-C_4H_9Li·TMEDA in isopentane as base, 0.32, 0.26. Unchanged using t-$C_4H_9O^-K^+$ in isopentane.

[f] Using n-C_4H_9Li·TMEDA in isopentane as base; unchanged with t-$C_4H_9O^-K^+$ in isopentane.

TABLE 38A[44b]

Dichlorocarbene

Cl_2C: from $LiCCl_3$ ($CHCl_3$ + n-C_4H_9Li, −115° to −110°), in tetrahydrofuran, at −73°.

Olefin structures	Relative reactivities
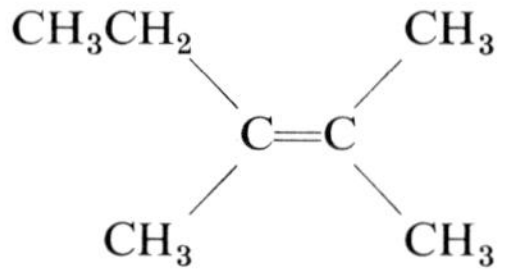 CH_3CH_2 CH_3 C=C CH_3 CH_3	23.4
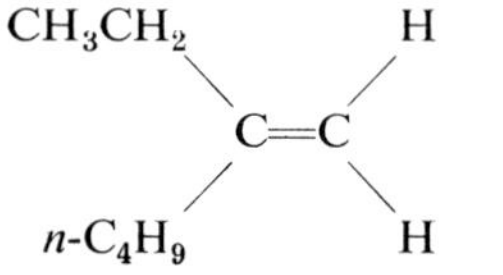 CH_3CH_2 H C=C n-C_4H_9 H	3.14
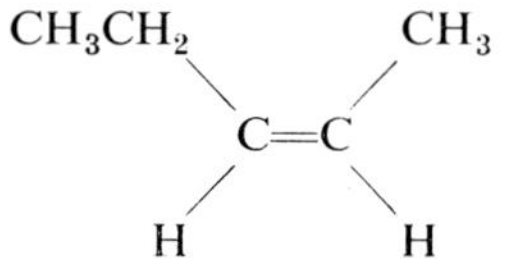 CH_3CH_2 CH_3 C=C H H	1.34
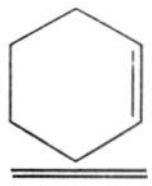	1.00
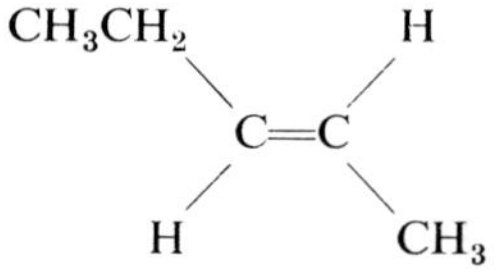 CH_3CH_2 H C=C H CH_3	0.79
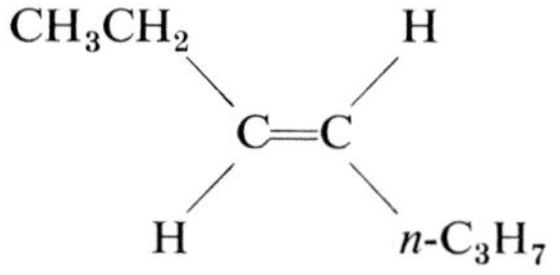 CH_3CH_2 H C=C H n-C_3H_7	0.40
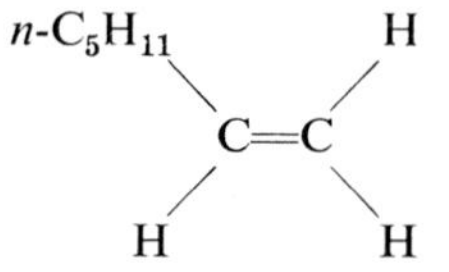 n-C_5H_{11} H C=C H H	0.12

TABLE 38B[44c]

Dichlorocarbene

Cl_2C: from $LiCCl_3$, in ether, at −100° and −15°.

Olefin structures	Relative reactivities[a] −100°	−15°
$(CH_3)_2C{=}C(CH_3)_2$	2.85	10.07
$(CH_3)_2C{=}CH(CH_3)$	2.38	3.73
$(CH_3)_2C{=}CH_2$	1.00	1.00
$CH_3CH{=}CHCH_3$ (cis)	0.27	0.29
Cyclohexene	0.099	0.12
n-$C_6H_{13}CH{=}CH_2$	0.0050	0.017

[a] Adjusted to an isobutene standard from data in the original.

TABLE 38C[44d]

Dichlorocarbene

Cl_2C: from $CCl_3COO^-Na^+$, in monoglyme.[a]

Olefin structures	Relative reactivities			
	k_{total}[b]	$k_{corr.}$[c]	k_{cis}[d]	k_{trans}[e]
cis-Cyclooctene[f]	0.90	0.90	0.90	—
cis-1-*cis*-5-Cyclooctadiene	2.60	1.30	1.30	—
cis-Cyclodecene	0.96	0.96	0.96	—
trans-Cyclodecene	4.00	4.00	—	4.00
1-*cis*-5-*trans*-Cyclodecadiene	4.68	2.34	1.15	3.53
cis-Cyclododecene	0.67	0.67	0.67	—
trans-Cyclododecene	1.00	1.00	—	1.00
1-*cis*-5-*trans*-Cyclododecadiene	1.72	0.86	0.66	1.06
1-*trans*-5-*trans*-Cyclododecadiene	2.28	1.14	—	1.14
1-*cis*-5-*trans*-9-*trans*-Cyclododecatriene	2.35	0.78	0.51	0.92
1-*trans*-5-*trans*-9-*trans*-Cyclododecatriene	3.06	1.02	—	1.02

[a] The exact temperature is not given, but it is presumably about 85°.

[b] Total reactivity, relative to *trans*-cyclododecene.

[c] Corrected reactivity (per bond basis).

[d] Reactivity of a *cis* double bond in the substrate, relative to *trans*-cyclododecene (per bond).

[e] Reactivity of a *trans* double bond in the substrate, relative to *trans*-cyclododecene (per bond).

[f] It seemed less confusing to give names, rather than structures, for these large cyclic substrates.

TABLE 39[7]

Dichlorocarbene

Cl_2C: from $CHCl_3 + K^{+-}O\text{-}t\text{-}C_4H_9 \cdot t\text{-}C_4H_9OH$, at 0°.

Olefin structures[a]	Relative rate constants
p-$(CH_3)_2N$	3.65
p-CH_3O	1.82
p-CH_3	1.29
p-CH_3CH_2	1.22
p-$(CH_3)_3C$	1.21
p-$(CH_3)_2CH$	1.12
m-CH_3	1.03
o-CH_3	0.11
H	1.00
p-F	0.99
m-CH_3O	0.97
o-CH_3O	0.33
p-Cl	0.78
p-Br	0.78
m-Br	0.53
p-O_2N	0.46
p-$CH_2{=}C(CH_3)$	1.06[b]
p-(1-methyl-2,2-dichlorocyclopropyl): *p*- CH_3, Cl, Cl 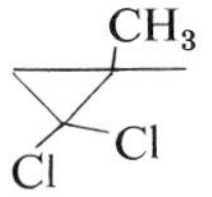	1.11
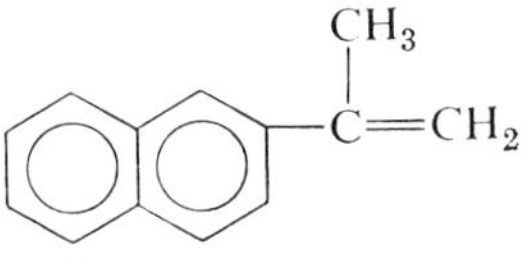 2-naphthyl–C(CH_3)$=CH_2$	1.02
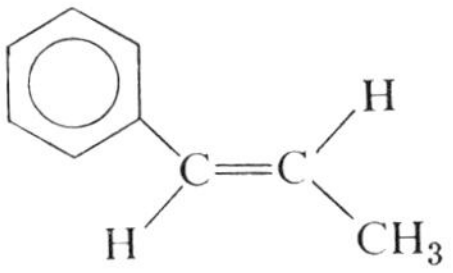 C_6H_5(H)C$=$C(H)CH_3	0.26
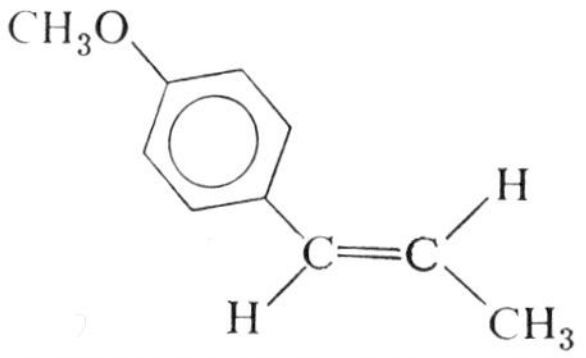 CH_3O-C_6H_4(H)C$=$C(H)CH_3	0.67

TABLE 39 (*Continued*)

Olefin structures[a]	Relative rate constants
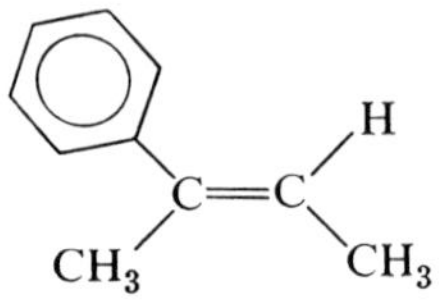 H C=C CH_3 CH_3	1.40
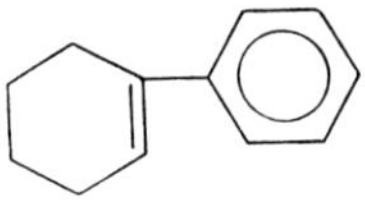	0.89[c]
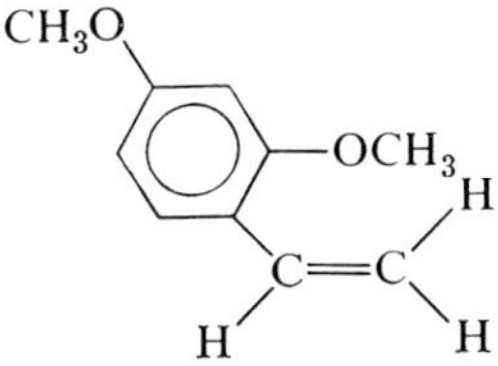 CH_3O OCH_3 H C=C H H	0.11
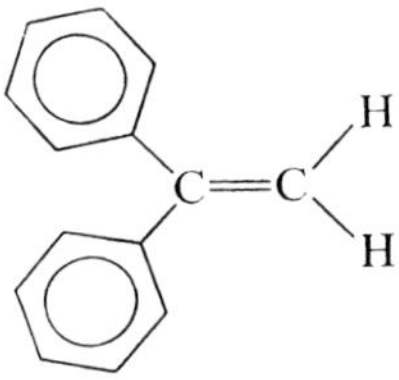 H C=C H	0.35
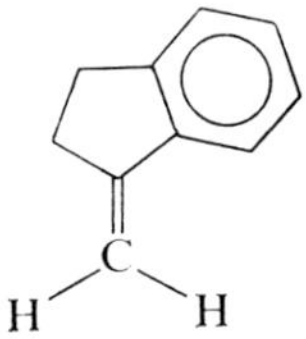 C H H	2.82
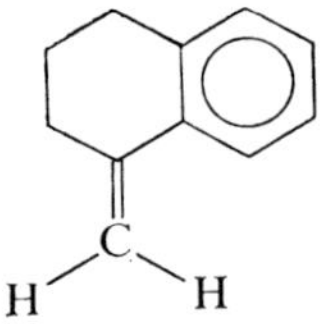 C H H	1.38

TABLE 39 (*Continued*)

Olefin structures	Relative rate constants
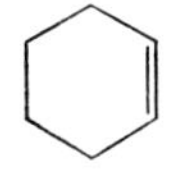	0.14[c]

[a] Substituents are R in R, $C{=}CH_2$, CH_3

[b] Reactivity is per double bond.

[c] These values appear in Table 42 where 1-phenylcyclohexene/cyclohexene is given as 6.19. The cyclohexene value in the present table is calculated from that figure. Multiplication by ~7.1 will put these data on the cyclohexene scale. It should be noted, however, that the values in Table 42 were determined at 15°. Moreover, the value for α-methylstyrene/cyclohexene (−10° to −15°) given in Table 35 is 2.2, very different from the 7.1 just suggested by the first comparison. This highlights the impossibility of trying to put much of the data on one scale.

TABLE 40[40,45]

Dichlorocarbene

Cl_2C: from (1), $C_6H_5HgCCl_3$ + NaI, in dimethoxyethane, at 80°;
(2), $C_6H_5HgCCl_2Br$, in benzene or dimethoxyethane, at 80°;
(3), $Cl_3CCOO^-Na^+$, in dimethoxyethane, at 80°.

Olefin structures	Relative reactivities		
	(1)	(2)	(3)
$(CH_3CH_2)(CH_3)C{=}C(CH_3)_2$	23.2	22.5	24.8
$(C_6H_5)(CH_3)C{=}CH_2$	—	7.25	7.35
$(CH_3CH_2)_2C{=}C(CH_3)H$	3.13	3.54	3.52
$(CH_3CH_2)(n\text{-}C_4H_9)C{=}CH_2$	2.30	2.31	—
$(CH_3CH_2)(n\text{-}C_3H_7)C{=}CH_2$	—	2.07	2.08
Vinylferrocene ($C_5H_5FeC_5H_4$–CH=CH_2)	—	—	1.60[a]

TABLE 40 (*Continued*)

Olefin structures	Relative reactivities		
	(1)	(2)	(3)
$C_6H_5(H)C{=}CH_2$ (C₆H₅ and H cis; H, H)	—	1.22	1.26
Cyclohexene	1.00	1.00	1.00
n-$C_3H_7(H)C{=}C(H)CH_2CH_3$ (cis)	0.835	0.828	0.800
n-$C_3H_7(H)C{=}C(H)CH_2CH_3$ (trans)	0.537	0.524	0.523
n-$C_5H_{11}(H)C{=}CH_2$	0.218	0.236	0.219
$Cl_2C{=}C(Cl)H$	—	0.0152	—

[a] Data of I. Moritani, monoglyme, 80°. Private communication from Professor Moritani.

TABLE 41[46]

Dichlorocarbene

Cl_2C: from $C_6H_5HgCCl_2Br$, in benzene, at 80.3°.

Olefin structures[a]	Relative reactivities
CH_3	1.52
H	1.00[b]
F	0.961
Cl	0.839
CF_3	0.453

[a] X in X—C₆H₄—CH=CH₂.

[b] To scale these values to a cyclohexene standard (for direct comparison with the data of Table 40) multiply by 1.22.

TABLE 42[47]

Dichlorocarbene

Cl_2C: from $CHCl_3 + K^{+-}O\text{-}t\text{-}C_4H_9$, at 15°.

Olefin structures	Relative rate constants[a]
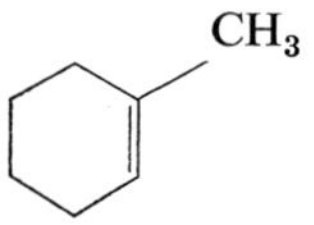 1-methylcyclohexene (CH_3)	8.43
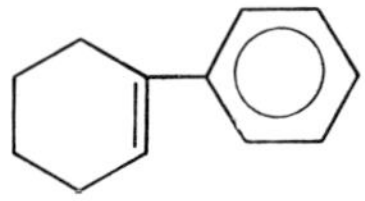 1-phenylcyclohexene	6.19
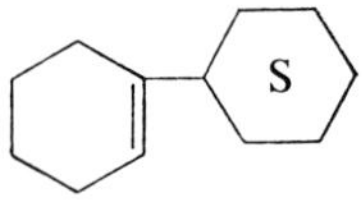 1-cyclohexylcyclohexene (S)	1.14
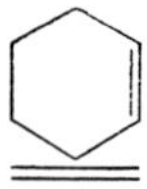 cyclohexene	1.00
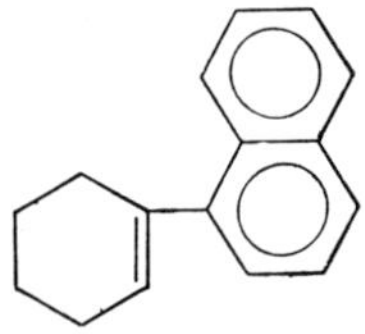 1-(1-naphthyl)cyclohexene	0.40

[a] Only values calculated by the Ingold expression[6] are given here.

TABLE 43[48]

Dichlorocarbene

Cl_2C: from (1), $CHCl_3 + K^{+-}O\text{-}t\text{-}C_4H_9$, at 15°;
(2), $Cl_3CCOOC_2H_5 + Na^{+-}OCH_3$, in benzene, at 2°;
(3), $CHCl_3$ + Li, in tetrahydrofuran, at 10°;
(4), $CHCl_3 + CH_3Li$, in ether, at −20°;
(5), $CBrCl_3 + CH_3Li$, in ether, at −60°;
(6), $Cl_3CCOO^-Na^+$, in monoglyme, at 80–83°;
(7), $C_6H_5HgCCl_3$, in benzene, at 82°.

Olefin structures	Relative rate constants[a]						
	(1)	(2)	(3)	(4)	(5)	(6)	(7)
CH_3	8.43	8.01	7.55	7.92	7.90	8.48	8.06
	1.00	1.00	1.00	1.00	1.00	1.00	1.00
Cl	0.45[b]	0.44	—	0.50	—	0.45[c]	—
O ‖ CCH_3	—	—	—	—	—	0.26	0.30

[a] From the Ingold expression.[6]
[b] −10° to −15°.
[c] 85°.

TABLE 44[49]

Dichlorocarbene

Cl_2C: from $CHCl_3 + K^{+-}O\text{-}t\text{-}C_4H_9$, $-8°$ to $-10°$.

Olefin structures[a]	Relative reactivities
$(CH_3)_2C=C=CH_2$	1.00
$(CH_3CH_2)(CH_3)C=C=CH_2$	0.64
$((CH_3)_2CH)(CH_3)C=C=CH_2$	0.11
$((CH_3)_3C)(CH_3)C=C=CH_2$	"0.0"
$((CH_3)_3CCH_2)(CH_3)C=C=CH_2$	0.28
cyclobutylidene$=C=CH_2$	1.67
cyclopentylidene$=C=CH_2$	1.4
cyclohexylidene$=C=CH_2$	0.67

[a] Only the more substituted double bond is attacked in the reaction.

TABLE 45[50]

Dichlorocarbene

Cl_2C: from $CHCl_3$ + K^{+} ^{-}O-t-C_4H_9, in t-butanol, at 0°–5°.

Olefin structures	Relative reactivities[a]
$(CH_3)_3CO(H)C{=}CH_2$	2.59
$(CH_3)_2CHO(H)C{=}CH_2$	1.66
$CH_3CH_2O(H)C{=}CH_2$	1.28
$CH_3O(H)C{=}CH_2$	1.10
$CH_3{-}CH(CH_3){-}CH_2O(H)C{=}CH_2$	1.06
cyclohexene	1.00
$ClCH_2CH_2O(H)C{=}CH_2$	0.51

[a] Rates adjusted to cyclohexene = 1.00 from data in the original.

TABLE 46[51]

Dichlorocarbene

Cl_2C: from $CHCl_3 + K^{+}\,{}^{-}O\text{-}t\text{-}C_4H_9$, in isopropyl- or *n*-butylcyclohexane, at $-30°$.

Olefin structures	Relative reactivities
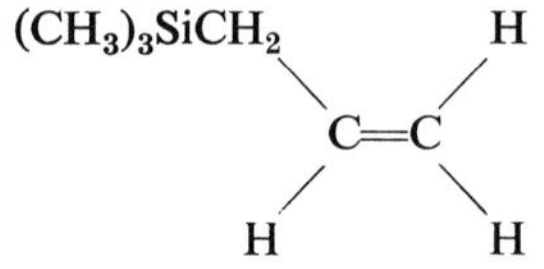 $(CH_3)_3SiCH_2(H)C{=}C(H)H$	5.00
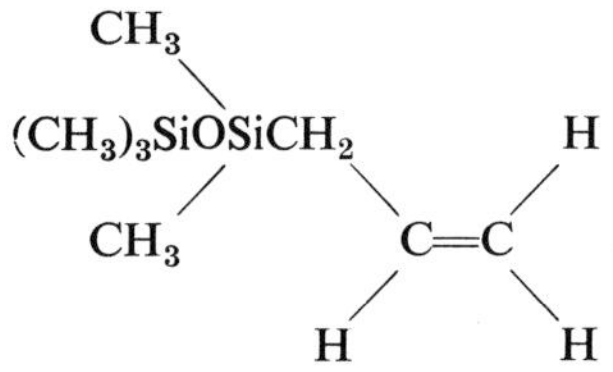 $(CH_3)_3SiOSi(CH_3)_2CH_2(H)C{=}C(H)H$	3.80
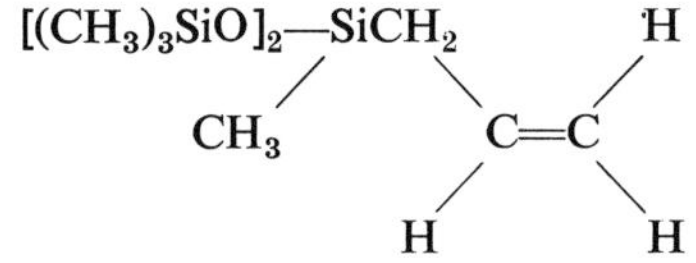 $[(CH_3)_3SiO]_2{-}Si(CH_3)CH_2(H)C{=}C(H)H$	2.50
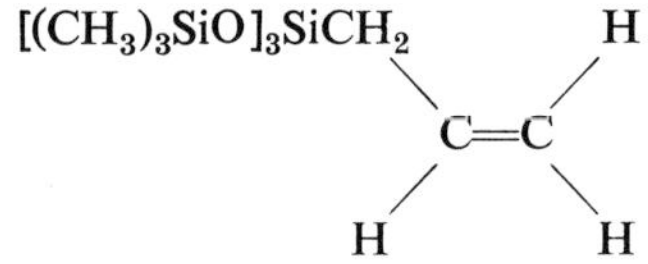 $[(CH_3)_3SiO]_3SiCH_2(H)C{=}C(H)H$	1.95
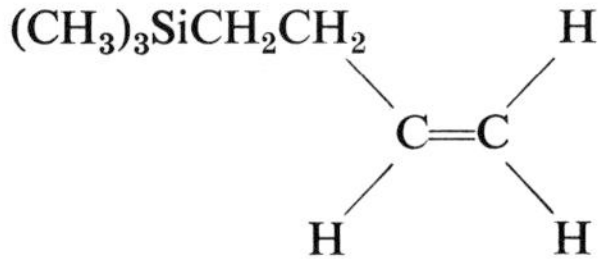 $(CH_3)_3SiCH_2CH_2(H)C{=}C(H)H$	1.20[a]

TABLE 46 (*Continued*)

Olefin structures	Relative reactivities
$n\text{-}C_5H_{11}CH=CH_2$	1.00
$(CH_3)_3SiCH=CH_2$	0.047
$(CH_3)_3SiOSi(CH_3)_2CH=CH_2$	0.033
$[(CH_3)_3SiO]_2Si(CH_3)CH=CH_2$	0.036
$[(CH_3)_3SiO]_3SiCH=CH_2$	0.036

[a] Reference 52.

TABLE 47[39]

Dichlorocarbene

Cl_2C: from $C_6H_5HgCCl_2Br$, in benzene, at 71° and 80°.

Olefin structures	Relative reactivities
$(CH_3)_3GeCH_2CH=CH_2$ (cis H, H; terminal $=CH_2$)	5.7[a]
$(CH_3)_3SiCH_2CH=CH_2$	4.2[a]
n-$C_5H_{11}CH=CH_2$	1.00[b,c]
$(CH_3)_3C—CH_2CH=CH_2$	0.78[a]
$CH_3CH_2—C(CH_3)_2—CH=CH_2$	0.043[b]

TABLE 47 (*Continued*)

Olefin structures	Relative reactivities
$(CH_3)(CH_3CH_2)(CH_3)Si\text{—}CH{=}CH_2$ (Si bearing CH_3, CH_3CH_2, CH_3; C=C with H, H, H)	0.069[b]
$(C_2H_5)_3Si\text{—}CH{=}CH_2$	0.048[b]
$(C_2H_5)_3Ge\text{—}CH{=}CH_2$	0.064[b]

[a] 71°.
[b] 80°.
[c] To adjust these data to a cyclohexene standard (80°) for comparison with the data of Table 40 multiply by 0.236.

TABLE 48[53,53a]

Dichlorocarbene

Cl_2C: from $Cl_3CCOOC_2H_5 + Na^{+}\,{}^{-}OCH_3$, temperature not given.

Olefin structures	Relative reactivities[a]
$(C_6H_5)(C_2H_5)C{=}CH_2$	60
$H_2C{=}C(C_6H_5)CH{=}CH_2$	41[b]
$C_6H_5CH{=}CHCH{=}CH_2$	21[c]
$C_6H_5CH{=}CH_2$	14
$C_6H_5CH_2CH_2CH{=}CH_2$	1

[a] Adjusted to a common standard from data in the original.

[b] Reaction at the 1,2 position.

[c] Reaction at the 3,4 position.

TABLE 48A[53b]

Dichlorocarbene

Cl_2C: from $Cl_3CCOOC_2H_5 + Na^{+-}OCH_3$, temperature not given.

Olefin structures	Relative reactivities
$C_6H_5CH=CH-CH=CH-CH=CH_2$ (all-cis structure drawn: C_6H_5(H)C=C(H)–(H)C=C(H)–(H)C=C(H)H)	18.5[a]
$C_6H_5CH=CH-(CH_2)_3CH_3$	11.9[b]
$C_6H_5CH_2CH_2CH=CHCH_2CH_3$	3.7[b]
$C_6H_5(CH_2)_4CH=CH_2$	1.0

[a] Reaction at the 5,6 position exceeds reaction at the 1,2 position by 13:1.

[b] Stereochemistry not given.

TABLE 48B[53c]

Dichlorocarbene

Cl_2C: from $Cl_3CCOOC_2H_5$ + $Na^{+-}OCH_3$, temperature not given.

Olefin structures	Relative reactivities[a]
$C_6H_5CH{=}CH{-}CH{=}CH{-}CH{=}CH{-}CH{=}CH_2$	17.6[b]
$C_6H_5CH{=}CH{-}CH{=}CH{-}CH{=}CH_2$	17.1[c]
$C_6H_5CH{=}CH{-}CH{=}CH_2$	15.6[d]
$C_6H_5CH{=}CH{-}n\text{-}C_6H_{13}$	9.
$C_6H_5(CH_2)_6CH{=}CH_2$	1.0

[a] Adjusted to a common standard from data in the original.

[b] Reaction at the 7,8 position exceeds reaction at the 1,2 position by 9:1.

[c] To adjust to the same standard as the data in Table 48A, multiply by 1.08.

[d] To adjust to the same standard as the data in Table 48, multiply by 1.35.

TABLE 49[54]

Dichlorocarbene

Cl_2C: from (1), C_6H_5-(2,2-dichlorocyclopropyl) + $h\nu$ (Vycor filter), at ~25°; from (2), $CHCl_3$ + $K^{+-}O$-t-C_4H_9, at −80°.

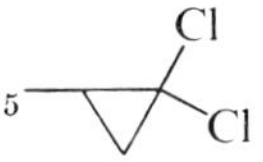

Olefin structures	Relative reactivities[a] (1)	(2)
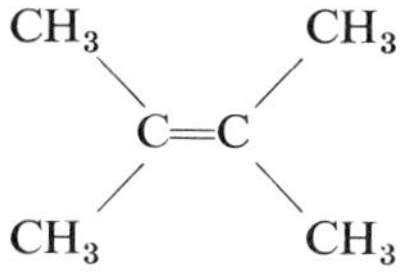 $(CH_3)_2C{=}C(CH_3)_2$	>44	60.8[b]
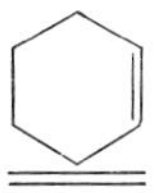 Cyclohexene	1.00	1.00
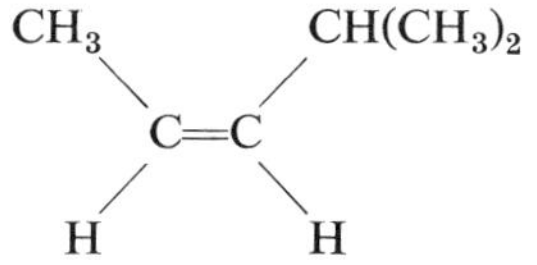 $CH_3CH{=}CHCH(CH_3)_2$ (cis)	0.45	0.55

[a] Adjusted to a cyclohexene standard from data in the original.

[b] Much faster than the comparable datum from Table 38, 5.25/0.14 = 37.5. The difference may be due to a solvent effect (hydrocarbon versus THF), but this would be a larger solvent effect than expected.

TABLE 50[3,4]

Dibromocarbene

Br_2C: from $CHBr_3 + K^{+-}O\text{-}t\text{-}C_4H_9$, (1) in $t\text{-}C_4H_9OH$ and pentane, at 0°–3°, and (2) at −15°.

Olefin structures	Relative reactivities (1)[a]	Relative reactivities (2)[b]
$(CH_3)_2C{=}C(CH_3)_2$	8.8	6.92
$(CH_3)_2C{=}CH(CH_3)$	8.0	7.42
$(CH_3)_2C{=}CH_2$	2.5	3.72
$(C_6H_5)_2C{=}CH_2$	2.0	—
Cyclopentene	1.2	—
Cyclohexene	1.0	1.00
$CH_3O\text{—}C_6H_4\text{—}CH{=}CHCH_3$	3.0	—
$C_6H_5CH{=}CH_2$	1.0	—

TABLE 50 (*Continued*)

Olefin structures	Relative reactivities	
	(1)[a]	(2)[b]
$n\text{-}C_4H_9(H)C{=}C(H)H$	0.18	0.19
$n\text{-}C_3H_7(H)C{=}C(H)H$	—	0.17
$C_6H_5CH_2(H)C{=}C(H)H$	0.05	—
$H_2C{=}CH{-}CH{=}CH_2$	1.2	—
$H_2C{=}C(Br)H$	"V. slow"	—

[a] Analysis by fractional distillation of competition reaction product mixtures. Relative rate constants calculated with Ingold equation.[6]

[b] Analysis by gas chromatography. Relative reactivities calculated by Doering approximation.[4]

TABLE 51[41]

Dibromocarbene

Br_2C: from $CHBr_3 + K^+{}^-O\text{-}t\text{-}C_4H_9$, at $-10°$ to $-15°$.

Olefin structures	Relative rate constants[a]
$C_6H_5(CH_3)C=CH_2$	3.3
$(C_6H_5)_2C=CH_2$	1.1[b]
cyclohexene	1.00
$C_6H_5CH=CH_2$	0.42[b]
$2,4,6\text{-}(CH_3)_3C_6H_2CH=CH_2$	0.13

[a] Calculated with the Ingold expression.[6]

[b] There are discrepancies with the corresponding values in Table 50. Solvent, temperature, and analytical differences may be responsible.

TABLE 52[49]

Dibromocarbene

Br_2C: from $CHBr_3$ + $K^{+-}O$-t-C_4H_9, at −8° to −10°.

Olefin structures[a]	Relative reactivities
$(CH_3)_2C{=}C{=}CH_2$	1.00
$(CH_3CH_2)(CH_3)C{=}C{=}CH_2$	0.56
$((CH_3)_2CH)(CH_3)C{=}C{=}CH_2$	0.21
$((CH_3)_3C)(CH_3)C{=}C{=}CH_2$	"0.0"
$((CH_3)_3C{-}CH_2)(CH_3)C{=}C{=}CH_2$	0.37
cyclobutylidene$=C{=}CH_2$	1.57
cyclopentylidene$=C{=}CH_2$	0.98
cyclohexylidene$=C{=}CH_2$	0.55

[a] Addition only to the more substituted double bond.

TABLE 53[42a,55]

Chlorofluorocarbene

$ClFC:$ from $(Cl_2FC)_2C{=}O + K^{+-}O\text{-}t\text{-}C_4H_9$, at $-12°$.

Olefin structures	Relative reactivities
$(CH_3)_2C{=}C(CH_3)_2$	31.
$(CH_3)_2C{=}CHCH_3$	6.5 (2.35)[a]
$(CH_3)_2C{=}CH_2$	1.00
cis-$CH_3CH{=}CHCH_3$	0.14 (3.08)[a]
trans-$CH_3CH{=}CHCH_3$	0.097
$CH_3CH_2CH{=}CH_2$	0.0087 (1.53)[a]

[a] Stereoselectivity, *syn*-chloro/*syn*-fluoro.

TABLE 54[56]

Chlorofluorocarbene

Cl(F)C: from $FCl_2CCOOCH_3 + Na^{+-}OCH_3$, at 35°.

Olefin structures	Relative reactivities[a]
$(CH_3)_2C{=}C(CH_3)_2$	103.
$C_6H_5(CH_3)C{=}CH_2$	3.0[b]
Cyclohexene	1.00 (1.2–1.5)
trans-$C_6H_5CH{=}CHCH_3$	0.81[b]
$C_6H_5CH{=}CH_2$	1.4 (1.2)
n-$C_4H_9CH{=}CH_2$	0.21[b]

[a] Data in parentheses are stereoselectivities, *syn*-chloro/*syn*-fluoro.
[b] No stereoselectivity data available.

TABLE 55[56]

Bromofluorocarbene

$Br(F)C:$ from $CHFBr_2 + K^{+-}O\text{-}t\text{-}C_4H_9$, at −13°.

Olefin structures	Relative reactivities[a]
$(CH_3)_2C{=}C(CH_3)_2$	66.
$(CH_3)_2C{=}CH(CH_3)$	42.0 (1.4–1.5)
$C_6H_5(CH_3)C{=}CH_2$	8.8[b]
Cyclohexene	1.00 (1.4–1.7)
$C_6H_5CH{=}CHCH_3$ (trans)	0.51[b]
$C_6H_5CH{=}CH_2$	0.75 (0.9)
$n\text{-}C_4H_9CH{=}CH_2$	0.26 (1.5)

[a] Data in parentheses are stereoselectivities, *syn*-bromo/*syn*-fluoro.

[b] No stereoselectivity given.

IV. COMMENTARY ON THE TABLES

A. Introduction

The early carbene-olefin addition work established that CBr_2[3] and CCl_2[4] reacted as electrophiles. Skell and Garner demonstrated a parallel of structure-reactivity behavior in the abilities of CBr_2 (T50*), bromine, and peracetic acid to discriminate between alkenes. A graph of log (k/k_0) for CBr_2 data versus the corresponding data for bromine gave a straight line with a slope of +2.1. CBr_2 was more discriminating than bromine.

Doering and Henderson's study of CCl_2 (T33) led to similar conclusions, though an attempt to correlate log (k/k_0) values for CCl_2 with those for CBr_2 showed that "the scatter about a hypothetical line (is) so large that no slope can be determined. One notices that relatively dibromocarbene reacts slower than dichlorocarbene, the greater the degree of substitution. This falling off may indicate the operation of a steric factor."[4,57]

Both reports suggested that the transition state for addition of CX_2 to an olefin could be represented by **1**. The olefinic centers become positive and the

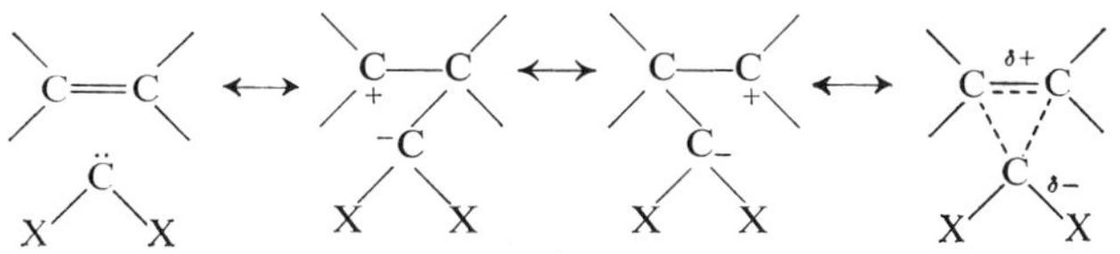

1

carbenic center becomes negative as progress is made along the reaction coordinate. Increasing alkylation of the olefinic carbons, which should stabilize positive charge and lower the energy of **1**, increased the rate of addition of CX_2, in the absence of overwhelming steric effects. Addition to isobutene was seen to be faster than addition to pentene-2. The more rapid reaction of the unsymmetrically disubstituted olefin was taken to mean that "contribution (to **1**) by one *t.* and one *p.* carbonium ion structure is more favorable than contribution by two *s.* structures."[4] The electrophilic selectivity of CBr_2, a dramatic contrast to the selectivity of trichloromethyl radical, was offered as support for a singlet (rather than a triplet) electronic state of the reactant carbene.[3]

In these initial reports, we see cited many of the principal themes which have since dominated research in carbene chemistry: electrophilic selectivity, electronic and steric control of selectivity, correlation of selectivities, and reactant multiplicity.

* Table numbers are identified by the letter T.

B. Effects Mainly Steric in Origin

1. Carbenes

Both steric and electronic effects must be considered in order to correlate relative addition rates with substrate structure.[4] Attempts at quantitative "separation" of these factors, however, have been limited.[57] It is possible to point out qualitative indications of steric hindrance to carbene addition; CCl_2 is a good starting point.*

A 1-cyclohexyl substituent is 7.4 times less effective than a 1-methyl substituent at promoting the addition of CCl_2 to cyclohexene (T42). Even phenyl, which should help electronically, is less effective than methyl, and a 1-α-naphthyl substitution retards the addition. These trends must be largely steric in origin. Comparison of T35 with T51 suggests that CBr_2 experiences greater steric hindrance to addition than does CCl_2.[41] Relative to cyclohexene, styrene and 2,4,6-trimethylstyrene are equally reactive toward CCl_2 (T35). Toward CBr_2, styrene is more than three times as reactive. The hindrance arises at the *ortho* substituents.

Sadler cites a related effect of *ortho* substituents.[7] The reactivity sequence for additions of CCl_2 to α-methylstyrene derivatives is unsubstituted $>$ 2-methoxy $>$ 2-methyl $\sim$ 2,2-dimethoxy (T39). A progressive twisting of the isopropenyl group from the plane of the aromatic ring is suggested. This steric inhibition of resonance decreases possible stabilization of the transition state. Direct hindrance to CCl_2 approach is also possible. The coplanarity of aryl and olefinic groups enforced on 1-methylenetetralin and 1-methylene-indane by their fused-ring structures maximizes favorable resonance interactions (presumably more important in transition states than in ground states). These olefins are more reactive than α-methylstyrene toward CCl_2.

Steric effects are indicated by the decreasing rates of addition of CCl_2 (T44) and CBr_2 (T52) to the 1,2-double bond of 1-alkyl-1-methylallenes as the alkyl group is made larger.[49] However, the greater steric discrimination of CBr_2 is not manifested here. Both CBr_2 and CCl_2 react with allenes of structure **2** in the order $n = 3 > 4 > 5$. The related greater reactivity of

$$(CH_2)_n\,C{=}C{=}CH_2$$

2

CCl_2 toward 1-methyleneindane as compared with 1-methylenetetralin is "attributed to relief of conformational strain upon formation of the transition state,"[7] though steric approach control may also be involved.

* The "stereoselectivity" of carbene additions will not be a major subject of this discussion; see reference 57 for a detailed consideration.

Steric hindrance is considered responsible for the decreasing rates of addition of CCl_2 to various alkenes at the *same* substitution level as the chain of a single substituent is lengthened (T34). For example, replacement of methyl by ethyl (in tetramethylethylene) halves the reactivity. Similar effects are seen on comparison of T33 with T34.

Addition of CCl_2 to 1-alkylethylenes is retarded by α- but not β-methylation (T37); rates for the series, alkyl = ethyl, isopropyl, *t*-butyl, correlate with E_s parameters, $\delta = 1.0$. A similar δ value is found upon application of the Taft polar-steric equation (Eq. 3) to the reactions of CCl_2 and simple alkenes.[58] $\Sigma\sigma^*$ and ΣE_s were used for each alkene.[44,59] The strong retarding

$$\log (k/k_0) = \rho^*\sigma^* + \delta E_s \tag{3}$$

effect of the *t*-alkyl substituent is also reported by Seyferth.[39] Although neopentylethylene is 0.78 times as reactive as 1-heptene toward CCl_2 at 80°, 3,3-dimethylpentene-1 is only 0.043 times as reactive, a rate reduction of 23 for the *t*-alkylethylene (T47). Gratifyingly, the smaller CF_2, expected to be less sterically demanding than CCl_2, experiences a retardation of only 8.3 in the same competition (T32).

We now consider steric effects in additions of other free carbenes. Singlet CH_2 (T1) does not show sufficient selectivity to characterize. On the other hand, the low rate of addition of 2,2-diphenylcyclopropylidene to tetramethylethylene (T7) is attributed to steric hindrance caused by opposition of the carbene's phenyl substituents and the olefinic alkyl groups.[14] *cis*-Butene and cyclohexene, which permit the phenyls to pair off with olefinic protons, react more rapidly than tetramethylethylene.

Dimethylethylidene carbene **3** (T8) is most interesting. If the carbenic center is *sp* hybridized, then the vacant *p* orbital is in the plane of the methyl

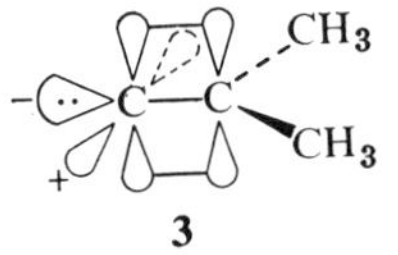

CH_3 CH_3 R_4 $\delta-$ $\delta+$ R_2 CH_3 CH_3 R_3 R_1

3 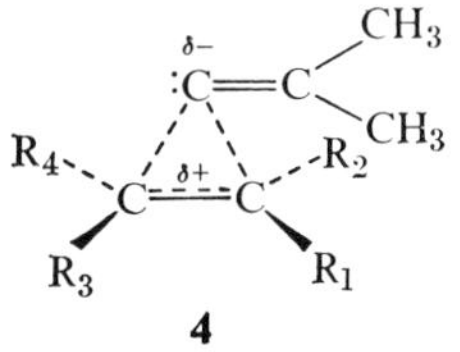 **4**

carbons. In a transition state, **4**, these methyl groups impinge on the olefinic substituents. This picture rationalizes the very slow rate of addition of **3** to tetramethylethylene. Here no arrangement approximating to **4** avoids costly steric interactions. The more facile addition of **3** to cyclopentene, compared with 1-methylcyclopentene, can be similarly explained.[15] However, it is not clear why addition of **3** to tetramethylallene is 45 times more rapid than addition of **3** to tetramethylethylene. Note that **3** attacks olefin **5** only at a peripheral double bond, in contrast to CCl_2, which preferentially attacks the more highly substituted central double bond (see Appendix).

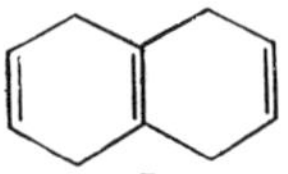

5

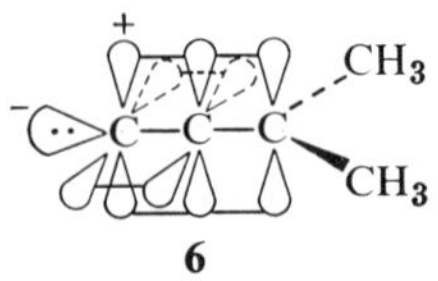

CH_3 CH_3

6

Studies of steric effects in the additions of *t*-butylmethylethylidene have also been reported,[15a] and a rather extreme interpretation involving two-step carbene additions with zwitterionic intermediates was offered. The evidence for such a mechanism is, in the reviewer's opinion, far from compelling, and unsupported by precedent. Further investigation in this area is clearly desirable.*

The vinylidene carbene **6**, in which the additional double bond permits the methyl carbons to occupy the plane perpendicular to the carbenic *p* orbital, does not display the steric selectivity of **3** (T9). This is especially dramatized by the behavior of diphenylvinylidene carbene **6a** (T9A) and of di-*t*-butylvinylidene carbene (T9B), which do not show marked evidence of steric

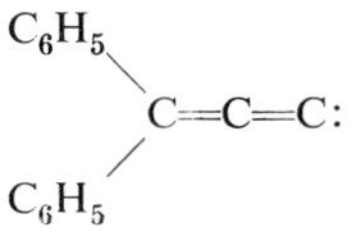

$$(C_6H_5)_2C{=}C{=}C:$$

6a

hindrance in addition to tetramethylethylene.[16d,16e] (Note also the similarity in reactivity of the base- and diazo-generated **6a**.) The data (T9) also show that the selectivity of dimethylvinylidene carbene is substantially independent of source, 1-chloro-3-methylbuta-1,2-diene or 3-chloro-3-methylbut-1-yne, suggesting the intermediacy of a free carbene. However, recent experiments of le Noble et al.[16c] suggest that a chloride ion could be associated with this carbene in an "anion-carbene pair."

Cyclopentadienylidene (T11) and 4,4-dimethylcyclohexadienylidene (T12) make an interesting comparison. The former, though highly reactive, shows a modest steric selectivity in that tri- and tetra-substituted alkenes react more slowly than *n*-butylethylene. No strong steric discrimination is observed in additions of the cyclohexadienylidene, but it is suggested that its similar reactivity toward mono- and disubstituted alkenes is due to a steric effect.[19] Unfortunately, selectivity differences between the carbenes are not very noticeable. It was hoped that these studies could detect the importance of

* See the correction reported by the authors in *J. Amer. Chem. Soc.*, **94**, 1793 (1972).

"aromatic" character in cyclopentadienylidene (**7**) as opposed to 4,4-dimethylcyclohexadienylidene (**8**), where such character should be absent.[18,19]

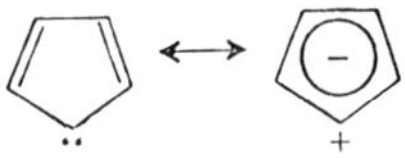

7 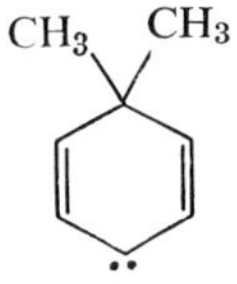 **8**

If **7** does feature an aromatic π system, its special properties are still undefined.

The similar reactivity of **8**, whether generated from a diazoalkane or a norcaradiene precursor (T12), suggests that a free carbene is involved in both reactions and tends to exclude the intermediacy of an excited state of the diazoalkane in the actual olefin addition. Evidence for or against diazoalkane intermediacy is hard to obtain, and is especially welcome here.

Photolytically generated arylcarbenes (T13) do not exhibit characteristic steric effects with simple alkenes. Diphenylcarbene, on the other hand, ought to manifest substantial steric selectivity, but the limited data available for this (triplet) species (T17, T17A) do not establish such a selectivity.

There appears to be a substantial overall selectivity difference between photolytically (T18) and thermally (T19) generated carboethoxycarbene. The former does not discriminate between *cis*- and *trans*-butene; the latter prefers *cis*-butene by a factor of 2.8. The role of steric effects here is not yet defined. Stereoselectivity data would be useful and should be obtained.

Dicarbomethoxycarbene singlet (T20) exhibits a modest steric selectivity, but its failure to discriminate between *n*-propylethylene and *t*-butylethylene is surprising, especially when compared with the behavior of CCl_2 (see above). CCl_2 presumably reacts via a tighter transition state, in which steric interactions are more strongly experienced.

Steric hindrance is assigned an important role in phenylthiocarbene addition reactions (T22A). Note the slower addition to 1-isopropyl-1-methylethylene relative to isobutene. It is suggested that the hindrance arises as the olefinic carbons rehybridize toward cyclopropyl carbons during the addition.[30a] Geminal and vicinal bond angle changes of 120–113° and 60–52.8°, respectively, lead to approach of geminal alkyl groups and concomitant strain, which is greater for the first olefin.

The free halocarbenes (T23, T24, T25) show little selectivity; steric effects are not isolable. Consideration of stereoselectivity in the addition reactions of phenylhalocarbenes (T26, T28, T30) allows some delineation of steric effects,[34,57] but overall relative reactivities are not very informative. This is also true of the moderately selective carbenes, chloromethylcarbene (T27) and bromocarboethoxycarbene (T29).

CF_2 (T31) is singular in its behavior toward butene-1.[38] The relative reactivities toward CF_2 of the other simple alkenes in T31 are correlated by calculated π electron energy differences between ground and transition states; butene-1 deviates substantially from the correlation. The Arrhenius pre-exponential term for this reaction is quite low, smaller by a factor of 162 than the corresponding datum for the propene-CF_2 reaction (T31). Therefore, the transition state for the butene-1-CF_2 addition is considered to be "appreciably different than that of the other olefins . . . ."[38]

The excellent linear relationship between log $(k/k_{\text{isobutene}})$ for CF_2 and CCl_2 shown in reference 38 does not extend to 1-butene. A parallel observation is reported for CFCl and CCl_2.[42a] However, a linear free energy correlation between CF_2 and CFCl *does* include 1-butene.[38] An explanation was offered, based on increasing steric hindrance ($CCl_2 > CFCl > CF_2$) in the "tight" transition states for additions of these highly selective carbenes to the poor substrate, 1-butene. According to this idea, CCl_2 should be most hindered and exhibit an abnormally low addition rate.[42a] Although accounting for the deviations in the free energy correlations, this suggestion is not consistent with the apparently normal differential preexponential factor which can be calculated for the CCl_2-1-butene reaction from the data in T38. Further study appears warranted.

Steric hindrance during additions of CBr_2 (T50, T51) has been noted previously. At $-15°$, CBr_2 adds 1.07 times faster to trimethylethylene than to tetramethylethylene (T50).

2. *Carbenoids*

The exact nature of the methylene transfer reagent derived from methylene iodide and zinc (Simmons-Smith reagent) is unclear;[9] a convenient formulation is bis(iodomethyl)zinc·zinc iodide, **9**.[60] Free CH_2 is not involved in these reactions,[9,60] and the reactive intermediate shows considerable steric discrimination, as reflected in the relative reactivity data (T2, T3, T4) and in

$$(ICH_2)_2Zn{\cdot}ZnI_2$$

9

studies of its selectivity toward complex substrates.[57] Note, in T2, that **9** adds more rapidly to trimethylethylene than to tetramethylethylene (factor, 1.69). Similarly, cyclohexene reacts more rapidly than 1,2-dimethylcyclohexene [factor, 1.72 (T2); 1.06 (T3)]. Note, too, that *n*-alkylethylenes react about 2.7 times faster than *t*-butylethylene with **9**. Reactivity similar to that of **9** has been reported for the carbenoid generated from CH_2Cl_2 and magnesium.[9a] This species is suggested to have the structure $(ClMg)_2CH_2$; it adds stereospecifically to *cis*- and *trans*-butene, and it affords the reactivity sequence *cis*-butene (1.1), *trans*-butene (1.00), 1-butene (0.56). Its steric demand has not been tested.

Rickborn and Chan have reported an extensive study of the reactions of **9** with cyclic olefins.[10,11] Alkylation of cyclopentene or cyclohexene at C-1 enhances reactivity, but an additional alkylation at C-2 of cyclohexene causes retardation (T3). A ring size effect on cycloalkene reactivity, $C_5 > C_7 > C_6$, is noted and tentatively attributed to "relief of torsional strain, with diminished bond angle strain also playing an appreciable role in cyclopentene";[10] analogy is made to diimide reductions.[61] Though 4-alkyl substitution on cyclohexene has little effect, 4,4-dialkylation, or 4,5-*cis*-dialkylation, halves the rate. Steric hindrance to carbene addition is implicated, originating at a pseudo-axial alkyl group. Note that 3-alkylation also has a retarding effect. Strikingly, although 1-phenylcyclohexene is 6.19 times more reactive than cyclohexene toward CCl_2 (T42), it is 3.3 times less reactive than cyclohexene toward **9**. This seems to be a clear example of electronic versus steric control of carbenic addition rates.

Oxygen-containing substrate functional groups are well known to affect the rate and stereochemistry of addition of **9**.[57] Of special note (T4) is the rate difference of 3.35 between *cis*- and *trans*-5-methyl-2-cyclohexenols. Conformational analysis suggests that "stereospecific methylenation of allylic cyclohexenols occurs through the quasi-equitorial hydroxyl conformation."[11] Methylenation assisted by an equitorial, allylic hydroxyl group is also postulated for larger cycloalkenols.[57,62] Note the substantial rate decrease which attends movement of the hydroxyl from C-3 to C-4. The homoallylic cycloalkenols may add **9** with participation of an axial hydroxyl group.[11] Addition of **9** to all of the olefins of T4 (excepting 1-hydroxymethyl-3-cyclohexene) involves oxygen participation; nevertheless, the sensitivity of rate to hydroxyl location is great.

Perhaps related is the curious observation that the CH_2 carbenoid generated from methylene bromide and a zinc-copper couple adds about 100 times faster to the vinyl aluminum compound **9a** than to octene-1.[62a] Since, in a simple

$$\begin{array}{ccc} n\text{-}C_4H_9 & & H \\ & C{=}C & \\ H & & Al(i\text{-}C_4H_9)_2 \end{array}$$

9a

resonance interpretation, deactivation of the double bond by aluminum would be expected, it seems likely that the aluminum is directly involved in the carbenoid addition, perhaps as a center for nucleophilic attack of the carbenoid. Further work in this area would be welcome.

Methylene carbenoids have also been generated from methylene dihalides and alkyllithiums,[11a,11b] from ethylaluminum dichloride and diazomethane,[11a] and from methyl chloride and phenylsodium.[11b] Competition data appear in T4A and T4B, and comparison with the Simmons-Smith reagent (T2, T3) shows an even greater steric selectivity on the part of the new reagents (T4A). The alkyllithium-methylene dihalide system shows an unusually high reactivity toward monosubstituted alkenes. This is further underscored by its exocyclic/endocyclic selectivity (1.10–1.15) toward the double bonds of 4-vinylcyclohexene.[11a] The Simmons-Smith reagent, however, shows typical electronic selectivity toward mono- and disubstituted alkenes. (See T2 and T3. With 4-vinylcyclohexene, its exocyclic/endocyclic selectivity is 0.50. Other zinc carbenoids are similar in this selectivity.[11a]) In contrast to the carbenoids derived from methylene dihalides, the ethylaluminum dichloride-diazomethane species shows the expected diminution in rate with monoalkylethylenes (T4A). Its exocyclic/endocyclic selectivity toward 4-vinylcyclohexene is 0.11.[11a]

Although the "methylenes" of T4A and T4B are clearly carbenoids and exhibit a halide precursor-selectivity dependence, a substantial question attends their organometallic dependency. Huisgen and Burger have argued that LiCHX is the methylene transfer reagent in CH_2XY-RLi systems, and that no dependence of selectivity on the kind of alkyllithium (*n*-butyl or *t*-butyllithium) exists.[11a] But Friedman et al.[11b] present clear evidence of such a dependence (T4B), the steric selectivity of the carbenoid being greater when neopentyllithium is used in place of *n*-butyllithium as the generative agent. More work will be needed to reconcile these divergent findings.

Little of the steric discrimination manifested by **9** is observed in the methylene transfer reactions of bis(bromomethyl)mercury (T5). The reaction involves a carbenoid (kinetic dependence on olefin concentration) and transition state **10** is suggested.[12] Solvation of this carbenoid is considered unimportant in benzene and part of the greater steric demand of **9** is attributed to its ether solvation (cf. transition state **11**[12]).

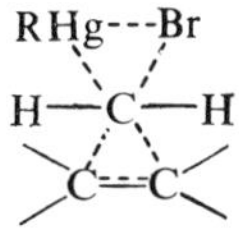

10 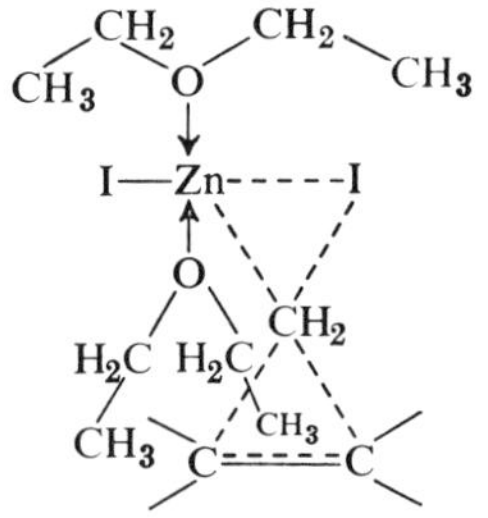 **11**

The selectivity of the methylene transfer reagent of the diethylzinc-methylene iodide system (the carbenoid is formulated as $C_2H_5ZnCH_2I$) resembles that of bis(bromomethyl)mercury more than that of the Simmons-Smith intermediate (compare T5A with T5 and T2). The $C_2H_5ZnCH_2I$ shows typical electrophilic selectivity in ether, pentane, and benzene solvents, and little of the steric selectivity manifested by **9** is apparent. The tetramethylethylene/cyclohexene rate ratio falls from 16.2 to 8.82 when the $C_2H_5ZnCH_2I$ reactant is transferred from pentane to ether. This may represent an increase in steric selectivity by the carbenoid, due to coordination of ether. Note that the carbenoid is *more* reactive toward cyclohexene (relative to heptene-1) in ether than in pentane; with the less sterically demanding olefins the effect of ether coordination is to enhance rather than retard the expression of electrophilic selectivity. Presumably this represents a stabilization of the carbenoid by ether. $C_2H_5ZnCH_2I$ in ether is not nearly as sterically selective as **9**, however, for which tetramethylethylene/cyclohexene is only 1.29 (T2).

Steric selectivity of *p*-tolylcarbenoid (T16) is considered here; a discussion of aryl carbenes and aryl carbenoids (T13–T15) follows later (Section IV.C.5). The selectivity of *p*-tolylcarbenoid, as generated from *p*-methylbenzal halide and alkyllithium, depends on solvent and halide, but the substrate reactivity sequence ethylethylene > *i*-propylethylene > *t*-butylethylene is invariant. Stereoselectivity data can be used to partition the relative reactivities (T16) into separate rates for *syn* and *anti* additions of the carbenoid to olefins. *Syn* addition is found to be more sensitive than *anti* addition to the size of the alkyl group.[22] For example, relative reactivities (pentane, iodide precursor) were ethylethylene, 0.99; *i*-propylethylene, 0.46; and *t*-butylethylene, 0.071 (*syn* mode). Corresponding *anti* mode values were 0.37, 0.24, and 0.17, respectively. Note, also, that *p*-tolylcarbenoid is somewhat more sterically discriminating than the Simmons-Smith reagent, based on the *n*-alkylethylene/*t*-butylethylene competition (T2, T16). Detailed discussion of this subject and illustrations of possible transition states appear elsewhere.[57]

A caveat emerges from T18, in which the data indicate little difference in discrimination by photolytically or catalytically generated carboethoxycarbene. There are, however, large differences in stereoselectivity;[57] the catalytically generated species is certainly a carbenoid.[63] The scope of T18 is too limited to differentiate the two intermediates. It is, nevertheless, peculiar that a substantial reactivity difference between tetramethylethylene and trimethylethylene is not observed toward the catalytically generated species.

Alkoxy and aryloxycarbenes have been extensively investigated with regard to stereoselectivity.[57] Less is known about their interalkene discrimination, although phenoxycarbenoid generated from chloromethylphenyl ether (T22) shows pronounced selectivity. Electronic factors are important, as shown by the sixteenfold greater reactivity of ketene diethylacetal over isobutene. However, trimethylethylene reacts 2.44 times faster than tetramethylethylene, suggesting that steric effects are also involved. Combined stereoselectivity and relative reactivity data indicate that phenoxycarbenoid adds to trimethylethylene (*syn*-phenoxy mode) at about the same rate as it adds to tetramethylethylene.[30] Moreover, the faster addition to trimethylethylene than to isobutene is, under this analysis, due to an increase in the rate of addition in the *anti*-phenoxy mode. A suggested transition state is **12**, in which a lithium atom is held responsible for steric hindrance toward olefinic alkyl groups *syn* to it. (The lithium is considered part of an oligomeric species.) The observa-

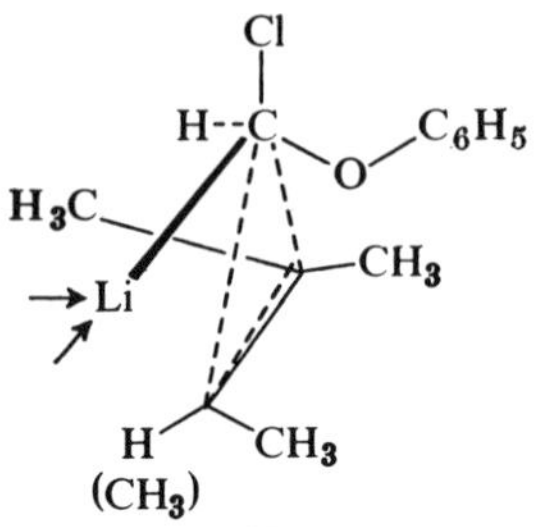

12

tion that k_{anti} (trimethylethylene) $\sim k$(tetramethylethylene) $< k_{syn}$ (trimethylethylene) is thus attributed to steric hindrance toward addition in the *anti*-phenoxy mode, originating at *syn*-lithium.

The foregoing differs from usual analyses of carbenoid additions, in which the leaving group (e.g., LiX) is not considered to play a direct steric role.[57] It does not explain the *p*-tolylcarbenoid data (T16), which shows *syn*-tolyl rather than *anti*-tolyl addition to be more strongly retarded by increasing bulk of an olefinic alkyl group.[22] It is, of course, conceivable that phenoxycarbenoid additions proceed over transition states dissimilar to those of arylcarbenoid additions. A study of phenoxycarbenoid discrimination and stereoselectivity toward $RCH{=}CH_2$, with R varied from ethyl to *t*-butyl, might help decide between **12** and other transition states. In this regard, it would be helpful to know the behavior of *free* alkoxy or aryloxycarbenes toward these olefins. Two methods reported to generate free methoxycarbene, however, employ radically different conditions; they yield different results in the single comparison presently available, stereoselectivity toward *cis*-butene.[64,65]

Chlorocarbenoid (T24) shows no dominant steric selectivity, although it is far more selective than the free carbene.[33] Its selectivity can be described as "electrophilic," and it is presumably governed largely by electronic effects. Combined stereoselectivity[32a] and relative reactivity data[32] indicate that *anti*-chloro mode additions to cyclohexene and *cis*-butene are about equally facile, whereas *syn*-chloro mode addition to *cis*-butene is more rapid (factor of 1.7). Addition to tetramethylethylene is twice as fast as *anti*-chloro mode addition to trimethylethylene, and faster than *syn*-chloro mode additions to trimethylethylene and *cis*-butene. The results are well explained by prevailing theories,[57] but they are not in accord with a transition state analogous to **12**.

Steric hindrance to addition is suggested by the phenylchlorocarbene and phenylbromocarbene data (T28, T30). The species generated from benzal halides and potassium *t*-butoxide show relatively little ability to discriminate between tetramethylethylene and trimethylethylene.[35–37] The photochemically generated carbenes do not show this behavior. The intermediates of the former reactions are probably carbenoids; the fall-off in reactivity toward the tetra-substituted alkene can be ascribed to steric hindrance in an "S_N2-like" transition state.

C. Effects Mainly Electronic in Origin

1. Olefinic Substituents

Carbene addition to an olefin has been regarded as an electrophilic process since the initial reports (see Section IV.A). A theoretical analysis of the addition of $^1CH_2$ to ethylene suggests initiation of the reaction as in **13,** a "π approach" in which the vacant *p* orbital of the carbene begins to overlap with the π system. As the reactants move along the reaction coordinate toward product geometry the π approach goes over to **14**, a "σ approach."[66]

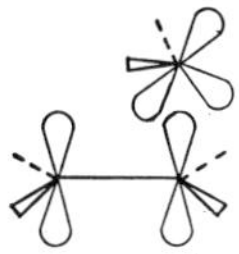

π approach
13

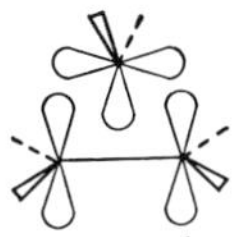

σ approach
14

The important point is the nature of the initial stage, the π approach, also suggested by previous investigators.[3,4,20,67] Transfer of electron density from the olefin's π system to the carbene's *p* orbital occurs; this is an *electrophilic* addition.

More recent calculations suggest that π approach **13a** (R = H) is energetically preferable to π approach **13b** (R = H), in the addition of $^1CH_2$ to ethylene.[66a] Addition of CF_2 is calculated to follow a similar course, with

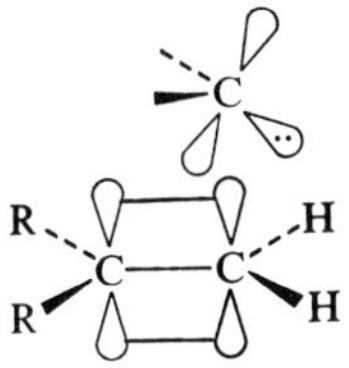

13a

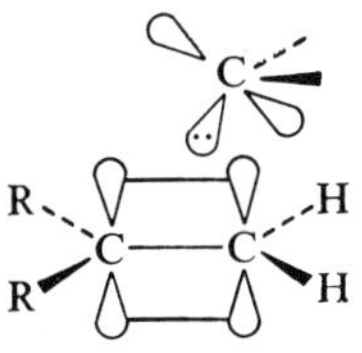

13b

less charge transfer from olefin to carbene at the transition state. Related calculations for additions of CH_2 and CF_2 to isobutene again suggest that **13a** (R = CH_3) is preferred to **13b** (R = CH_3), but only by a small energy difference, <2 kcal/mole.[66a] Results of these calculations must be treated cautiously, as guides to thought rather than as firm conclusions.

Experimentally, the charge separation in generalized transition state **15** can be probed by variation of either olefinic substituents, R_1–R_4, or carbenic

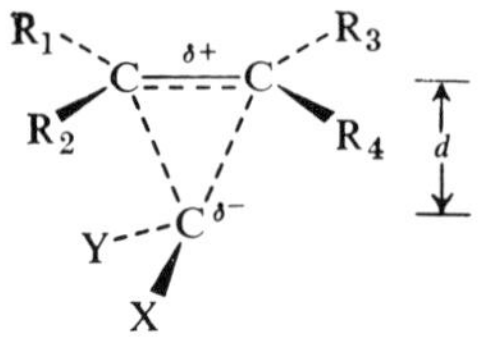

15

substituents, X and Y. This procedure, however, can alter *both* steric and electronic factors; the separation *d* changes in response to altered carbenic or olefinic reactivity. In the following discussion, therefore, the observations are not reflections of isolated electronic effects.

We note in passing that electrophilic reactivity of carbenes is not restricted to olefinic substrates. Insertions of CCl_2 into the Si—H bonds of **16** give relative reactivity data which can be correlated by a Hammett σ-ρ treatment, $\rho = -0.632$.[68]* Hammett studies of the analogous reactions of **16** with carbenoids have also appeared. With the methylene carbenoid derived from

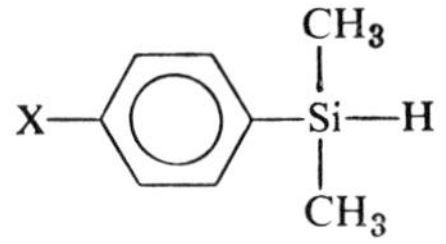

16

diethylzinc and methylene iodide, $\rho = -11.1$ (σ°),[68a] whereas the methylene transfer reagent $Hg(CH_2Br)_2$ afforded relative insertion rate data which could be correlated by $\rho = -1.31$ (σ°).[68b] Similarly, the methylcarbenoid derived from diethylzinc and ethylidene iodide inserted into the Si—H bond of **16** giving data correlated by $\rho = -1.19$ (σ).[68a]

Note that the three carbenoids all show similar negative ρ values for insertion into the Si—H of **16** and that $|\rho|$ is about twice the value for the corresponding insertion of CCl_2. Transition states for the former reaction are then somewhat more polar, and the carbenoids appear to be more selective than CCl_2. Free CH_2 or CH_3CH would be expected to be far less selective than CCl_2 in reaction with **16**. A structure such as **16a** has been

* CCl_2 insertion into the analogous *cumene* C—H bond gives $\rho = -0.942$ (σ^+) and -1.223 (σ). It would appear that a larger partial positive charge is transmitted from C to the aryl group in the transition states of the cumene insertions than in the corresponding silicon cases. (D. Seyferth and Y. M. Cheng, *forthcoming*.)

suggested for the transition state of the carbenoid insertions.[68a] Reactions of

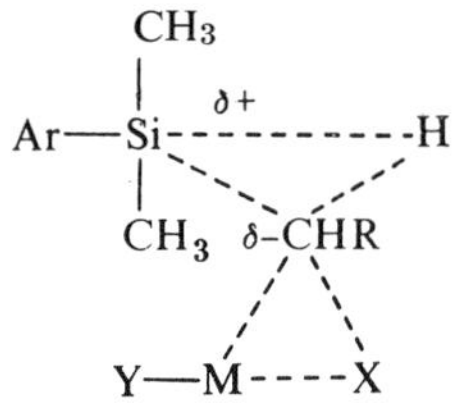

$$\mathrm{Ar{-}Si(CH_3)_2^{\delta+}\cdots H\cdots {}^{\delta-}CHR\cdots M(Y)\cdots X}$$

16a

CCl_2 with ylids **17** give relative reactivities of 1.20, 1.06, 1.00, and 0.89, for

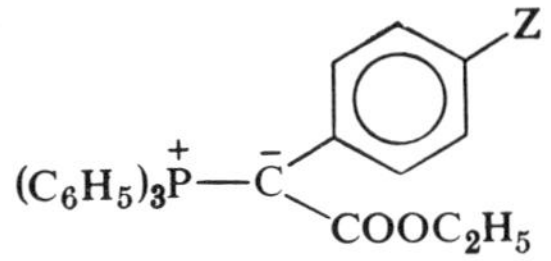

$$(C_6H_5)_3\overset{+}{P}{-}\overset{-}{C}(C_6H_4Z){-}COOC_2H_5$$

17

Z = CH_3O, CH_3, H, and Cl, respectively. In this reaction, CCl_2 is believed to attack the ylid carbon in the rate-determining step; expulsion of triphenylphosphine from the resulting intermediate affords product [$Cl_2C{=}C(Ar)$-$COOC_2H_5$].[69]

Similarly, 1,2 aryl migrations in carbenes **18**,[70,71] **19**,[72] and **20**[72,73] afford migratory aptitudes for Ar, which can be fitted to Hammett relations. Reaction constants, ρ, are -0.68 (σ^+) for **18**, -0.824 (σ) for **19**, and -0.28 (σ^+) or -0.449 (σ) for **20**.[73,72] Aryl migrates to an electron-deficient center,

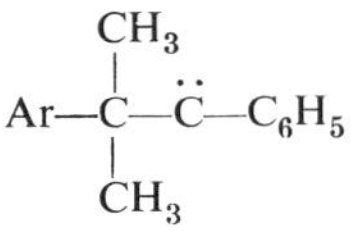

$$\mathrm{Ar{-}C(CH_3)_2{-}\ddot{C}{-}C_6H_5}$$

18

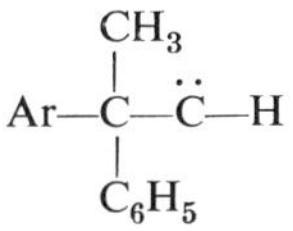

$$\mathrm{Ar{-}C(CH_3)(C_6H_5){-}\ddot{C}{-}H}$$

19

$$\mathrm{Ar{-}C(C_6H_5)_2{-}\ddot{C}{-}H}$$

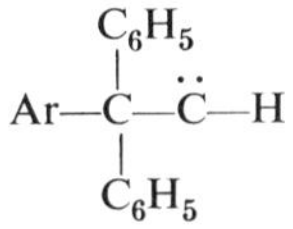

20

the carbenic p orbital; donating substituents on the aryl ring accelerate this process. The reaction may be viewed as intramolecular attack by the carbenic p orbital on the aryl π system at C-1.

The intermolecular reactions of thermally generated carboethoxycarbene with benzene derivatives are also correlated by the Hammett treatment, $\rho = -0.38$ (σ_p).[74] Here too the carbene acts as an electrophile, impinging on the aromatic π system. In a related process, the methylcarbenoid, derived from ethylidene iodide and diethylzinc, attacks alkylbenzenes to yield methylalkyltropylidenes. Substrate reactivity (relative to toluene) ranges from 0.40 (benzene) to 2.8 (*m*-xylene). Correlation of substrate reactivity with that observed in a Friedel-Crafts reaction leads to postulation of a carbenoid-aromatic σ complex as the key intermediate.[74a]

Returning to the discussion of **15**, we reiterate that there are two handles for qualitatively probing charge separation: olefinic substituents and carbenic substituents. Each is discussed in turn.

Expectations are clear for variation of olefinic substituents; olefin reactivity will parallel π electron availability. The substrate's ability to bear a positive charge in **15** will be paramount. We have seen that olefinic reactivity toward CBr_2 and CCl_2 parallels reactivity toward bromination and epoxidation reagents.[3,4] Cvetanović discusses these trends in connection with a study of olefin-(^3P) oxygen reactions.[75] Correlations of olefin reactivities toward these five reagents can be achieved with such olefin parameters as ionization potential (see also reference 44), excitation energy (spectroscopic and theoretical), heat of hydrogenation, and bond order.[75]

Linear free energy correlations are receiving much attention. The CCl_2 data of T38 fits a Taft polar-steric relation (Eq. 3, Section IV.B.1), $\rho^* \sim -4.3$. Landgrebe[75a] reports that the relative rates of addition of CCl_2 to $RCH_2CH{=}CH_2$ ($R = n\text{-}C_3H_7$, CH_3O, Cl, $C_6H_5CH_2$, C_6H_5O) at 0° can be correlated by σ^*, with $\rho^* = -0.74$. Note that ρ^* here is much smaller than ρ^* for CCl_2 additions to alkenes in which substituents are varied directly at the olefinic carbons as in T38.[44] In the former case, the olefinic centers are insulated from the varied substituent by a methylene unit.

Additions of CCl_2 (80°) to **21** (T41), **22** (0°, T39), and **23** (0°, T39) are correlated by σ^+; ρ values are -0.619, -0.53 (two points only), and -0.378, respectively. These ρ values are much smaller than the ρ^* observed in CCl_2

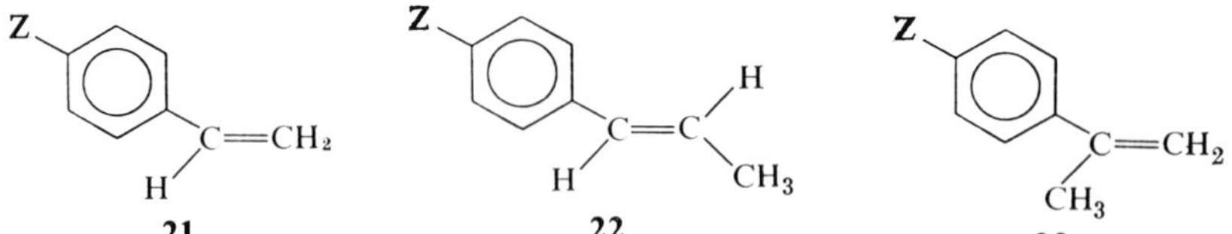

Z–C₆H₄–C(H)=CH₂ **21** Z–C₆H₄–C(H)=C(H)CH₃ **22** Z–C₆H₄–C(CH₃)=CH₂ **23**

additions to alkenes (T38). There, substituent variation is accomplished directly at the reaction center; response is great. For **21–23**, the effect of substituent variation on reactivity is attenuated because it is mediated by the aromatic π system. Note the order of increasingly negative ρ, **23** < **22** < **21**, which is the order of *decreasing* substrate reactivity. Substrate **21** should yield the most "advanced" transition state, with the largest charge separation;

the strongest response to substituent variation is expected. [There is, however, one report[15] of a large negative ρ for carbene addition to **21**; $(CH_3)_2C{=}C:$ (T8) gives $\rho = -3.4$ (3 points).]* After incorporating the concepts of Hoffmann,[66,66a] the transition state for CCl_2 addition to **23** is represented as **24**.[7]

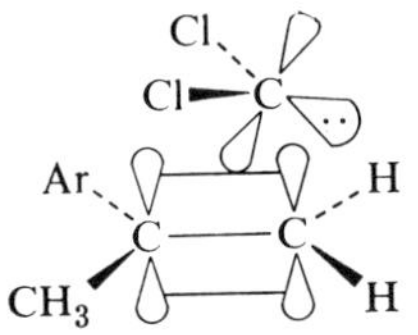

24

The methylene carbenoid $C_2H_5ZnCH_2I$ has also been the subject of a Hammett study with **21** (78.6°, T5C) and ρ was found to be -1.61 (σ). Note that this carbenoid is more selective toward **21** than is CCl_2 ($\rho = -0.619$, above). Transition state **24a** was offered;[12a] presumably $\delta+$ is larger here than in the corresponding transition state for CCl_2 addition. It will be

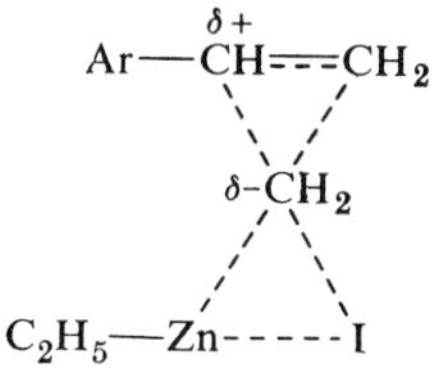

24a

recalled that this carbenoid also shows a significantly larger $|\rho|$ than CCl_2 in insertion into the Si—H bond of aryldimethylsilicon hydrides (see above, and structure **16a**).

CCl_2 also reacts with aryl azides, leading via the dichloroimines, $ArN{=}CCl_2$, to the corresponding tetrachloroaziridines, **24b**.[75b] Under

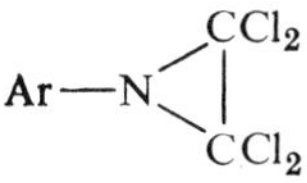

24b

comparable conditions (80°, CCl_2 from the thermal decomposition of excess $C_6H_5HgCCl_3$ in dimethoxyethane) the observed pseudo first-order rate constants for disappearance of azide were in the ratio 1.6:1.2:1 for Ar = *p*-methoxyphenyl, phenyl, and *p*-chlorophenyl, respectively, suggesting an electrophilic attack of the carbene on the azide.

* See Addenda, number 2.

Evidence strongly favors transition state **15**. Even the *number* of alkyl substituents on the olefinic π bond correlates, albeit crudely, the CCl_2 relative reactivity data, Figure 1.[44] Beyond these correlations, one may inquire more deeply into the origins of the relative reactivity trends. A useful

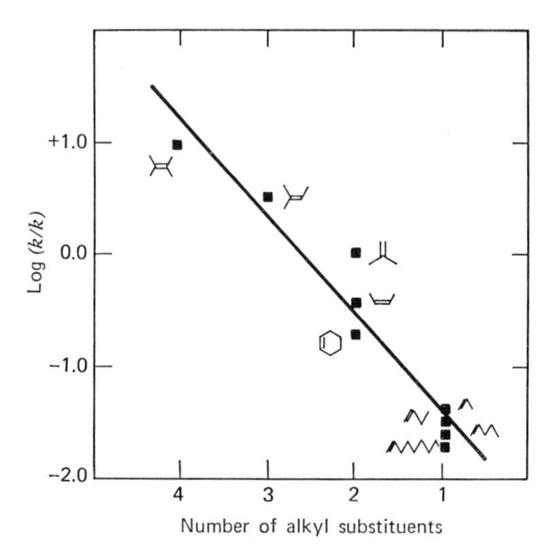

Fig. 1. Relative reactivity versus number of alkyl substituents in the additions of CCl_2 to olefins.[44]

approach is the extraction of relative activation parameters, and results are now at hand for CF_2 (T31), CCl_2 (T38, T36), and C_2O (T10). Such studies are certain to become more common; they will make important contributions.

Experimental differential activation energies for CF_2 additions (T31) correlate with calculated π electron energy differences ($\Delta E_\pi^\ddagger$) between

CH_3, CH_3, C, 1.46Å, CH_2, 80°, $\beta=0.8$, 1.52Å, $\beta=0.4$, CF_2

25

transition states and ground states. The transition state model employed resembles **15**; it uses the dimensions and overlap integrals shown in **25** (addition of CF_2 to isobutene). The reader is referred to reference 38 for further details; the correlation appears in Figure 2. Calculated $\Delta E_\pi^\ddagger$ values accord well with the observations, except for the 1-butene point (see Section IV.B.1.). CF_2 therefore seems guided in its selectivity mainly by activation

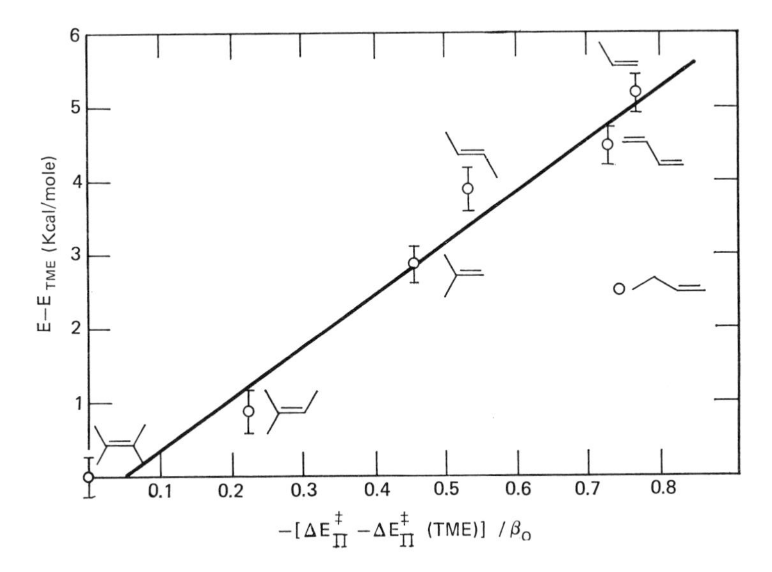

Fig. 2. Correlation of activation energy with calculated π electrons energy differences between transition and ground states in the additions of CF_2 to olefins.[38]

energy variations, which can be rationalized in terms of models **15** or **25**. Note (T31) that the preexponential terms are also important in setting olefinic reactivity. The more favorable A factor contributes substantially, for example, to the reactivity advantage enjoyed by isobutene over *cis*- and *trans*-butenes.

Crude relative activation data for CFCl indicate that it resembles CF_2 in that it, too, is largely guided in its selectivity by differential activation energies. For example, ΔE_a (tetramethylethylene/isobutene) is -2.37 ± 0.13 kcal/mole.[42b] The corresponding value for CF_2 is -2.88 kcal/mole.[38]

The case of CCl_2 is more abstruse. The relative reactivity data (T36, T38) are generally dominated by entropy factors. Although $\Delta\Delta F^{\ddagger}$ varies in the expected way with olefin structure (T38), $\Delta\Delta H^{\ddagger}$ is near zero for all substrates except the monoalkylethylenes. Indeed $\Delta H^{\ddagger}$ is *greater* for CCl_2 addition to, for example, tetramethylethylene than it is for CCl_2 addition to isobutene (see also T36). It is $\Delta\Delta S^{\ddagger}$ which parallels the rates and dominates. Only with the monoalkylethylenes is the expected $\Delta H^{\ddagger}$ behavior observed. Table 36 supports similar conclusions. A rationale has been offered,[44] but it is recognized that much of the discussion is "contingent on the ability of the Arrhenius equation to separate effectively $\Delta H^{\ddagger}$ and $\Delta S^{\ddagger}$ terms of the small magnitude encountered here."[44] This ability is in doubt, and the interpretation is tentative.

Much more work is needed along these lines. Our entire conception of the relative reactivity data could change in the light of extensive, carefully determined activation parameters.

2. *Carbenic Substituents*

a. Qualitative Studies. The second "handle" for probing **15** is variation of the carbenic substituents. The tables show that whatever the potential reactivity differences of the olefins, it takes a properly substituted carbene to "develop" them. $^1CH_2$ (T1), for example, shows little selectivity toward simple alkenes, with the exception of butadiene. (Note, however, that there is some question whether the $^1CH_2$ is free of $^3CH_2$.[8,8a]) Less than a factor of 2 separates tetramethylethylene from ethylene.

2,2-Diphenylcyclopropylidene (T7) and dimethylethylidene carbene (T8) show far more selectivity than CH_2; steric control is important here. The selectivity of dimethylvinylidene carbene (T9) is independent of precursor; it appears to lie between that of CCl_2 and CBr_2. Resonance stabilization of the carbene **26** is suggested to account for its selectivity; the vacant carbenic *p*

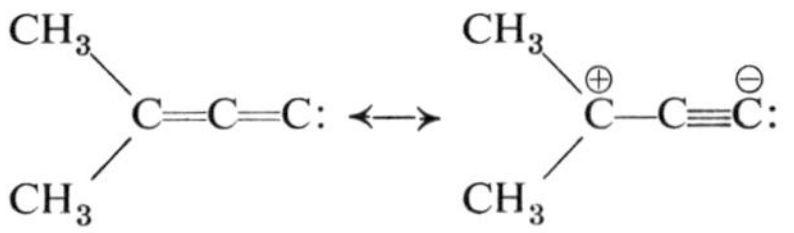

$$(CH_3)_2C{=}C{=}C: \longleftrightarrow (CH_3)_2\overset{\oplus}{C}{-}C{\equiv}\overset{\ominus}{C}:$$

26

orbital is part of an allylic cation system (**6**). Note that **26** should not be sterically demanding, in contrast to dimethylethylidene carbene (Section IV.B.1).

Singlet carbonylcarbene (T10) is perhaps more selective than $^1CH_2$, for it reacts 2.17 times faster with tetramethylethylene than with ethylene.[17a] The comparable factor for $^1CH_2$ is only 1.34 (T1). Toward the other alkenes of T10, singlet C_2O shows a "normal" but highly compressed selectivity sequence. As with $^1CH_2$, butadiene seems relatively reactive toward C_2O.

Cyclopentadienylidene (T11) and 4,4-dimethylcyclohexadienylidene (T12) have been discussed in connection with steric effects (Section IV.B.1.) The latter species shows, at most, a mild electrophilic selectivity sequence, tetramethylethylene/pentene-1 = 5.1. Butadiene is the most reactive olefin examined.[19]

The substituted phenylcarbenes also show only modest ability to distinguish between simple alkenes (T13). The isobutene/*trans*-butene reactivity ratio varies with carbenic aryl substituent in the order, *m*-Cl > *p*-Cl > H > *p*-CH_3 > *p*-OCH_3, which parallels the expected order of increasing carbene stability. Representing these carbenes as in **27**, the π system is seen to be

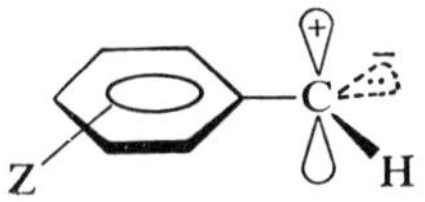

Z H C

27

isoelectronic with a benzylic cation. It follows that **27**, Z = OCH_3, should be the most stabilized and selective species.

The variation of $k_{\text{isobutene}}/k_{trans\text{-butene}}$ with substituent could be correlated with σ constants, $\rho = +0.3$.[20] This value is interpreted as the $\Delta\rho$ for additions to isobutene versus *trans*-butene; ρ is more positive for addition to isobutene, in agreement with greater negative charge on the carbenic carbon in the transition state for this addition. The idea is reasonable, because isobutene can more easily accept the conjugate partial positive charge in the transition state.[20]

Free carboethoxycarbene has been generated by photolytic (T18) and thermal (T19) means from ethyl diazoacetate. The photolytic intermediate shows almost no selectivity over the usual substrate set; the thermal intermediate shows some selectivity, but its reactivity toward butene-1 seems too high. The reactivity ratio isobutene/*trans*-butene is 1.11 for the former and 5.18 for the latter species.[25,27] Despite the 125° temperature difference between the two reactions, we believe a significant difference would still persist if the comparison were to be made at an identical temperature. Further investigation is called for.

The thermal reaction (T19) can be studied as a function of solvent. Relative reactivity ratios, for example, isobutene/*trans*-butene, in the presence and absence of a given solvent, define an influence factor f for that solvent (Eq. 4).

$$f = \frac{(k_{\text{iso}}/k_{trans})_{\text{solv}}}{(k_{\text{iso}}/k_{trans})_{\text{neat}}} \tag{4}$$

Observed effects are small, the largest being $f = 3\text{–}4$ for solvent benzophenone. Larger solvent effects are encountered for insertion versus addition reactions of carboethoxycarbene. Solvent-carbene "complexes," more selective than the free carbene, are postulated.[27]

The copper carbenoid derived from ethyl diazoacetate cannot be regarded as a free carbene, as originally suggested.[26] The counterevidence, discussed at length,[57] stems from studies of stereoselectivity. Therefore, note the similarity in the interolefinic discrimination of the carbenoid and the photo-generated carbene (T18). Presumably, both species add via transition states which lack substantial polarization of the olefinic carbon atoms.

Compare carboethoxycarbene [reaction (1), T18], dicarbomethoxycarbene [reaction (1), T20, T21], and bromocarboethoxycarbene (T29). The greater selectivity of the latter, compared with carboethoxycarbene, suggests stabilization as in **28**,[25] in addition to stabilization due to the carboethoxy substituent. The latter factor (cf. **29**) is presumably responsible for stabilizing carboethoxycarbene relative to CH_2; that is, it has somewhat greater selectivity in the C—H insertion reaction.[76] Little of this selectivity difference

$$:\ddot{Br}-\ddot{C}-COOC_2H_5 \longleftrightarrow :\overset{+}{Br}=\bar{C}-COOC_2H_5$$

28

$$H-\ddot{C}-C(=O)-OC_2H_5 \longleftrightarrow H-\overset{+}{C}=C(-\ddot{O}:^{-})-OC_2H_5$$

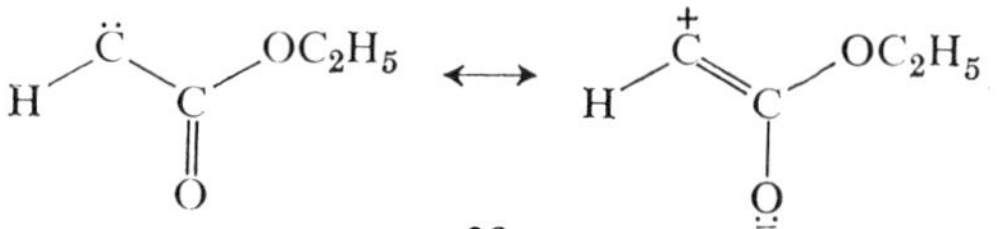

29

appears in the addition reactions, however (compare T1 and T18). Dicarbomethoxycarbene is more selective than either CH_2 or carboethoxycarbene toward the C—H bond;[76] it is also more selective in addition reactions (T21). The added selectivity presumably corresponds to extension of the stabilization depicted in **29** by the second carboalkoxy group.

Halocarbene selectivity (T23–T55) has been extensively investigated. Substitution of halogen for hydrogen (CHX versus CH_2) does not greatly enhance discrimination toward olefins (T1, T23, T24, T25). For the competition tetramethylethylene/*cis*-butene, the maximum reactivity ratios are CH_2, 1.57 (gas phase, 24°); CHBr, 1.16 (−30°); CHCl, 1.21 (−30°); CTF, 1.95 (gas phase, 22°). The expected stabilization of the halocarbenes **30** is not well expressed in addition selectivity, but the order of increasing selectivity,

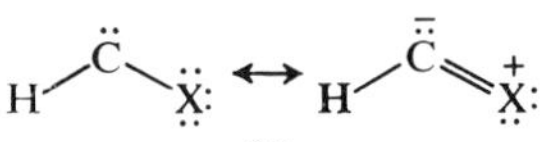

$$H-\ddot{C}-\ddot{X}: \longleftrightarrow H-\bar{C}=\overset{+}{X}:$$

30

CHBr < CHCl < CHF, does correlate with the expected increasing ability of X to donate an electron pair (*p-p* overlap) to the carbenic *p* orbital (cf. 30).* Indeed, Rowland concludes that "the fluorine atom has stabilized [CTF] by more than 25 kcal/mole relative to the unsubstituted carbene."[31]

Skell provides a valuable discussion of the relation of carbene selectivity and carbene stabilization via forms like **30**:

"The concept of electrophilicity or degree of electrophilicity is in common parlance without a precise definition. The concept relates in principle to the freeness of a vacant *p* orbital, and relates to experiment by an increasing rate of reaction with increasing nucleophilicity of a series of reagents. For the

* For a discussion of resonance stabilization in halocarbenes, see reference 77, Chapter 3.

present only relative rates are available somewhat clouding the use of the concept. From theory, CH_2 has a freer *p* orbital than CCl_2, and thereby CH_2 would be classed the more electrophilic. With olefins of increasing alkyl substitution, as test reagents, it is found that CCl_2 is more selective (a greater spread of rates). This inverse relation between selectivity and electrophilicity is rationalized by a decrease in $\Delta\Delta H^{\ddagger}$ with a decrease in $\Delta H^{\ddagger}$ . . . . A strongly electrophilic reagent does not spread the olefin rates and thus the feature which distinguishes electrophiles from nonelectrophiles tends to vanish with increasing electrophilicity of the reagent."[44]

A similar treatment of "electrophilicity" has been provided by Harrison,[44f] who offers the series of increasingly electrophilic carbenes: $CF_2 < CHF < CH_2$, and $CF_2 < CCl_2 < CBr_2 < CI_2 < CH_2$.

In a sense, then, there is a "competition" between the substituents X and Y and the olefinic π bond over donation of electrons to the carbenic *p* orbital; the balance of this competition changes during the course of addition of **31**

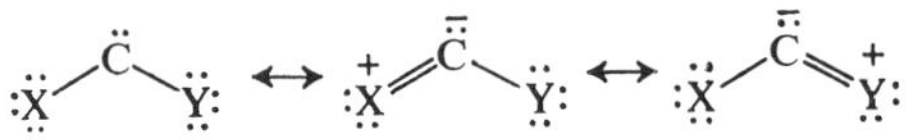

31

to an olefin. The more strongly X and Y interact with the carbenic center, the higher will be the activation energy for addition to a π bond. The transition state will be further advanced along the reaction coordinate, and more important will be the ability of the olefinic carbon atoms to support a positive charge. Thus strong resonance interaction in **31** will "develop" the olefin structure-reactivity sequence. Doering recognized this long ago when he spoke of relative carbene selectivity as reflecting the "internal stabilization" of the carbene.[4]

Recent calculations suggest significantly higher activation energies for CF_2 versus CH_2 addition reactions, but they indicate *less* rather than more charge transfer from olefin to carbene in the CF_2 addition transition state.[66a] If the calculations correctly mirror reality, then the selectivity difference of CF_2 and CH_2 is not attributable to the relative magnitude of charge transfer in the respective transition states but to the greater inequality with which partial positive charge is imposed on the olefinic carbon atoms during CF_2 addition.

CH_3CCl is more selective than CHCl [T27; reaction (2), T24], and C_6H_5CCl is yet more selective [reaction (2), T28]. The selectivity "spreads" (defined as the reactivity ratio tetramethylethylene/*trans*-butene) are 25.5, 7.45, and 1.10. The results indicate extended stabilization by (Y =) CH_3 or C_6H_5 of **31** (X = Cl), compare to **30**. C_6H_5CCl is also more selective than C_6H_5CH (T13).

Additional comparisons of T23 with T26, and of T25 with T30 [reaction (2)], show that C_6H_5CF is more selective than CTF (spreads: 27 versus 1.50), and that C_6H_5CBr is more selective than CHBr (spreads: 17.6 versus 1.07). Phenylhalocarbene selectivity increases in the order Br < Cl < F, as does monohalocarbene selectivity. Again, however, the range is small.

The C_6H_5CCl and C_6H_5CBr results show that α-elimination reactions of benzal halides do not give the same (presumably free) carbenic intermediate as is obtained by photolysis of phenylhalodiazirines (T28, T30). This clouds interpretation of the C_6H_5CF results, for which only a benzal halide precursor has been studied. Nevertheless, the increasing similarity in selectivity of base and light generated C_6H_5CX, as X is changed from Br to Cl, suggests that the selectivity of a free C_6H_5CF would be similar to that shown in T26.

Dihalocarbenes are more selective than either monohalocarbenes or phenylhalocarbenes. Reactivity spreads are 177 (36°, gas phase, T31) and 49.8 (25°, T36) for CF_2 and CCl_2, respectively. Not only is CX_2 selectivity greater than that of CHX, but the effect on selectivity of successive halogenation in the series CH_2, XCH, and CX_2 is unequally additive. The major enhancement comes with the *second* halogen.[34] This effect can be observed for X = F, Cl, or Br, on comparisons of the proper tables. Moreover, sensitivity to halogen identity is greater with dihalocarbenes than with either phenylhalocarbenes or monohalocarbenes. Compare, for example, the response of "spread" to the change of Cl to F in ClCX, as opposed to HCX. The spread increases from 49.8 (25°, T36) to 320 (−12°, T53) in the first case, but only from 1.10 (−30°, T24) to 1.50 (gas phase, 22°, T23) in the second case. It is believed that the differing reaction conditions do not invalidate this comparison. In terms of our previous discussion, only with the dihalocarbene additions has $\Delta F^{\ddagger}$ become large enough so that $\Delta\Delta F^{\ddagger}$ terms are substantial functions of carbenic and olefinic substituents.

Most of the tables of CCl_2 data manifest the steric and electronic effects we have discussed. Some demonstrate the independence of CCl_2 reactivity from CCl_2 source. Table 45 indicates the high reactivity of vinyl ethers toward CCl_2. Thus from T45 and T38, pentene-1 is ~0.15 times as reactive as cyclohexene toward CCl_2, but ethoxyethylene (the corresponding vinyl ether) is 1.28 times as reactive. A possible transition state for the CCl_2-vinyl ether addition is **32**, in which the contribution of oxygen electrons is emphasized.[50]

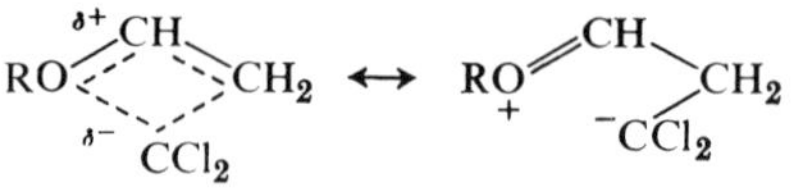

$$\mathrm{RO}\overset{\delta+}{\cdots}\mathrm{CH}\cdots\mathrm{CH_2}\cdots\overset{\delta-}{\mathrm{CCl_2}} \longleftrightarrow \overset{+}{\mathrm{RO}}{=}\mathrm{CH}{-}\mathrm{CH_2}{-}\overset{-}{\mathrm{C}}\mathrm{Cl_2}$$

32

Note that substitution of a chlorine atom for a β proton in ethoxyethylene cuts the reactivity by a factor of 2.5; the β-C—Cl dipole raises the energy of

32. The reactivity ordering in T45 parallels the expected basicity of the substrate oxygen atoms.[50]

The data of T45 have been compared to those of T5B, which summarizes the selectivity of the carbenoid $C_2H_5ZnCH_2I$ toward alkenyl ethers. Parallel behavior is seen, but the selectivity of this carbenoid along the series $ROCH{=}CH_2$ (R = t-C_4H_9, i-C_3H_7, C_2H_5) is much greater than that of CCl_2. Indeed if the (log) reactivities for the two carbenic species (relative to cyclohexene) are used to construct a linear free energy graph ($C_2H_5ZnCH_2I$ data as the ordinate), the slope of the correlation line is about 3.3. The greater electrophilic selectivity of this methylene carbenoid, as compared to CCl_2; has previously been noted in additions to substituted styrenes, and also in insertions into Si—H bonds (see above).

There are also steric effects visible in the carbenoid data of T5B. Note that substrate reactivity toward $C_2H_5ZnCH_2I$ is in the order R = CH_3 > C_2H_5 > i-C_3H_7 in *cis*-$RCH{=}CHOC_2H_5$. These results correspond to a decrease in reactivity as the substrate is alkylated at an α carbon atom. Similar behavior has been noted in the reactions of CCl_2 with 1-alkenes (see above). The general selectivity of $C_2H_5ZnCH_2I$ toward $ROCR'{=}CR''R'''$ is in accord with the picture of an electrophilic addition. Note the typical behavior in T5B: additional alkylation of the olefin enhances addition, and a *cis*-disubstituted substrate is more reactive than its *trans* isomer. Note, too, that the monosubstituted, oxygen-isolated olefin $ROCH_2CH{=}CH_2$ (R = i-C_4H_9) is *more* reactive than heptene-1 (factor of 10.4), suggesting the Zn—O interaction depicted in **32a**.[12a]

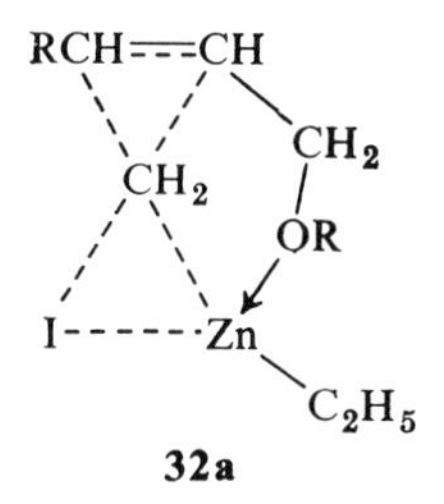

32a

The data of T46 and T47 suggest $(CH_3)_3Ge > (CH_3)_3Si > (CH_3)_3C$ as the order of inductively releasing α substituents, capable of increasing alkene reactivity toward CCl_2. Substitution of silyl groups directly on the olefinic carbons decreases substrate reactivity; d_π–p_π overlap drains π electron density toward the silicon.[39,51] The analysis of the data is complicated by the substituents' strongly competing steric, inductive, and resonance effects on π bond accessibility.

Data for CBr_2 are less extensive than those for CCl_2. Taking note, however, of the greater importance of adverse steric effects during CBr_2 additions,[4] the reactivity trends are similar for both carbenes. Comparison of common data for BrCH (T25), $BrCCOOC_2H_5$ (T29), $BrCC_6H_5$ (T30), BrCBr (T50), and BrCF (T55) shows that selectivity increases in this order, as expected from a consideration of the changing substituents' carbene-stabilizing potentials. The observations parallel those for ClCX (see above). The greatest selectivity enhancement occurs with the change from CBr_2 to CBrF, as expected, although steric effects may contribute to this result.

b. More Quantitative Studies. Linear free energy relations may be used to measure the relative selectivities of several reagents toward a standard set of olefinic substrates. A plot of log (k/k_0) for CBr_2 additions [reaction (1), T50] against bromination data (Br_2 in CH_2Cl_2) gave a straight line with slope of +2.1, indicating that CBr_2 was the more selective reagent under the particular reaction conditions.*[3] The linearity indicates mechanistic similarity of the two reactions.

Doering and Henderson report failure in an attempted correlation of CCl_2 and CBr_2 additions, indicative of a differential steric effect.[4] Lithium chlorocarbenoid [reaction (1), T24] can be correlated with CCl_2 (T33). The original discussion[32] needs modification, in view of later changes in assignment of product configurations for the chlorocarbenoid reaction.[32a] The slope of log $(k/k_0)_{\text{"CHCl"}}$ (−35°) versus log $(k/k_0)_{CCl_2}$ (−10°) is now found to be 0.40, showing CCl_2 to be more selective under these conditions; this agrees with the original conclusion.[32] Hartzler compares dimethylvinylidene carbene (−10°, T9) with CCl_2 (−10° to −20°, T33), and the correlation slope (not reported in the original) is 0.71.

Moss and Gerstl compare CFCl (−12°, T53) with CCl_2 (−10°, T36, reference 42a). Over the olefin set tetramethylethylene to *trans*-butene, the slope is 1.53; CFCl is more selective. Inclusion of data for butene-1 causes a poorer fit and lowers the slope to 1.28. Analyzed a different way, over the partial set tetramethylethylene to isobutene, the slope is 1.84, but over the partial set *cis*-butene to butene-1, it is only 0.92. CFCl appears more selective than CCl_2 toward the more reactive olefins; the selectivities are reversed toward the less reactive olefins. An explanation cites differing steric demand of CFCl and CCl_2, which becomes increasingly important as substrate reactivity is decreased[42a] (see Section IV.B.1).

Mitsch and Rodgers present linear free energy plots for CF_2 (adjusted to −15°, T31) versus CCl_2 (−15°, T33) and for CF_2 (adjusted to −10°, T31) versus CFCl (−12°, T53).[38] The slopes are 1.87 and 1.12, respectively. The authors note that, "stabilizing effects are apparently not a linear function of substituents, as chlorofluorocarbene is nearly as selective as difluorocarbene . . . ".[38] Other aspects of these correlations have been discussed previously (Section IV.B.1).

* In discussing these correlations, the Y reagent will be taken as the first-named reagent.

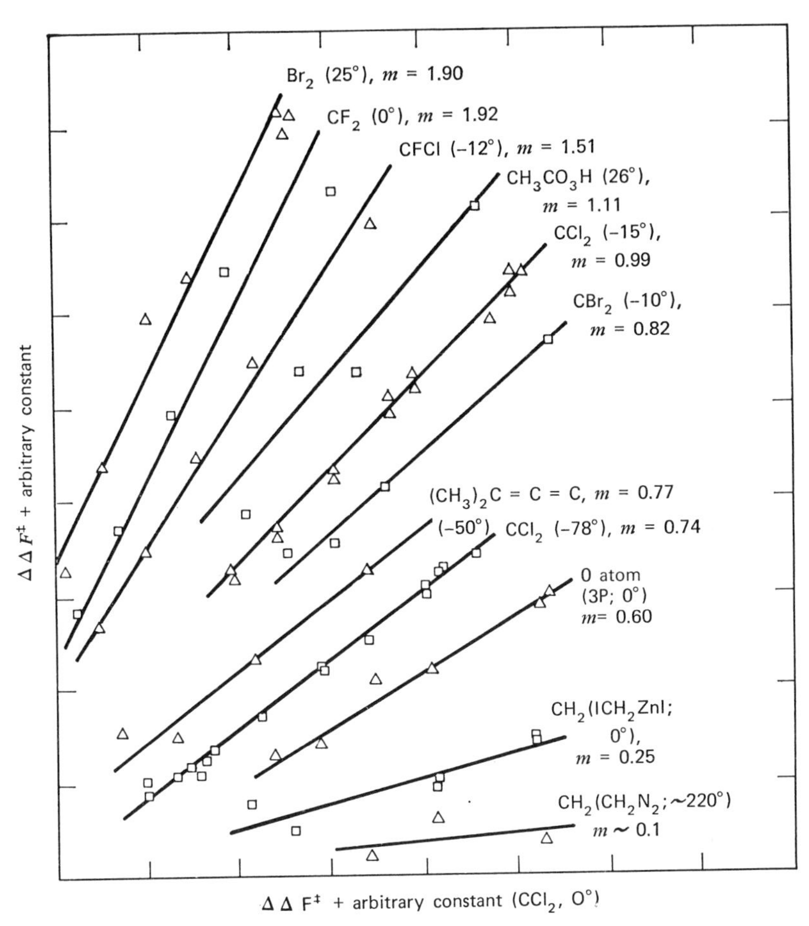

Fig. 3. Linear free energy graphs; m refers to slope of the correlation line.[44]

Skell and Cholod[44] present a composite of earlier and new correlations (Figure 3). The data permit an ordering of reagent selectivity toward olefins: $CH_2 < O(^3P) < CBr_2 < (CH_3)_2C{=}C{=}C < CCl_2 < CH_3CO_3H < CFCl < CF_2 \sim Br_2$ (in CH_3OH). The order of increasing selectivity is taken to be the order of decreasing electrophilicity; "electrophilicity" is defined, as earlier, in terms of "freeness" of a vacant p orbital.

Moss and Mamantov[34a] present linear free energy plots for XCCl versus CCl_2. The following slopes are obtained, as a function of X: $X = CH_3$, $m = 0.50$ (25°–30°, T27); $X = C_6H_5$, $m = 0.83$ [25°, reaction (1), T28]; $X = F$, $m = 1.48$ (−12°, T53). The CCl_2 data are measured at −10° (T36), but use of CCl_2 data measured at 25° does not change the m values for $X = CH_3$ or C_6H_5. Adding $m = 1.00$ for the CCl_2 versus CCl_2 correlation

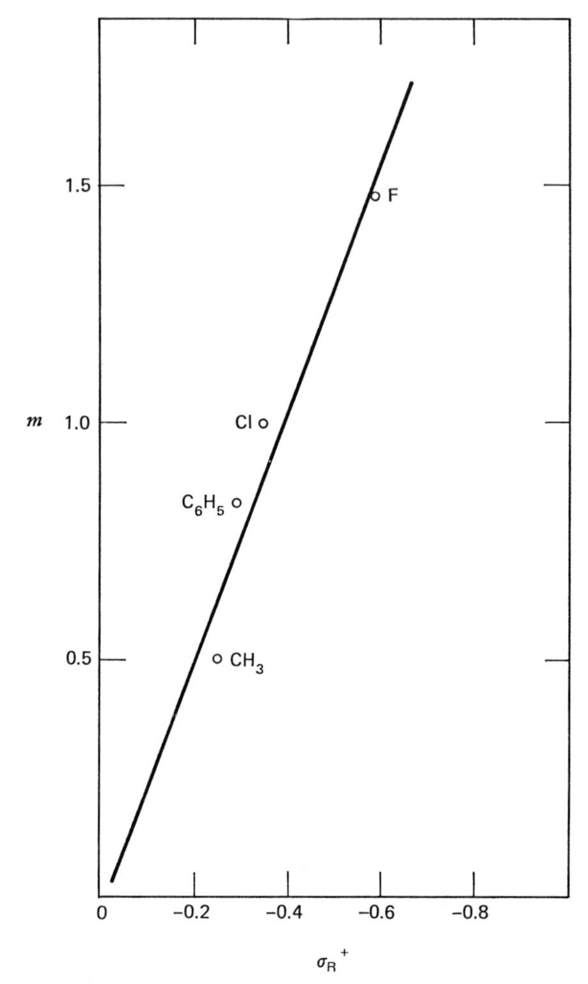

Fig. 4. Slopes (m) of $\log (k/k_0)_{XCCl}$ versus $\log (k/k_0)_{CCl_2}$, correlated with $\sigma_R{}^+$ (X); the correlation coefficient is 0.96.[34a]

gives four m values which can be correlated with $\sigma_R{}^+$ for X of XCCl (Figure 4). The relation manifests the connection between electron-donating ability of carbenic substituents and carbene selectivity (see Section IV.C.2a). The possibility that the correlation of Figure 4 can be extended to other carbenes is being explored.

3. *Highly Stabilized Carbenes*

If a carbene can be made less electrophilic by overlap of its vacant p orbital with electron-donating substituents, then we expect continued increase of the donating power of such groups to lead ultimately to singlet carbenes which are not electrophilic at all toward olefins. Alternatively, the carbenic center could be incorporated in an aromatic π system such as cyclopropenylidene (**33**). Several examples of these types are considered here.

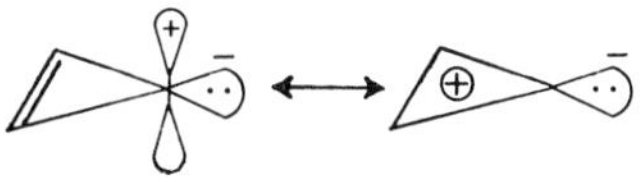

33

Theoretical considerations are available for both approaches.[78,79] With total suppression of its "vacant" p orbital, a carbene can behave as a nucleophile via the carbenic lone pair. See reference 80 for a brief review and leading references to the chemistry of highly nucleophilic carbenes.

Although dimethoxycarbene **34** prefers to dimerize rather than add to simple alkenes,[81] the sulfur analog **35** does add to such electron-rich alkenes as ketene diethylacetal. In cyclohexene solution, however, **35** dimerizes and

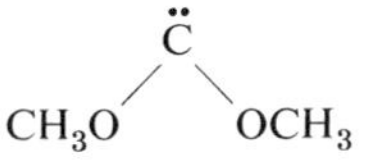

$CH_3O-\ddot{C}-OCH_3$ **34** $CH_3S-\ddot{C}-SCH_3$ **35**

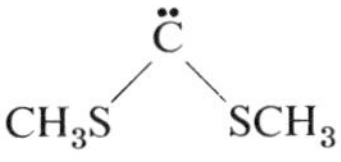

also affords an orthothioformate.[82] Other dialkylthiocarbenes[82,83] can be trapped with triphenylphosphine but not with cyclohexene.[83] Two alkylthio substituents thus bring a carbene to the border of its structure-selectivity relation toward alkenes; freeness of the p orbital is now so minimal that only the most electron-rich olefins are suitable substrates. Two alkoxy substituents further decrease p orbital availability, and the carbene-olefin addition reaction has not yet been reported. (Additions of **34** to ketene acetals, however, should be studied.) An expanded ordering of CX_2 selectivity is thus $CBr_2 < CCl_2 < CF_2 < C(SR)_2 < C(OR)_2$.

There is evidence that R_2N is a more effective carbene-stabilizing substituent than RO.[83a] This would be anticipated on the basis of the σ_R^+ relationship of Moss and Mamantov (see above). Recent work also suggests that siloxycarbene **35a** is a nucleophilic intermediate, since it adds to diethyl fumarate (giving **35b**) but not to cyclohexene, tetramethylethylene, or even ketene dimethylacetal.[83b] It is not clear why **35a** does not add to the acetal, since the dialkylthiocarbenes do. Moreover, it is also odd that **35a** does not add to tetracyanoethylene.

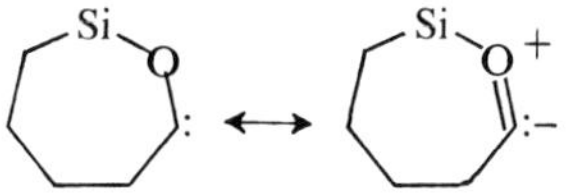

35a 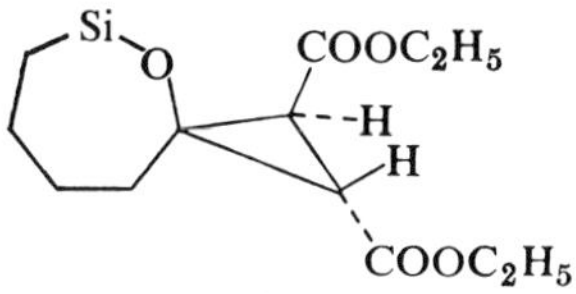   **35b** ($COOC_2H_5$, H, H, $COOC_2H_5$)

Krebs and Kimling[83c] have reported the cycloaddition of isocyanide **35c** to the thiocycloheptyne **35d**. This may represent an extremely rare example in

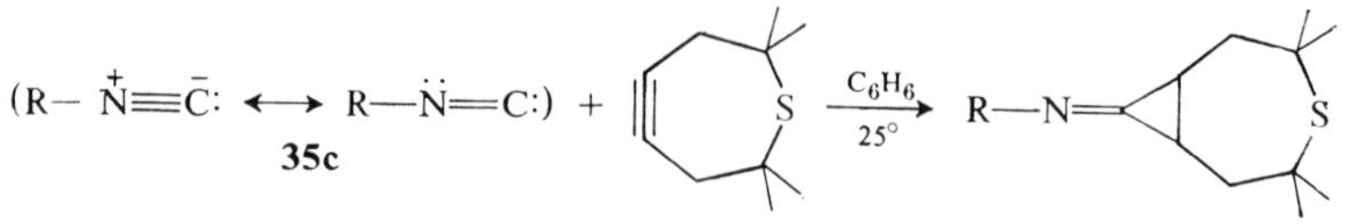

(R—$\overset{+}{N}\equiv\overset{-}{C}$: ⟷ R—$\ddot{N}$=C:) + **35d** $\xrightarrow[25°]{C_6H_6}$ R—N=

35c **35d**

which an isocyanide behaves as an azaethylidene carbene in an olefin addition reaction (cf. the right-hand canonical form of **35c**). In this regard, it is suggestive that the reaction is most rapid when R = *p*-nitrophenyl, of intermediate rate when R = phenyl, and slowest when R = cyclohexyl. The azaethylidene form would be expected to make the greatest contribution to **35c** in the R = *p*-nitrophenyl case. It is also noteworthy that dimethylacetylene, tolane, and cyclooctyne did not react with **35c**. The acetylene must be extremely reactive for the cycloaddition to proceed.

Monocyclic systems are reported, in which the carbenic center is part of a potentially aromatic π system; examples are diphenylcyclopropenylidene **36**[84–86] and cycloheptatrienylidene **37**.[87–90] Carbene **36** apparently forms from

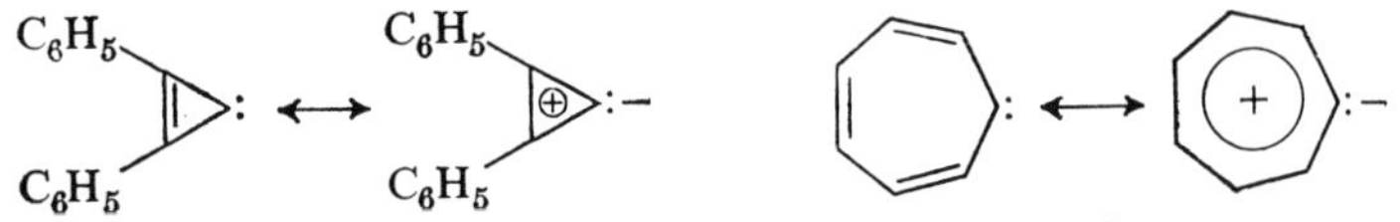

C_6H_5 C_6H_5 C_6H_5 C_6H_5

36 **37**

38, upon loss of a proton and carbon dioxide.[85] The intermediate is trapped by addition to the excess trityl cation used to generate **38**. Added dimethyl fumarate did not compete successfully with trityl cation for **36**. More practicably, **39** apparently eliminates the elements of *N*,*N*-dimethylcarbamic acid in basic media, affording **36**.[84,84a,86] Addition to dimethyl fumarate occurs (70%, maximum), giving a labile spiropentene, probably **40**. The same

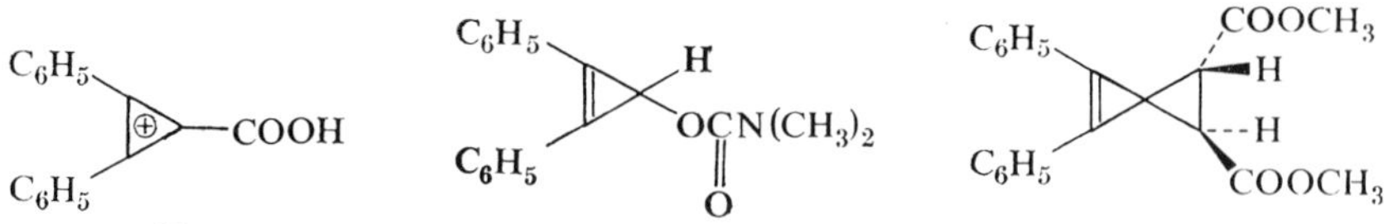

C_6H_5 C_6H_5 —COOH

C_6H_5 C_6H_5 H $OCN(CH_3)_2$ O

C_6H_5 C_6H_5 $COOCH_3$ H H $COOCH_3$

38 **39** **40**

product is obtained from **36** and dimethyl maleate. There is now no explanation for the lack of stereospecificity; its significance is not clear.[86] Carbene **36** also adds to fumaronitrile, diethyl fumarate, and *N*,*N*,*N*′,*N*′-tetramethylfumaramide, but no addition occurs with cyclohexene, *trans*-stilbene, tetrachloroethylene, diphenylacetylene, or dimethyl acetylenedicarboxylate.[86]

Photolysis[87] or thermolysis[88] of tropone tosylhydrazone (sodium salt) gives cycloheptatrienylidene **37**. Products include the formal carbene dimer, heptafulvaene, and, in the presence of dimethyl fumarate, adduct **41**. Addition to dimethyl maleate also gives **41**, but this apparently nonstereospecific

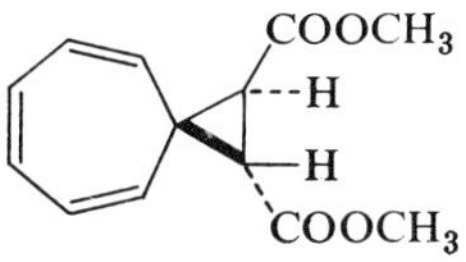

COOCH$_3$
H
H
COOCH$_3$

41

reaction is probably due to instability of *cis*-**41** under the experimental conditions.[89] Carbene **37**, generated photolytically in tetrahydrofuran, adds stereospecifically to fumaronitrile and maleinitrile.[89] Approximate relative reactivities toward **37** of fumaronitrile, maleinitrile, dimethyl fumarate, and dimethyl maleate are 1:1:0.5:0.007, respectively;[89] steric effects appear to hinder addition to the maleate ester. The stereospecific additions to the nitriles, which persist in the presence of heavy atom solvents (e.g., CH_2Br_2), suggest a singlet ground state for **37**. The carbene does not add to *cis*-butene, cyclohexene, *N*-piperidylcyclohexene, *trans*-1,2-dibromoethylene, methyl cinnamate, or β-chloromethyl acrylate; heptafulvaene (dimer of **37**) forms in these reactions.

In summary, **36** and **37** behave as nucleophilic species, adding only to electron-poor alkenes. This supports extensive contributions of the delocalized forms in resonance hybrids **36** and **37**. The results are clouded, however, by the possibility that the precursors of the carbenes, or pyrazolines derived from 1,3-dipolar additions of these precursors, are the progenitors of the products. Definitive evidence for **36** and **37** as the ultimate intermediates is still lacking; it would be most welcome.*

4. Triplet Carbenes

In addition to the stereochemical test for *triplet* carbene addition,[91] it is suggested that they should show a "radical-like" selectivity toward olefins.[3] In particular, because such addition should be a two-step procedure, involving a triplet 1,3-biradical intermediate, triplet carbenes are expected to show exalted reactivity toward such substrates as 1,3-dienes, styrene, and 1,1-diphenylethylene. Here the intermediates are resonance stabilized. Singlet carbenes, which add to olefins in one step, are not expected to show unusual reactivity toward these substrates. Therefore a sensitive indicator for a triplet carbene might be the reactivity ratio, 1,3-alkadiene/alkene-1. ^3R_2C

* See Addenda, number 3.

would afford the stabilized diradical **42** on addition to the diene; corresponding stabilization is not available on addition to the 1-alkene. Examination of the tables permits some test of this idea; the stereochemical aspects of triplet carbene additions are reviewed elsewhere.[92]

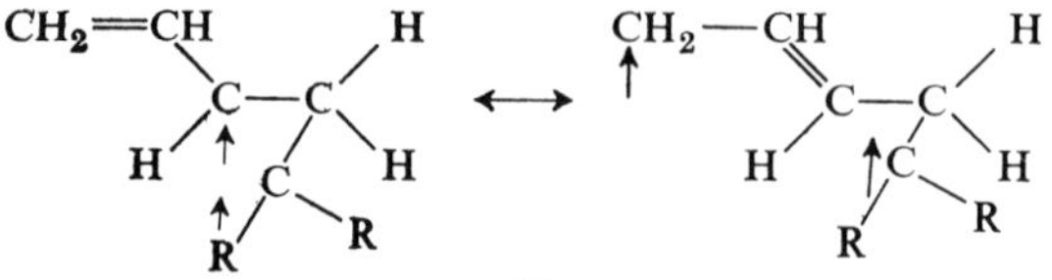

42

$^3CH_2$ shows some enhanced selectivity in the 1,3-butadiene/butene-1 competition, as compared with $^1CH_2$(T1); relative reactivities are 11.8 and 4.92, respectively. Neither methylene species is particularly selective toward alkenes and the purity of the $^3CH_2$ from $^1CH_2$ is still unclear;[8a,8b,8c,8d] the data may need revision.

Triplet propargylene **43** and methyl propargylene are reported to select in

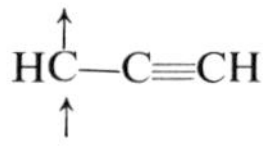

HC—C≡CH

43

favor of dienes, as opposed to monoenes;[93] singlet propargylene shows little selectivity in additions to olefins.[93a]

Triplet carbonylcarbene (T10) is more selective toward alkenes than is the corresponding singlet. It is also more selective than $^3CH_2$ (T1). The selectivity of $^3(C_2O)$ is controlled largely by differences in activation energies. The activation energy for addition of this species to ethylene is ~6 kcal/mole, which would make the corresponding datum for tetramethylethylene ~10.4 kcal/mole.[17a] A decrease in preexponential factor with increasing substrate alkylation indicates sensitivity to steric effects. Note that such sensitivity is not shown by (^3P) oxygen atoms, for which $A_{tetramethylethylene}/A_{ethylene}$ is 1.24.[75] The 1,3-butadiene/butene-1 reactivity ratio is 30.4 (29°) for $^3(C_2O)$. Relative to a propene standard, the ratio is 35.0. For comparison the latter ratio is only 2.06 (31°) for $^1(C_2O)$. Note, too, that alkenes which can add $^3(C_2O)$ so as to give diradical intermediates with *tert* character at a substrate carbon are much more reactive than alkenes that cannot, for example, isobutene versus the 2-butenes. Calculations indicate that $^3(C_2O)$ should add with an electrophilic selectivity, and that the diradical character of the transition state should be "between (that of) the olefin-O atom transition state and the olefin-propargylene transition state, probably closer to the former."[17c]

Diphenylcarbene (T17, T17A) shows marked radical-like selectivity. Particularly impressive is the relative reactivity ratio, 1,3-butadiene/hexene-1, which is >100. The 1,1-diphenylethylene/isobutene competition is noteworthy (T17A). Ferrocenylphenylcarbene, a closely related species, is reported to add to 1,1-diphenylethylene but not to cyclohexene or ethylvinyl ether. A triplet multiplicity was assigned, partly on this basis.[94]

Triplet dicarbomethoxycarbene shows a mild radical-like selectivity, relative to the singlet; the reactivity ratios, 2,3-dimethylbutadiene-1,3/pentene-1, are 9.6 and 2.8, respectively (T20). Neither species is very discriminating, and both encounter steric hindrance in addition to tetramethylethylene. A discussion of the reactions of singlet and triplet carboethoxycarbene and dicarbomethoxycarbene with allyl halides and allyl sulfides is given in the Appendix (Section V).

The available data, though sparse, suggests that there are real differences in the abilities of singlet and triplet carbenes to discriminate between dienes and alkenes. These differences can be used diagnostically, but only with care, for their magnitudes depend strongly on the carbenic substituents.

5. Carbenoids

Carbenoids are "intermediates which exhibit reactions qualitatively similar to those of carbenes without necessarily being free divalent carbon species."[20] Here they are discussed only in relation to the tables and only if they have not been discussed earlier.

Arylcarbenoids generated from benzal halides and alkyllithium (T14) differ in selectivity from the corresponding free carbenes (T13). In particular, the carbenoids react significantly faster with *cis*-butene than with *trans*-butene; the free arylcarbenes display a slight preference for the *trans*-butene. Examination of the *combined* carbenoid relative reactivity-stereoselectivity data reveals that the enhanced selectivity toward *cis*-butene is largely a product of preferred *syn*-arylcarbenoid addition to this olefin (cf. transition state **44**).[20]

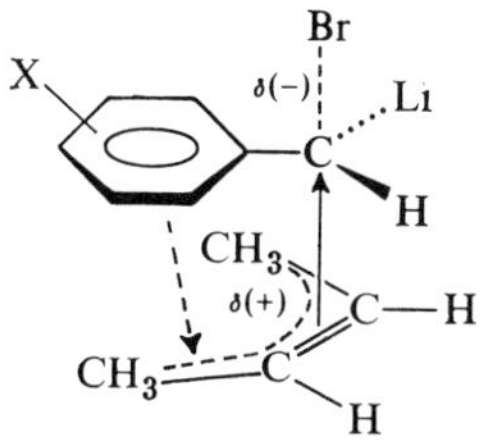

Note also (T13 versus T14) that the arylcarbenoids react more rapidly with isobutene than with *trans*-butene, but that the free arylcarbenes do not exhibit this trend. The greater ability of the carbenoid to distinguish between unsymmetrically and symmetrically disubstituted olefins is attributed to greater bond-making character in the carbenoid transition state; a similar conclusion follows from the stereoselectivity.[20,57] The arylcarbenoids fail to show the Hammett relation between selectivity and *p*-substituent which is shown by the free arylcarbenes (Section IV.C.2a); this is in agreement with an S_N2-like transition state for the carbenoid addition.

Lithium carbenoids derived from either diphenyldibromomethane[96] or benzal bromides[20] can add to olefins, but the lithium carbenoids derived from **45** and **46** fail to add.[97] A modified S_N2-like transition state is suggested, in

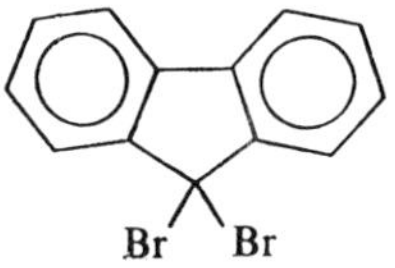

Br Br
45

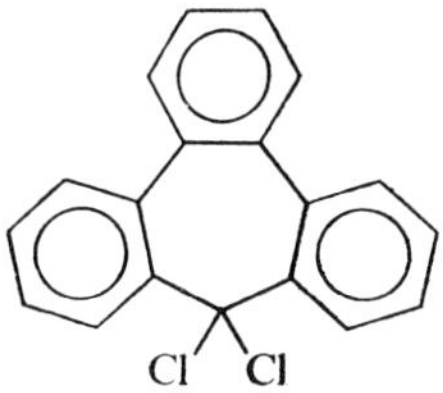

Cl Cl
46

which steric problems prevent the required attack of the olefinic π bond on the back side of the carbenic center.[97]

Comparative selectivities of the *p*-tolylcarbenoids derived from *p*-tolyldiazomethane and various metal halides are displayed in T15. Note that the species derived from *p*-methylbenzal bromide and methyllithium differs in selectivity from that derived from *p*-methylphenyldiazomethane and lithium bromide; the latter is more selective. Differing states of carbenoid aggregation could be responsible.

Approximate lifetimes of the intermediates were obtained by means of continuous-flow experiments, in which the diazoalkane and ethereal zinc chloride were mixed, passed at a variable rate through a capillary tube, and drowned in cyclohexene. At 20°, halflives were $\sim$0.2 and 10 seconds, respectively, for *p*-$CH_3C_6H_5CH{\cdot}ZnCl_2$ and $C_6H_5CH{\cdot}ZnCl_2$.[21] Both species afford comparable yields of cyclopropanes in olefinic solution, and the half life data therefore mean that the *p*-tolylcarbenoid addition is the faster reaction. It is not yet clear whether the substituent-rate of addition effect is mainly dominated by alteration of transition state or ground state energies. Both energies probably are important functions of the aryl substituent.

Table 15 indicates that arylcarbenoid selectivity in the ZnX_2-catalyzed reactions increases in the order X = I > Br > Cl. This is attributed to "increasing electron deficiency on the carbenoid carbon reflecting greater carbon-halogen bond breaking in the transition state with increasing atomic number of the halide ion."[21] Greater carbenoid electron deficiency provokes greater transition state polarization and a stronger dependence of rate on substrate structure. The likely intermediate is formulated as **47**.[21]

```
       ZnX
        |
  Ar—C—X
        |
        H
```

47

Chlorocarbenoid (T24) and the phenylhalocarbenic species, derived from benzal halides and potassium *t*-butoxide (T28–T30), are differentiated from their photochemically generated analogs, which are presumed to be free carbenes. In contrast to the arylcarbenoids, the halocarbenoids do show clear carbenelike selectivity.

The practically invariant selectivity of CCl_2, as generated under various conditions (T34, T40, T43, T49), is taken as evidence that free CCl_2 is the intermediate in each case. A similar conclusion follows from studies of relative addition activation parameters as a function of CCl_2 precursor.[44a] The published data indicate that (above −80°) CCl_2, from chloroform and various bases, and free CCl_2 (chloroform pyrolysate, −125°) possess temperature-dependent olefin selectivities, which are correlated by the same Arrhenius expression.[44,44a]

In the temperature range −70 to −100°, there is independent evidence for the existence of the carbenoid $LiCCl_3$.[98] The following scheme must be considered:

$$LiCCl_3 \underset{k_{-1}}{\overset{k_1}{\rightleftharpoons}} LiCl + :CCl_2$$

$:CCl_2$ + olefin $\xrightarrow{k_2}$ 1,1-dichlorocyclopropane (Cl, Cl)

$:CCl_2 \xrightarrow{k_3} Cl_2C{=}CCl_2$ + other products

$LiCCl_3$ + olefin $\xrightarrow[-LiCl]{k_d}$ 1,1-dichlorocyclopropane (Cl, Cl) (5)

It was originally believed that cyclohexene could react directly with $LiCCl_3$ at −72° ($k_d > k_1$, Eq. 5).[98] The data of T38A suggest, however, that only free CCl_2 is involved in this reaction.[44b] Thus the selectivity of CCl_2 from chloroform and *n*-butyllithium (−73°, tetrahydrofuran, T38A) is nearly identical to that of CCl_2 from phenyl(chlorodibromomethyl)mercury or sodium trichloroacetate (80°, T40). Similar results are obtained when CCl_2 is generated from the mercurial and sodium iodide or from chloroform and potassium *t*-butoxide (−15°, T34).

Moreover, though the rate of decomposition of $LiCCl_3$ in a mixture of cyclohexene and methylcyclohexane at $-73°$ increases with the cyclohexene content, the selectivity of the ultimate intermediate is that of CCl_2 and presumably not that of the carbenoid.[44b] $LiCCl_3$ is believed to be in rapid equilibrium with LiCl and CCl_2; it is the destruction of CCl_2, by olefin capture, dimerization, or other reactions, which constitutes the rate-limiting stage of Eq. 5.[44a,44b] Lodoen argues, however, that direct reaction can occur between $LiCCl_3$ and olefin at $-100°$.[44c] The argument rests on the different selectivities of the reactive intermediates at $-100°$ (T38B) and $-78°$ (T38); the difference is greater than expected for the temperature effect. Moreover, though LiBr-free $LiCCl_3$ is unstable in the presence of olefin at $-100°$, it is stable to excess methyllithium. This is considered inconsistent with the presence of substantial free CCl_2 at $-100°$.[44c] The low solubility of $LiCCl_3$ under these conditions, however, clouds the results.

A reaction scheme similar to Eq. 5 has been proposed for CCl_2 generation from phenyl(chlorodibromomethyl)mercury (cf. Eq. 6).

$$C_6H_5HgCCl_2Br \underset{k_{-1}}{\overset{k_1}{\rightleftharpoons}} C_6H_5HgBr + :CCl_2 \tag{6}$$

$$:CCl_2 + \text{>C=C<} \xrightarrow{k_2} \text{(1,1-dichlorocyclopropane)}$$

The reaction of the mercurial with olefins is zero order in olefin, but the initial rate is higher with more reactive olefins, and added phenylmercuric bromide retards the rate with the least reactive substrates.[99] In Eq. 6, k_1, the *formation* of CCl_2 is the slow step.[99] This contrasts to Eq. 5, in which the *capture* of CCl_2 is believed to be rate determining.[44b]

V. APPENDIX

Here we briefly review some examples of carbene attack on bifunctional or multifunctional alkenes. These substrates can be regarded as intramolecularly competitive, and we find that the simple ideas that have been put forward help systematize their chemistry. The discussion is illustrative, not exhaustive.

Interesting results are reported for olefins **48–52**. Triene **48** shows a high preference for addition of CCl_2 to the central, tetrasubstituted double bond, as opposed to the peripheral, disubstituted double bonds.[100] A recent investigation gives this preference as 85:15, which, corrected for the statistical advantage of the peripheral double bonds, indicates a per bond selectivity of about 11:1.[101] (A preference for attack on the central bond of **48** is also shown by a modified Simmons-Smith reagent.[102] This selectivity parallels the intermolecular selectivity. The selectivity, central/peripheral (c/p), must be

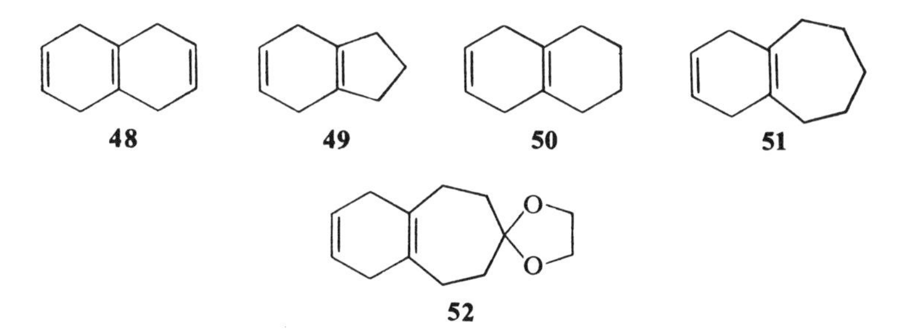

even higher when **49** is the substrate, since both CCl_2[101] and CBr_2[103] apparently yield only cyclopropanes corresponding to central addition. The *c/p* selectivity of CCl_2 decreases, however, along the series **49**, **50**, and **51**, being only 4:1 with **50**[101] and near unity with **51**.[104] An important factor in this ordering of *c/p* could be a developing steric hindrance to central bond attack afforded by the atoms of the saturated ring, which can increasingly project near or over the central double bond, as this ring is made larger and more flexible.[101] A really detailed understanding would also require analysis of changing ground and transition state strain energies along the series.

In comparison to CCl_2, (lithium) chlorocarbenoid adds with much lower *c/p* selectivity to **48**, 2.0 (per bond).[101] This is in keeping with the lower intermolecular olefin discrimination of the chlorocarbenoid versus CCl_2.[32]

It is interesting that, whereas the *c/p* selectivity is near unity with diene **51**, it increases to about 3:1 with ketal-diene **52**.[104] It was suggested that with **52** a CCl_2-oxygen atom complex might "deliver" CCl_2 to the central bond, once again enhancing its reactivity relative to the peripheral bond. It would be very helpful to have intermolecular competition experiments for CCl_2 performed with various combinations of the foregoing olefins.

In this context, we note that 3-methoxycyclohexene **53** reacts with CCl_2 at 0.18 the rate of cyclohexene, and that the addition occurs very markedly on

OCH_3

53

the face of the ring *trans* to methoxy.[104a] There is no evidence here for a directive influence in the addition; on the contrary, a deactivating influence is clearly seen.

This idea receives confirmation through the careful work of Seyferth and Mai, who studied the additions of CCl_2 to cycloalkene-3-ols.[104b] C_7, C_8, and C_9 cycloalkenols afford only *trans*-hydroxydichlorobicyclo[*n*.1.0]alkanes.

Additions of CCl_2 to 3-methoxycycloalkenes (C_6–C_9), to 3-acetoxycyloalkenes (C_6–C_9), and to 3-trimethylsiloxycyclooctene also yield very predominantly one cyclopropane isomer, which is of the *trans* configuration. Finally, a competition of cyclooctene and cyclooctene-3-ol for CCl_2 shows only a minor preference (~1.2:1) for addition to the latter. In all, no compelling evidence for an intermediate CCl_2-oxygen atom complex was indicated.* This lack of facilitation of CCl_2 addition by the substrate hydroxy group should be contrasted to the analogous reactions of the Simmons-Smith reagent, where such facilitation is very marked (see above, T4, and references 11, 57, and 62).

Indeed, we note that facilitation of Simmons-Smith methylenation involves coordination of the zinc atom to the substrate oxygen, followed by attack of the olefinic π electrons on the still-electrophilic carbenoid center. On the other hand, complexation of CCl_2 to oxygen, of the type suggested to operate in dichlorocyclopropanation of **52**, would presumably require transfer of a dichloromethide type within an ylid complex, **52a**. It is not clear that such a transfer to a tetraalkylated (i.e., nucleophilic) double bond should readily occur, for ylidic cyclopropanation usually requires that the double bond be activated toward nucleophilic addition by an electron-withdrawing substituent.[104c] Note that the enhanced reactivity of the central bond of **52** can

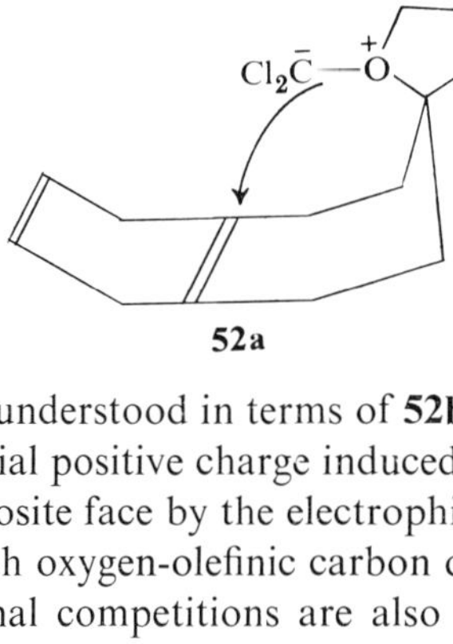

$Cl_2\bar{C}$—$\overset{+}{O}$ O $\ddot{C}Cl_2$ $\ddot{O}$ O

52a 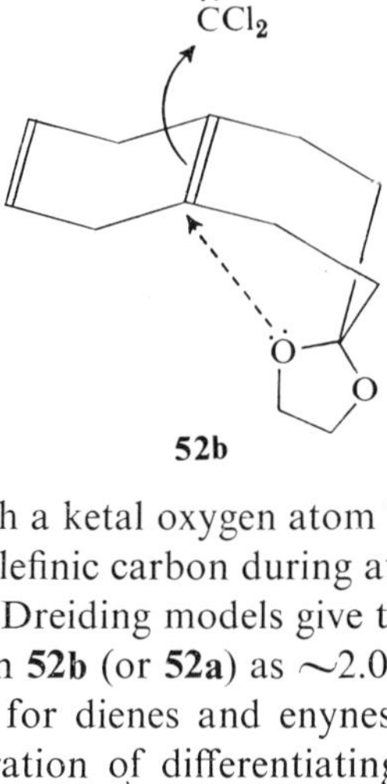 **52b**

also be understood in terms of **52b**, in which a ketal oxygen atom "solvates" the partial positive charge induced on the olefinic carbon during attack from the opposite face by the electrophilic CCl_2. Dreiding models give the closest-approach oxygen-olefinic carbon distance in **52b** (or **52a**) as ~2.0 Å.

Internal competitions are also reported for dienes and enynes. In some cases the results clearly point to the operation of differentiating steric or electronic effects. Addition of CCl_2 or CBr_2 to allenes **54** occurs mainly at

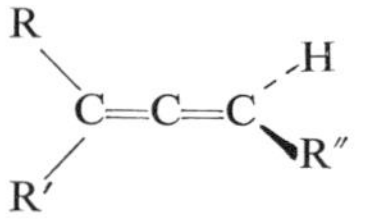

R, R′ C=C=C H, R″

54

* See Addenda, number 4.

the more substituted double bond, as expected for simple electrophilic discrimination.[105] The alkyl groups are all unbranched; steric effects do not dominate.

Related results have been reported for mono- and 1,1-disubstituted allenes with both CBr_2 and CCl_2.[105a] The direction of approach of CX_2 to the disubstituted double bond of a 1,1,3-trisubstituted allene is such as to minimize steric interaction with the 3-alkyl substituent. Analogous reactions with carboethoxycarbene (Cu carbenoid) followed a similar selectivity pattern toward di- and trisubstituted allenes; monosubstituted allenes gave mainly spiropentane derivatives. The application of the Simmons-Smith reaction to simple alkylallenes[105b] and to allenyl alcohols[105c] is interesting from a synthetic point of view, but the results cannot yet be clearly interpreted on a selectivity basis because the reactions were carried out so as to maximize spiropentane formation. The reaction of the Simmons-Smith reagent with the cyclopropylidene allene **54a**, however, occurred mainly at the terminal double bond, as expected on the basis of steric and electronic effects.[105d]

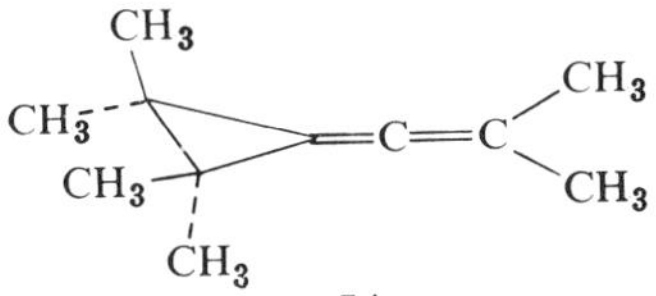

CH_3 CH_3 CH_3 CH_3 (tetramethylcyclopropylidene)$=C=C(CH_3)_2$

54a

It is interesting that CBr_2 attacked 1,2,6-cyclononatriene mainly at the allene linkage.[105e] Both steric accessibility and high ground state energy combine to augment the statistically favored preference for allenic attack.

With 2-substituted-1,3-dienes such as isoprene and chloroprene (**55**), the more highly substituted double bond is usually the principal site of carbene

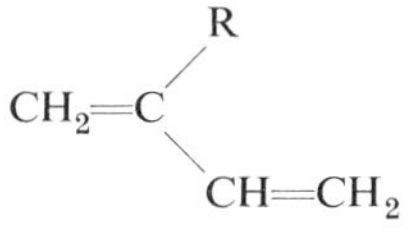

$CH_2{=}C(R){-}CH{=}CH_2$

$(R = CH_3, Cl)$

55

attack. CCl_2, for example, attacks isoprene mainly at the methyl-substituted double bond;[106,107] the selectivity is reported to be as high as 49:1.[106] CFCl shows analogous selectivity, >9:1.[107] CBr_2,[106] acetylcarbene,[108] and benzoylcarbene[108a] (as copper carbenoids) also prefer this site, as does carboethoxycarbene (see reference 108). Chloroprene is also attacked by CCl_2, mainly at the more substituted double bond; selectivity, 19:1.[109] Acetylcarbene (copper carbenoid) behaves similarly.[108] These results are in accord with a dominant electronic effect in the addition transition states.

With 1,3-pentadiene (**56**), the situation is more complex. Both CCl_2 and CBr_2 are reported to attack the less substituted double bond; selectivities

$$CH_3CH{=}CH{-}CH{=}CH_2 \qquad RCH{=}CH{-}CH{=}CH_2$$

56 **57**

58:42 and 60:40, respectively.[106] However, a reversed selectivity (1:3) is also reported for CCl_2.[107] These reactions have been carried out with different mixtures of the geometric isomers of **56**; the reactions should be studied with the pure isomeric substrates. Acetylcarbene (copper carbenoid) is reported to attack **56** with a preference for the more substituted double bond,[108] but benzoylcarbene (copper carbenoid) attacks only at the other double bond.[108a] A steric effect, dominant with the bulkier benzoyl species, could account for the turnabout.

The $CuSO_4$-catalyzed additions of carboethoxycarbene to both *trans*- and *cis*-1,3-pentadiene appear to follow the pattern of the benzoylcarbenoid. The preferences for attack at the less substituted double bond, relative to the more substituted double bond, are 89:11 and 80:20, respectively.[108b]

The electronic and steric effects of the terminal methyl group of **56** are clearly not overwhelming, and it is therefore of interest that with **57**, where the terminal substituent is chloro* or phenyl, CCl_2 adds very selectively at the less substituted double bond.[53,110] An explanation might consider the extent of conjugation in the alternative transition states for electrophilic addition of the carbene, but inductive effects may also help to deactivate the more substituted double bond.[110] Note (T48) that addition of CCl_2 to **57** ($R = C_6H_5$) is more rapid than addition to styrene or to 4-phenylbutene-1.

Additions of CCl_2 and CBr_2 to 2,4-dimethyl-1,3-pentadiene (**58**) occur mainly at the isoprene-like double bond; the CCl_2 selectivity is 88:12.[106]

$$(CH_3)_2C{=}CH{-}\underset{}{\overset{\displaystyle CH_3}{\overset{|}{C}}}{=}CH_2$$

58

The data in T48, T48A, and T48B can be understood in terms of the extent of conjugation in the alternative transition states for electrophilic addition of CCl_2 at either end of the various polyene systems. In such an analysis, it is assumed that the CCl_2 adds via an unsymmetrical transition state, involving more advanced C—C bonding at the terminal methylene carbon of the polyene, and a partial positive charge at the adjacent substrate carbon. Note in T48B that the triene and tetraene are not very different in reactivity from the diene. It is the fact of conjugation, rather than its extent, which generates

* See Addenda, number 5.

the major reactivity increase, relative to a simple alkene. Indeed the relative reactivities of 4-phenylbutene-1, styrene, 1-phenylbutadiene, 1-phenylhexatriene, and 1-phenyloctatetraene fall in the order 1:14:21:23:24.*

Duck, Locke, and Wallis[106a] have made an extensive study of the addition of CCl_2 to various trienes. The carbene was generated from chloroform with NaOH in glyme or diglyme. From a consideration of the monoadduct product distributions, they derived a set of approximate relative reactivities (at 60°): *CH_2=CH*—C=C, 19; —*CH*=*CH*—C=C, 8; *cis*-CH=CH—, 4.5; *trans*-CH=CH—, 2.2; CH_2=CH—, 1.0. The preference for CCl_2 attack on a terminal conjugated double bond, as opposed to an internal conjugated double bond, was here attributed to steric effects (see, however, above). Note the activating effect of the vinyl substituent relative to an alkyl substituent.

There has been substantial investigation of selective carbene additions to such substrates as *cis,trans,trans*-1,5,9-cyclododecatriene (**59**) and to 1-*cis*-5-*trans*-cyclodecadiene (**60**).[111] The qualitative data are reviewed else-

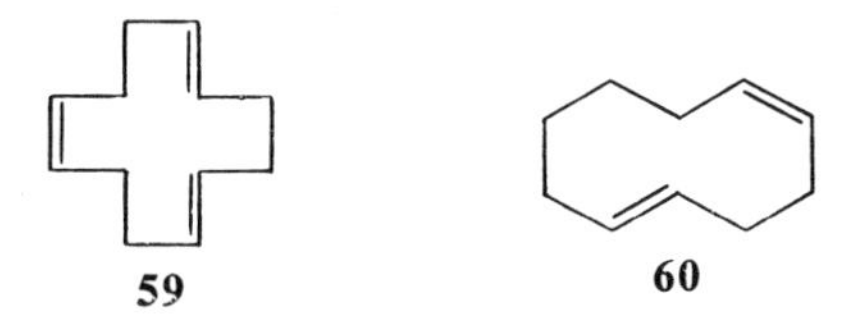

where,[57] but we note here that CCl_2, CBr_2, and CH_2 (Simmons-Smith) each give mainly monoadducts derived from attack at a *trans* double bond. (Selectivities toward **59** are greater than expected on a statistical basis.) Addition of a second mole of these carbenic reagents to the *trans*-monocyclopropanes derived from **59** involves preferential attack at the *cis* double bond.

A more quantitative study of these reactions has now appeared; data are displayed in T38C.[44d] It is clear that CCl_2 adds more rapidly to *trans*-cycloalkenes than to *cis*-cycloalkenes in the C_{10} and C_{12} systems. This is true, whether the competitions are conducted intermolecularly or intramolecularly, and the *trans*/*cis* ratios are similar in both kinds of competition. *Trans*/*cis* is substantially larger in the C_{10} series than in the C_{12} series. The specific *trans*/*cis* ratios, as derived from T38C, are 4.17 (cyclodecene), 3.07 (1-*cis*-5-*trans*-cyclodecadiene), 1.49 (cyclododecene), 1.61 (1-*cis*-5-*trans*-cyclododecadiene), and 1.80 (1-*cis*-5-*trans*-9-*trans*-cyclododecatriene). Note, in this context, that CBr_2 reacts about three times faster with *trans*-cyclooctene than with *cis*-cyclooctene.[112]

Reactions of *cis*- and *trans*-cyclododecene, of 1-*cis*-5-*trans*-cyclododecadiene, and of 1-*cis*-5-*trans*-9-*trans*-cyclododecatriene with chlorocarbenoid (from CH_2Cl_2 and methyllithium) have recently been reported. A

* See Addenda, number 6.

pattern similar to that of the CCl_2 additions was followed, although the stereoselectivity of "CHCl" had to be considered. The carbenoid added preferentially to *trans* double bonds in these systems, the *trans*/*cis* preferences were 1.51, 1.59, and 1.47 for the monoene, diene, and triene, respectively. *Endo*-chloro addition was very marked in additions to the *cis* double bonds. With 1-*cis*-5-*trans*-cyclodecadiene, an even higher *trans*/*cis* preference, 4.25, was observed.[44c]

These results contrast dramatically with the behavior of CCl_2, chlorocarbenoid, and CFCl toward acyclic alkenes, where the *cis* olefins react more rapidly than their *trans* isomers (T24, T34, T36, T38A, T40, T53). CF_2, however, reacts more rapidly with *trans*-butene than with *cis*-butene (T31). Data do not appear to be available for CBr_2.

A complete rationale for these effects must *inter alia* consider the ground-state energy difference between the *cis* and *trans* double bond, as well as how much this inherent difference is altered in the transition state. (The latter will vary with the selectivity of the carbene.) Other factors, such as the electronic interactions which control stereoselectivity,[57] are also important. It is possible that relief of sp^2–sp^2 torsional strain may at least partly account for the greater reactivity toward CCl_2 of *trans* middle-sized cycloalkenes, as opposed to their *cis* isomers.[44d] Such an explanation appears to hold for related epoxidation reactions.[112a]

Relief of strain may also be a factor in acyclic alkene *cis*, *trans* competitions; the reactivity ordering here varies with the selectivity of the carbene. Poorly selective species show slight preferences for *trans* isomers: CH_2 (T1), ArCH (T13) and HCX (T23–T25). More selective carbenes kinetically prefer the more energetic *cis* isomers: C_6H_5CX (T26, T28, T30), CH_3CCl (T27), CCl_2, and CFCl (see above). The most selective carbene tabulated, CF_2, shows a slight preference for *trans*-butene over *cis*-butene (T31). These trends suggest a possible correlation of *cis*/*trans* relative reactivity and the extent of bond-making in the transition state. Finally, carbenoids invariably select *cis* alkenes (acyclic) over their *trans* isomers. Certain other carbenes, in which specific steric effects attend the addition reaction, have been discussed previously.

Enynes such as **61** add CCl_2 mainly at the triple bond,[113] but **62**,[113] **63**,[114] **64**,[114] and **65**[113] are attacked mainly at the double bond. A steric effect could be operative in **61**, favoring triple bond attack.

A more recent account of these and related reactions generalizes that with conjugated enynes carrying a substituent in the 2-position, two or there substituents on the double bond, or a free acetylenic hydrogen, addition to the double bond predominates.[115] With conjugated enynes carrying one or no substituents on the 1-position, and having an internal acetylenic bond, addition to the triple bond predominates, but isolated yields are poor. With

$R = C_6H_5, (CH_3)_3C$

61

$CH_3C{\equiv}C{-}C(CH_3){=}CH_2$

62

$HC{\equiv}C{-}C(CH_3){=}CHCH_3$

63

$HC{\equiv}C{-}C(C_2H_5){=}CH_2$

64

$C_6H_5{-}C{\equiv}C$–(1-cyclohexenyl)

65

nonconjugated, internal triple bonds, the reactivity toward CCl_2 is slightly higher than that of a terminal double bond. Further details and earlier literature citations can be found in reference 115. CCl_2 adds to the conjugated ene-allene system of 2-methylocta-1,3,4-triene at the ene position (50%); attack on the allene sites amounts to no more than 10%.[115a]

With the conjugated *cis*-hex-2-ene-4-yne, carboethoxycarbene (Cu carbenoid) prefers attack at the triple bond by a factor of 1.6 (T19A); with **62**, however, the double bond appears to be more reactive.[27b] These selectivities are in line with the foregoing general analysis of the additions of CCl_2 to conjugated eneynes.[115]

Toward catalytically (Cu) generated carboethoxycarbenoid, **66** reacts

$CH_3C{\equiv}C{-}CH_2{-}C(CH_3){=}CH_2$

66

mainly at the double bond; selectivity 3:1.[116] (CCl_2 shows an even greater selectivity in the same direction.[115]) A recent report indicates that free carboethoxycarbene (photogenerated from ethyl diazoacetate at 15°) adds more rapidly to *cis*-2-hexene than to 2-hexyne, and more rapidly to *trans*-4-octene than to 4-octyne (factors: 1.4 and 7.7, respectively). However, the corresponding carbenoid (Cu or $CuSO_4$, 80°) selects in favor of the 4-octyne (factor: 3.0) and also adds more rapidly to 2-hexyne than to *trans*-2-hexene (factor: 5.6).[117] Similar preference for triple bond attack is shown by carboethoxycarbene (Cu carbenoid) in T19A, addition to 2-hexyne being more rapid than additions to cyclohexene or *cis*-2-hexene.

The carbenoid is a bulkier species than the free carbene and is also less reactive, suggesting a tighter and more sterically demanding addition transition state. Therefore its preference for the triple bond over the double bond could be steric in origin. In the absence of dominant steric effects, one would expect an electronic preference for a selective carbene to attack a double bond more rapidly than a triple bond, since the transition state for the latter reaction would have unfavorable vinyl cation character **67**.

$$\underset{\underset{\overline{}}{CX_2}}{R-\overset{+}{C}=C-R} \longleftrightarrow R-C=\overset{+}{C}-R \quad (CX_2^-)$$

67

There are also reports of carbene attack on olefins which carry heteroatomic substituents. In these reactions, the double bond comes into competition with the lone electron pairs of the heteroatom. Addition to the former affords a cyclopropane; addition to the latter affords an ylid, which can then undergo further reactions. Examples of these reactions include CH_2 (copper carbenoid) with allyl halides,[118] CH_2 (copper carbenoid) and CCl_2 with allyl amines and related substrates,[119,120] CH_2 (copper carbenoid) with allyl sulfides, allyl ethers, and vinyl oxirane,[121,122] and CCl_2 with allyl sulfides.[123] There are also several studies of intermolecular competitions between olefins and heteroatomic substrates. Thus it is reported that tetramethylethylene and triphenylphosphine are of comparable reactivity toward CFCl.[124] Dicarbomethoxycarbene (singlet, photolytically generated) reacts more rapidly with the sulfur atom of dimethyl sulfide than with the cyclohexene double bond (factor of ~4); other olefins behave similarly.[125] This reaction has also been studied with a sole allyl sulfide substrate. The results indicate that the singlet carbene prefers to attack the sulfur atom over the double bond; the copper carbenoid exclusively attacks the sulfur atom; and the triplet carbene prefers addition to the double bond over attack on sulfur.[126] A related study is reported with allylic halide substrates. Singlet dicarbomethoxycarbene and carboethoxycarbene prefer attack on halogen over addition to the double bond. This behavior is reversed for the triplet carbenes. Some quantitative data are presented in Table 56.[127] The data are derived from competitions of pairs of allylic compounds. The singlet carbene attacks allylic Br and S about twice as fast as it attacks allylic Cl. However, both singlet and triplet carbenes add to the olefinic bonds at rates which are nearly independent of the allylic substituent.

A detailed study of the allyl chloride dicarbomethoxycarbene reaction has been made as a function of added solvents such as cyclohexene, dichloromethane, dibromomethane, and diiodomethane.[128] These diluents enhance addition to the double bond, relative to attack on chlorine, presumably because they catalyze the singlet to triplet interconversion of the carbene. A "heavy atom" effect may be operative in the cases of the dibromo and diiodomethane dilutions.

TABLE 56

Competitive Reactions of Singlet and Triplet Dicarbomethoxycarbene with Allyl Compounds

Allyl compound	Singlet	Triplet
CH_2=CH—CH_2Cl (addition)	1.00[a]	1.00[a]
(insertion)	2.3	0.07
CH_2=CH—CH_2Br (addition)	1.0	1.0
(insertion)	8.5	0.3
CH_2=CH—$CH_2SC_2H_5$ (addition)	1.2	1.0
(insertion)	6.3	0.5

[a] Standard, defined as 1.00.

A study of the reaction of carboethoxycarbene with allylic sulfides, ethers, and chlorides has also appeared.[129] The conclusions, in general, were similar to those reached in the dicarbomethoxycarbene studies. Of interest was the observation that the tendency for singlet carboethoxycarbene attack on the heteroatom, as opposed to the double bond, increased in the order heteroatom = Cl < O < S. This might suggest that heteroatom nucleophilicity was important, and that the carbene's attack on the heteroatom was that of an electrophile.*

More detailed discussion of the heteroatom-double bond competition is beyond the scope of this chapter; the reader is referred to the citations for amplification and further references.

VI. ADDENDA

The following noteworthy results have been added in proof:

1. Bromocarbenoids can be generated by the action of metal bis-trimethylsilylamides on methylene bromide and added to olefins in fair yields to give bromocyclopropanes. Relative reactivities are reported for the additions of $NaCHBr_2$ (and $LiCHBr_2$) to the following: trimethylethylene, 2.3 (3.0); *cis*-2-pentene, 1.20 (1.1); cyclooctene, 1.65; **cyclohexene, 1.00**; and 1-pentene, 0.68 (0.45). *Syn*-bromo stereoselectivity was noted for the sodium carbenoid's addition to *cis*-2-pentene (3.0:1) and cyclohexene (1.5:1). The selectivity behavior of these bromocarbenoids, compared with that of bromocarbene (T25) is similar to that of chlorocarbenoid, as compared with chlorocarbene (T24). [B. Martel and J. M. Hiriart, *Angew. Chem., Int. Ed.*, **11**, 326 (1972); *Synthesis*, 201 (1972).]

* See Addenda, number 7.

2. Further relative reactivity studies have now appeared concerning the additions of dimethylethylidene and dimethyl vinylidene carbenes to *p*-substituted styrenes. $(CH_3)_2C{=}C{=}C$: was generated by the action of potassium *t*-butoxide on 1,1-dimethyl-3-bromoallene (−10°, *t*-butanol). $(CH_3)_2C{=}C$: was generated either by the action of ethereal methyllithium on $(CH_3)_2C{=}CBr_2$ (−40°) or of potassium *t*-butoxide (*t*-butanol, −10°) on $(CH_3)_2C{=}CHBr$. Substrates included styrene, its *p*-chloro and *p*-methyl derivatives; ρ values (σ^+) were −0.95, −4.3 and −4.3, respectively. These are the largest ρ values reported for carbenic additions to styrenes and bespeak electrophilic additions proceeding through highly polarized transition states. Even allowing for temperature effects on ρ, these additions appear "more electrophilic" than those of CCl_2, which, in the case of the apparently unstabilized dimethyl ethylidene carbene, is most surprising. One cannot easily dismiss this observation by claiming that the species is a carbenoid of undetermined structure, which adds via a highly polarized transition state, for it shows a substantial independence of generative source. Nevertheless, this explanation is not totally excluded, because even the generation of dimethylethylidene carbene from 5,5-dimethyl-*N*-nitrosooxazolidone (ρ = −3.4)[15] involves lithium cations, and in some cases, lithium and other metal carbenoids can display large electrophilic selectivities[21]. Note that the electrophilic behavior of $(CH_3)_2C{=}C$: toward styrenes substantiates the idea that its low reactivity toward tetraalkylated olefins (see T8) is a specific steric effect. [T. B. Patrick, E. C. Haynie, and W. J. Probst, *J. Org. Chem.*, **37,** 1553 (1972).]

3. Wishes sometimes come true. Cycloheptatrienylidene has now been added to styrene and derivatives (photolysis of the sodium salt of tropone tosylhydrazone in THF-olefin) affording the following relative reactivities: *p*-methoxy (styrene), 0.51; *p*-methyl, 0.57; **styrene, 1.00;** *p*-chloro, 1.59; *p*-bromo, 1.68; *m*-bromo, 2.22. A Hammett relation (σ) gave

$$\rho = +1.05 \pm 0.05$$

(correlation coefficient 0.982). This first positive ρ for a carbene addition is powerful evidence that cycloheptatrienylidene is indeed the key intermediate and that it is *nucleophilic* in selectivity (cf. structure **37**). [L. W. Christensen, E. E. Waali, and W. M. Jones, *J. Amer. Chem. Soc.*, **94,** 2118 (1972).]

4. Further relative reactivity data pertaining to this question have now been obtained. For addition of CCl_2 (from $C_6H_5HgCCl_2Br$, 80–85°) we find: **cyclohexene, 1.00;** cyclohexene-4-one ethylene acetal, 0.44; 1-carboethoxycyclohexene, 0.11; 2-carboethoxycyclohexene-4-one ethylene acetal, 0.11; 1-carboethoxycyclohexene-4-one ethylene acetal, 0.028; cyclohexene-3-one ethylene acetal, 0.012. No evidence is found here for cooperative or synergistic addition of CCl_2 to these olefins, mediated by the ethylene acetal

function; but the data can be analyzed in terms of substituent effects that operate mainly on an highly unsymmetrical transition state for CCl_2 addition. [R. A. Moss, *J. Amer. Chem. Soc.*, **94,** 6004 (1972).]

5. A similar effect has been reported for CCl_2 addition to **57** (R = C_2H_5O and C_2H_5S). [J-P. Gouesnard, *Compt. Rend. Acad. Sci.* (C), **274,** 1407 (1972).]

6. Kostikov et al., have now issued a somewhat augmented and unified study of the relative reactivities for CCl_2 additions to polyenes and related molecules (cf. T48, T48A, T48B). CCl_2 was generated from ethyl trichloroacetate and sodium methoxide, in hexane, at 0°. The reactivities, relative to styrene, are the following: all *trans*-1-phenylocta-1,3,5,7-tetraene, 2.02 (7,8-bond) and 0.20 (1,2-bond); all *trans*-1-phenylhexa-1,3,5-triene, 2.02 (5,6-bond) and 0.16 (1,2-bond); *trans*-1-phenyl-1,3-butadiene, 1.87 (3,4-bond) and 0.12 (1,2-bond); *trans*-1-phenylpropene, 1.53; *trans*-1-phenyl-1-butene, 1.39; *trans*-1-phenyl-1-hexene, 1.17; *trans*-1-phenyl-1-octene, 1.07; **styrene, 1.00;** *trans*-1-phenyl-3-hexene, 0.40; 8-phenyl-1-octene, 0.113; 6-phenyl-1-hexene, 0.109; 3-phenyl-1-butene, 0.095; 3-phenyl-1-propene, 0.067. Trends include those previously noted (see above). Moreover, there is a regular increase in C=C reactivity with an increase in calculated free valence indices or with decreases in calculated bond orders and localization energies. These observations lend quantitative support to analyses that simply examine the extent of conjugation in the alternative transition states for CCl_2 additions at various sites of the polyenes. An inductive deactivating effect of a non-conjugated phenyl group is to be noted with some of the substrates. [R. R. Kostikov, A. P. Molchanov, and I. A. D'yakonov, *J. Org. Chem. USSR*, **7,** 2297 (1971).]

7. Additional studies of the reactions of dicarbomethoxycarbene with allylic sulfides, ethers, amines, and halides, and with alkyl sulfides and disulfides have now appeared. The results are supportive of the previously reported trends. [W. Ando et al., *J. Org. Chem.*, **37,** 1721 (1972); *J. Amer. Chem. Soc.*, **94,** 3870 (1972).]

References

1. J. Hine, *Divalent Carbon*, Ronald Press, New York, 1964.
2. W. von E. Doering and A. K. Hoffmann, *J. Am. Chem. Soc.*, **76,** 6162 (1954).
3. P. S. Skell and A. Y. Garner, *J. Am. Chem. Soc.*, **78,** 5430 (1956).
4. W. von E. Doering and W. A. Henderson, Jr., *J. Am. Chem. Soc.*, **80,** 5274 (1958).
5. W. J. R. Tyerman, *Trans. Faraday Soc.*, **65,** 1188 (1969), and references therein.

5a. Y. Yamamoto, I. Moritani, Y. Maeda, and S. Murahashi, *Tetrahedron*, **26,** 251 (1970).

6. C. K. Ingold and M. S. Smith, *J. Chem. Soc.*, **1938,** 905.
7. I. H. Sadler, *J. Chem. Soc.* (*B*), **1969,** 1024.
8. S. Krzyzanowski and R. J. Cvetanović, *Can. J. Chem.*, **45,** 665 (1967).

8a. W. Braun, A. M. Bass, and M. Pilling, *J. Chem. Phys.*, **52,** 5131 (1970).
8b. C. McKnight, P. S. T. Lee, and F. S. Rowland, *J. Am. Chem. Soc.*, **89,** 6802 (1967).
8c. H. M. Frey and R. Walsh, *Chem. Commun.*, **1969,** 158.
8d. D. C. Montague and F. S. Rowland, *J. Phys. Chem.*, **72,** 3705 (1968).
9. E. P. Blanchard and H. E. Simmons, *J. Am. Chem. Soc.*, **86,** 1337 (1964).
9a. C. Fauveau, Y. Gault, and F. G. Gault, *Tetrahedron Lett.*, **1967,** 3149.
10. B. Rickborn and J. H-H. Chan, *J. Org. Chem.*, **32,** 3576 (1967).
11. J. H-H. Chan and B. Rickborn, *J. Am. Chem. Soc.*, **90,** 6406 (1968).
11a. U. Burger and R. Huisgen, *Tetrahedron Lett.*, **1970,** 3057; see also U. Burger and R. Huisgen, *Tetrahedron Lett.*, **1970,** 3049, and R. Huisgen and U. Burger, *Tetrahedron Lett.*, **1970,** 3053.
11b. L. Friedman, R. S. Honour, and J. G. Berger, *J. Am. Chem. Soc.*, **92,** 4640 (1970).
12. D. Seyferth, R. M. Turkel, M. A. Eisert, and L. J. Todd, *J. Am. Chem. Soc.*, **91,** 5027 (1969).
12a. J. Nishimura, J. Furukawa, N. Kawabata, and M. Kitayama, *Tetrahedron*, **27,** 1799 (1971).
12b. H. Nozaki, H. Takaya, S. Moriuti, and R. Noyori, *Tetrahedron*, **24,** 3655 (1968).
13a. P. S. Skell and R. R. Engel, *J. Am. Chem. Soc.*, **88,** 3749 (1966).
13b. P. S. Skell and R. R. Engel, *J. Am. Chem. Soc.*, **87,** 2493 (1965).
13c. P. S. Skell, J. E. Villaume, J. H. Plonka, and F. A. Fagone, *J. Am. Chem. Soc.*, **93,** 2699 (1971).
14. W. M. Jones, M. H. Grasley, and W. S. Brey, Jr., *J. Am. Chem. Soc.*, **85,** 2754 (1963).
15. M. S. Newman and T. B. Patrick, *J. Am. Chem. Soc.*, **91,** 6461 (1969).
15a. M. S. Newman and T. B. Patrick, *J. Am. Chem. Soc.*, **92,** 4312 (1970).
16a. H. D. Hartzler, *J. Org. Chem.*, **29,** 1311 (1964).
16b. H. D. Hartzler, *J. Am. Chem. Soc.*, **83,** 4997 (1961).
16c. W. J. le Noble, Y. Tatsukami, and H. F. Morris, *J. Am. Chem. Soc.*, **92,** 5681 (1970).
16d. D. J. Northington and W. M. Jones, *Tetrahedron Lett.*, **1971,** 317.
16e. H. D. Hartzler, *J. Am. Chem. Soc.*, **93,** 4527 (1971).
17a. D. G. Williamson and K. D. Bayes, *J. Am. Chem. Soc.*, **90,** 1957 (1968).
17b. C. Devillers, *Comp. Rendu*, **262C,** 1485 (1966).
17c. J. F. Olsen and L. Burnelle, *Tetrahedron*, **25,** 5451 (1969).
18. R. A. Moss, *J. Org. Chem.*, **31,** 3296 (1966).
19. M. Jones, Jr., A. M. Harrison, and K. R. Rettig, *J. Am. Chem. Soc.*, **91,** 7462 (1969).
19a. R. H. Levin and M. Jones, Jr., *Tetrahedron*, **27,** 2031 (1971).
20. G. L. Closs and R. A. Moss, *J. Am. Chem. Soc.*, **86,** 4042 (1964); R. A. Moss, Ph.D. thesis, University of Chicago, 1963. Data are adapted from the latter.
21. S. H. Goh, L. E. Closs, and G. L. Closs, *J. Org. Chem.*, **34,** 25 (1969).
22. R. A. Moss, *J. Org. Chem.*, **30,** 3261 (1965).
23. Prof. I. Moritani, private communication.
24. R. M. Etter, H. S. Skovronek, and P. S. Skell, *J. Am. Chem. Soc.*, **81,** 1008 (1959).
25. U. Schöllkopf and M. Reetz, *Tetrahedron Lett.*, **1969,** 1541.
26. P. S. Skell and R. M. Etter, *Chem. Ind.*, **1958,** 624.
26a. R. M. Etter, Ph.D. thesis, Pennsylvania State University, 1959.
27. T. C. Neil, Ph.D. thesis, Pennsylvania State University, 1964.
27a. R. N. Gmyzina, I. A. D'yakonov, and L. P. Danilkina, *J. Org. Chem. USSR*, **6,** 2168 (1970).
27b. I. A. D'yakonov and L. P. Danilkina, *Zh. Obsch. Khim.*, **34,** 738 (1964).
28. M. Jones, Jr., A. Kulczycki, Jr., and K. F. Hummel, *Tetrahedron Lett.*, **1967,** 183.
29. M. Jones, Jr., W. Ando, and A. Kulczycki, Jr., *Tetrahedron Lett.*, **1967,** 1391.

30. U. Schöllkopf and H. Görth, *Ann.*, **709,** 97 (1967).
30a. K. Riedel, Dissertation, University of Göttingen, 1968.
31. Y-N. Tang and F. S. Rowland, *J. Am. Chem. Soc.*, **89,** 6420 (1967).
32. G. L. Closs and G. M. Schwartz, *J. Am. Chem. Soc.*, **82,** 5729 (1960).
32a. G. L. Closs and L. E. Closs, *J. Am. Chem. Soc.*, **82,** 5723 (1960); G. L. Closs, R. A. Moss, and J. J. Coyle, *J. Am. Chem. Soc.*, **84,** 4985 (1962).
33. G. L. Closs and J. J. Coyle, *J. Am. Chem. Soc.*, **87,** 4270 (1965).
34. R. A. Moss and J. R. Przybyla, *Tetrahedron*, **25,** 647 (1969).
34a. R. A. Moss and A. Mamantov, *J. Am. Chem. Soc.*, **92,** 6951 (1970).
34b. J. M. Bollinger, private communication, December 9, 1970; J. M. Bollinger, J. Brinich, and G. A. Olah, *J. Am. Chem. Soc.*, **92,** 4025 (1970); G. A. Olah and J. M. Bollinger, *J. Am. Chem. Soc.*, **90,** 6082 (1968).
35. R. A. Moss, J. R. Whittle, and P. Freidenreich, *J. Org. Chem.*, **34,** 2220 (1969).
36. R. A. Moss and R. Gerstl, *Tetrahedron*, **22,** 2637 (1966).
37. R. A. Moss, *Tetrahedron Lett.*, **1967,** 4905.
38. R. A. Mitsch and A. S. Rodgers, *Int. J. Chem. Kinet.*, **1,** 439 (1969).
39. D. Seyferth and H. Dertouzos, *J. Organometal. Chem.*, **11,** 263 (1968).
40. D. Seyferth, M. E. Gordon, J. Y-P. Mui, and J. M. Burlitch, *J. Am. Chem. Soc.*, **89,** 959 (1967).
41. O. M. Nefedov and R. N. Shafran, *Zh. Obsch. Khim.*, **37,** 1561 (1967).
42a. R. A. Moss and R. Gerstl, *J. Org. Chem.*, **32,** 2268 (1967).
42b. R. A. Moss and A. Mamantov, unpublished work.
43. R. A. Moss and A. Mamantov, *Tetrahedron Lett.*, **1968,** 3425.
43a. C. M. Starks, *J. Am. Chem. Soc.*, **93,** 195 (1971).
44. P. S. Skell and M. S. Cholod, *J. Am. Chem. Soc.*, **91,** 7131 (1969).
44a. P. S. Skell and M. S. Cholod, *J. Am. Chem. Soc.*, **91,** 6035 (1969).
44b. G. Köbrich, H. Büttner, and E. Wagner, *Angew. Chem., Int. Ed.*, **9,** 169 (1970).
44c. G. A. Lodoen, Ph.D. thesis, Cornell University, 1969; *Diss. Abst.*, **30(B),** 4050 (1970).
44d. J. Graefe, M. Mühlstädt, and P. Kuhl, *Z. Chem.*, **10,** 192 (1970).
44e. J. Graefe, M. Mühlstädt, B. Bayerl, and W. Engewald, *Tetrahedron*, **26,** 4199 (1970).
44f. J. F. Harrison, *J. Am. Chem. Soc.*, **93,** 4112 (1971).
45. D. Seyferth and J. M. Burlitch, *J. Am. Chem. Soc.*, **86,** 2730 (1964).
46. D. Seyferth, J. Y-P. Mui, and R. Damrauer, *J. Am. Chem. Soc.*, **90,** 6182 (1968).
47. O. M. Nefedov, M. N. Manakov, and A. A. Ivashenko, *Izv. Akad. Nauk. SSSR. Otd. Khim. Nauk.*, **1962,** 1242.
48. O. M. Nefedov and R. N. Shafran, *Izv. Akad. Nauk. SSSR. Ser. Khim.*, **1965,** 538.
49. A. Bezaguet and M. Bertrand, *Comp. Rend., Sér. C*, **262,** 428 (1966).
50. A. Ledwith and H. J. Woods, *J. Chem. Soc. (B)*, **1967,** 973.
51. J. Cudlín and V. Chvalovský, *Coll. Czech. Chem. Comm.*, **28,** 3088 (1963).
52. J. Cudlín and V. Chvalovský, *Coll. Czech. Chem. Comm.*, **27,** 1658 (1962).
53. V. S. Aksenov, I. A. D'yakonov, and R. R. Kostikov, *J. Org. Chem. USSR*, **4,** 1680 (1968).
53a. R. R. Kostikov and A. Ya. Bespalov, *J. Org. Chem. USSR*, **6,** 629 (1970).
53b. R. R. Kostikov and A. P. Molchanov, *J. Org. Chem. USSR*, **6,** 628 (1970).
53c. R. R. Kostikov and A. P. Molchanov, *J. Org. Chem. USSR*, **7,** 415 (1971).
54. M. Jones, Jr., W. H. Sachs, A. Kulczycki, Jr., and F. J. Waller, *J. Am. Chem. Soc.*, **88,** 3167 (1966).
55. R. A. Moss and R. Gerstl, *Tetrahedron*, **23,** 2549 (1967).
56. W. Funasaka, T. Ando, H. Yamanaka, H. Kanehira, and Y. Shimokawa, *Symposium on Organic Halogen Compounds*, Tokyo, Nov. 29, 1967, *Abstracts*, pp. 25ff.

57. For a review of "Steric Selectivity in the Addition of Carbenes to Olefins," see R. A. Moss in *Selective Organic Transformations*, Vol. 1, B. S. Thyagarajan, Ed., Interscience, New York, 1970, pp. 35ff.
58. R. W. Taft, Jr., in *Steric Effects in Organic Chemistry*, M. S. Newman, Ed., Wiley, New York, 1956, pp. 556ff.
59. For discussion of this method, see G. Mouvier and J. E. DuBois, *Bull. Soc. Chim. France*, **1968,** 1441.
60. H. E. Simmons, E. P. Blanchard, and R. D. Smith, *J. Am. Chem. Soc.*, **86,** 1347 (1964).
61. E. W. Garbisch, Jr., S. M. Schildcrout, D. B. Patterson, and C. M. Sprecher, *J. Am. Chem. Soc.*, **87,** 2932 (1965).
62. C. D. Poulter, E. C. Friedrich, and S. Winstein, *J. Am. Chem. Soc.*, **91,** 6892 (1969).
62a. G. Zweifel, G. M. Clark, and C. C. Whitney, *J. Am. Chem. Soc.*, **93,** 1305 (1971).
63. W. R. Moser, *J. Am. Chem. Soc.*, **91,** 1135, 1141 (1969), and references therein.
64. W. H. Atwell, D. R. Weyenberg, and J. G. Uhlmann, *J. Am. Chem. Soc.*, **91,** 2025 (1969).
65. P. S. Skell and J. H. Plonka, *J. Am. Chem. Soc.*, **92,** 2160 (1970).
66. R. Hoffmann, *J. Am. Chem. Soc.*, **90,** 1475 (1968).
66a. R. Hoffmann, D. M. Hayes, and P. S. Skell, *J. Phys. Chem.*, **76,** 664 (1972).
67. W. R. Moore, W. R. Moser, and J. E. LaPrade, *J. Org. Chem.*, **28,** 2200 (1963).
68. D. Seyferth, R. Damrauer, J. Y-P. Mui, and T. F. Jula, *J. Am. Chem. Soc.*, **90,** 2944 (1968).
68a. J. Nishimura, J. Furukawa, and N. Kawabata, *J. Organomet. Chem.*, **29,** 237 (1971).
68b. D. Seyferth, R. Damrauer, R. M. Turkel, and L. J. Todd, *J. Organomet. Chem.*, **17,** 367 (1969).
69. Y. Ito, M. Okano, and R. Oda, *Tetrahedron*, **22,** 2615 (1966).
70. A. G. Kirk, *Diss. Abst.*, **28,** 3222B (1968).
71. J. A. Landgrebe and A. G. Kirk, *J. Org. Chem.*, **32,** 3499 (1967).
72. C. G. Moseley, *Diss. Abst.*, **28,** 2351B (1967).
73. P. B. Sargent and H. Schechter, *Tetrahedron Lett.*, **1964,** 3957.
74. J. E. Baldwin and R. A. Smith, *J. Am. Chem. Soc.*, **89,** 1886 (1967).
74a. J. Nishimura, J. Furukawa, N. Kawabata, and T. Fujita, *Tetrahedron*, **26,** 2229 (1970).
75. R. J. Cvetanović, *Can. J. Chem.*, **38,** 1678 (1960), and references therein.
75a. E. V. Couch and J. A. Landgrebe, *J. Org. Chem.* **37,** 1251 (1972).
75b. H. H. Gibson, Jr., J. R. Cast, J. Henderson, C. W. Jones, B. F. Cook, and J. B. Hunt, *Tetrahedron Lett.*, **1971,** 1827.
76. W. von E. Doering and L. H. Knox, *J. Am. Chem. Soc.*, **83,** 1989 (1961).
77. J. Hine, *Divalent Carbon*, Ronald Press, New York, 1964.
78. R. Hoffmann, G. D. Zeiss, and G. W. Van Dine, *J. Am. Chem. Soc.*, **90,** 1485 (1968).
79. R. Gleiter and R. Hoffmann, *J. Am. Chem. Soc.*, **90,** 5457 (1968).
80. W. Kirmse, *Carbene, Carbenoide und Carbenanaloge*, Verlag Chemie, GmbH, Weinheim, Germany, 1969, pp. 78–81.
81. D. M. Demal, E. P. Gosselink, and S. D. McGregor, *J. Am. Chem. Soc.*, **88,** 582 (1966), and references therein; see also R. W. Hoffmann, *Angew. Chem. Int. Ed.*, **10,** 529 (1971).
82. U. Schöllkopf and E. Wiskott, *Ann.*, **694,** 44 (1966); *Angew. Chem. Int. Ed.*, **2,** 485 (1963).
83. D. M. Lemal and E. H. Banitt, *Tetrahedron Lett.*, **1964,** 245.
83a. J. M. Brown and B. D. Place, *Chem. Commun.*, **1971,** 533.
83b. A. G. Brook, H. W. Kucera, and R. Pearce, *Can. J. Chem.*, **49,** 1618 (1971).

83c. A. Krebs and H. Kimling, *Angew. Chem. Int. Ed.*, **10,** 409 (1971).
84. W. M. Jones and J. M. Denham, *J. Am. Chem. Soc.*, **86,** 944 (1964).
84a. W. M. Jones and M. E. Stowe, *Tetrahedron Lett.*, **1964,** 3459.
85. S. D. McGregor and W. M. Jones, *J. Am. Chem. Soc.*, **90,** 123 (1968).
86. W. M. Jones, M. E. Stowe, E. E. Wells, Jr., and E. W. Lester, *J. Am. Chem. Soc.*, **90,** 1849 (1968).
87. W. M. Jones and C. L. Ennis, *J. Am. Chem. Soc.*, **89,** 3069 (1967).
88. T. Mukai, T. Nakazawa, and K. Isobe, *Tetrahedron Lett.*, **1968,** 565.
89. W. M. Jones, B. N. Hamon, R. C. Joines, and C. L. Ennis, *Tetrahedron Lett.*, **1969,** 3909.
90. W. M. Jones and C. L. Ennis, *J. Am. Chem. Soc.*, **91,** 6391 (1969).
91. P. S. Skell and R. C. Woodworth, *J. Am. Chem. Soc.*, **78,** 4496 (1956).
92. G. L. Closs, in *Topics in Stereochemistry*, Vol. 3, E. L. Eliel and N. L. Allinger, Eds., Interscience, New York, 1968, pp. 193ff; P. P. Gaspar and G. S. Hammond, in *Carbene Chemistry*, W. Kirmse, Academic Press, New York, 1964, pp. 235ff.
93. J. V. Gramas, *Diss. Abst.*, **26,** 4235 (1966).
93a. R. A. Bernheim, R. J. Kempf, J. V. Gramas, and P. S. Skell, *J. Chem. Phys.*, **43,** 196 (1965), and references therein.
94. A. Sonoda, I. Moritani, T. Saraie, and T. Wada, *Tetrahedron Lett.*, **1969,** 2943.
95. G. Köbrich, *Angew. Chem. Int. Ed.*, **6,** 41 (1967); W. Kirmse, *Angew. Chem. Int. Ed.*, **4,** 1 (1965).
96. G. L. Closs and L. E. Closs, *Angew. Chem.*, **74,** 431 (1962).
97. S. Murahashi and I. Moritani, *Tetrahedron*, **23,** 3631 (1967); I. Moritani, S-I., Murahashi, K. Yoshinaga, and H. Ashitaka, *Bull. Chem. Soc. Japan*, **40,** 1506 (1967).
98. G. Köbrich, K. Flory, and R. H. Fischer, *Chem. Ber.*, **99,** 1793 (1966).
99. D. Seyferth, J. Y-P. Mui, and J. M. Burlitch, *J. Am. Chem. Soc.*, **89,** 4953 (1967).
100. E. Vogel and H. D. Roth, *Angew. Chem.*, **76,** 145 (1964).
101. J. J. Sims and V. K. Honwad, *J. Org. Chem.*, **34,** 496 (1969).
102. P. H. Nelson and K. G. Untch, *Tetrahedron Lett.*, **1969,** 4475.
103. E. Vogel, W. Wiedemann, H. Kiefer, and W. F. Harrison, *Tetrahedron Lett.*, **1963,** 673.
104. E. Vogel, private communication to P. S. Skell, quoted in reference 44a, footnote 22; see also W. Grimme, J. Reisdorff, W. Jünemann, and E. Vogel, *J. Am. Chem. Soc.*, **92,** 6335 (1970).
104a. M. A. Tobias and B. E. Johnson, *Tetrahedron Lett.*, **1970,** 2703.
104b. D. Seyferth and V. A. Mai, *J. Am. Chem. Soc.*, **92,** 7412 (1970).
104c. R. S. Matthews and T. E. Meteyer, *Chem. Commun.*, **1971,** 1576.
105. W. J. Ball, S. R. Landor, and N. Punja, *J. Chem. Soc.* (*C*), **1967,** 194.
105a. P. Battioni, L. Vo-Quang, and Y. Vo-Quang, *Bull. Soc. Chim. France*, **1970,** 3938.
105b. P. Battioni, L. Vo-Quang, and Y. Vo-Quang, *Bull. Soc. Chim. France*, **1970,** 3942.
105c. R. Martin and M. Bertrand, *Bull. Soc. Chim. France*, **1970,** 2261.
105d. D. R. Paulson, J. K. Crandall, and C. A. Bunnell, *J. Org. Chem.*, **35,** 3708 (1970).
105e. E. H. Dehmlow and G. C. Ezimora, *Tetrahedron Lett.*, **1970,** 4047.
106. L. Skattebøl, *J. Org. Chem.*, **29,** 2951 (1964).
106a. E. W. Duck, J. M. Locke, and S. R. Wallis, *J. Chem. Soc.* (*C*), **1970,** 2000.
107. P. Weyerstahl, D. Klamann, M. Figge, C. Finger, F. Nerdel, and J. Buddrus, *Ann.*, **710,** 17 (1968).
108. V. A. Kalinina and Yu. I. Kheruze, *J. Org. Chem. USSR*, **4,** 1347 (1968).
108a. V. A. Kalinina, Yu. I. Kheruze, and A. A. Petrov, *J. Org. Chem. USSR*, **3,** 637 (1967).
108b. I. S. Lishanskii, A. M. Guliev, V. I. Pomerantsev, L. D. Turkova, and A. S. Khachaturov, *J. Org. Chem. USSR*, **6,** 918 (1970).

109. I. A. D'yakonov, T. A. Kornilova, and T. V. Nizovkina, *J. Org. Chem. USSR*, **3,** 272 (1967).
110. I. A. D'yakonov and T. A. Kornilova, *J. Org. Chem. USSR.*, **5,** 178 (1969).
111. J. Graefe and M. Mühlstädt, *Tetrahedron Lett.*, **1969,** 3431.
112. A. C. Cope, W. R. Moore, R. D. Bach, and H. J. S. Winkler, *J. Am. Chem. Soc.*, **92,** 1243 (1970), note 28.
112a. F. H. Allen, E. D. Brown, D. Rogers, and J. K. Sutherland, *Chem. Commun.*, **1967,** 1116.
113. E. V. Dehmlow, *Tetrahedron Lett.*, **1966,** 3763.
114. L. P. Danilkina, I. A. D'yakonov, and G. I. Roslovtseva, *J. Org. Chem. USSR*, **1,** 465 (1965).
115. E. V. Dehmlow, *Chem. Ber.*, **101,** 427 (1968).
115a. Ya. M. Slobodin and I. Z. Égenburg, *J. Org. Chem. USSR*, **6,** 188 (1970).
116. I. A. D'yakonov, L. N. Danilkina, and R. N. Gmyzina, *J. Org. Chem. USSR*, **5,** 1026 (1969).
117. I. A. D'yakonov, M. T. Komendantov, L. P. Danilkina, R. N. Gmyzina, T. S. Smirnov, and A. G. Vitenberg, *J. Org. Chem. USSR*, **5,** 383 (1969).
118. W. Kirmse and M. Kapps, *Angew. Chem. Int. Ed.*, **4,** 691 (1965).
119. W. Kirmse and H. Arold, *Chem. Ber.*, **101,** 1008 (1968).
120. W. E. Parham and J. R. Potoski, *J. Org.. Chem.*, **32,** 275, 278 (1967).
121. W. Kirmse and M. Kapps, *Chem. Ber.*, **101,** 994 (1968).
122. M. Kapps and W. Kirmse, *Angew. Chem., Int. Ed.*, **8,** 75 (1969).
123. W. E. Parham and S. H. Groen, *J. Org. Chem.*, **31,** 1694 (1966).
124. D. J. Burton and H. C. Krutzsch, *Tetrahedron Lett.*, **1968,** 71; *J. Org. Chem.*, **35,** 2125 (1970).
125. W. Ando, T. Yagihara, S. Tozune and T. Migita, *J. Am. Chem. Soc.*, **91,** 2786 (1969).
126. W. Ando, K. Nakayama, K. Ichibori, and T. Migita, *J. Am. Chem. Soc.*, **91,** 5164 (1969).
127. W. Ando, S. Kondo, and T. Migita, *J. Am. Chem. Soc.*, **91,** 6516 (1969).
128. W. Ando, S. Kondo, and T. Migita, *Bull. Chem. Soc. Japan*, **44,** 571 (1971).
129. W. Ando, T. Yagihara, S. Kondo, H. Nakayama, H. Yamato, S. Nakaido, and T. Migita, *J. Org. Chem.*, **36,** 1732 (1971).

CHAPTER 3

Generation of Carbenes by Photochemical Cycloelimination Reactions*

GARY W. GRIFFIN and NOELIE R. BERTONIERE
Department of Chemistry, Louisiana State University in New Orleans, New Orleans, Louisiana

I. INTRODUCTION

It has become increasingly apparent that divalent carbon intermediates play an important role in the photochemistry of a variety of diverse organic substrates. In many cases the reactions have synthetic utility which has not been exploited fully and which may complement existing conventional techniques for the generation of carbenes. It is our intent in this chapter to summarize those examples of photoinduced cycloeliminations that lead to carbene formation. In accordance with the precedent established in recent reviews[1a,1b] on photolytic and thermolytic cycloeliminations, we shall employ

* A shorter version of this chapter has been published as a review article in *Angewandte Chemie*.

the convention suggested by Huisgen[1c] in designating the mode of cycloelimination. Such terms as "photofragmentation" and "photocleavage" are also employed in this context.

To the extent that the reactions described here are concerted, they are subject to orbital symmetry constraints.[1d] For example, the [3 → 2 + 1] cycloelimination reaction of a cyclopropane to give a carbene if entirely concerted may be classified as a cheletropic reaction (i.e., a reaction in which two sigma bonds terminating at a single atom are broken in concert) and the orbital symmetry theory and selection rules for such processes have been advanced and discussed elsewhere.[1d,1e] Although compelling evidence is available that intermediates are implicated in many of the cycloeliminations reviewed, constraints are still imposed on the bond-breaking process leading to the intermediate(s) which control the stereochemical mode of ring opening.[1d,1e] Where in the authors' opinion the data available merit discussion of such theoretical considerations, the data are included for the reader's benefit.

II. [3 → 2 + 1] PHOTOCYCLOELIMINATIONS

A. Carbenes from Cyclopropanes

Among the most efficient methods of synthesizing cyclopropanes is one involving addition of carbenes to alkenes.[1f] The [3 → 2 + 1] cycloeliminations of interest in this context are, at least formally speaking, the reverse of this well-known reaction, that is, the production of a carbene and alkene by fragmentation of a cyclopropane.

$$\text{cyclopropane (R, R')} \xrightarrow[\text{gas phase}]{h\nu} \underset{\mathbf{2}}{:CH_2} + CH_2{=}CH_2$$

1a, R = R′ = H
1b, R = H; R′ = CH_3
1c, R = R′ = CH_3

Gas phase vacuum ultraviolet photolytic studies of the parent hydrocarbon, cyclopropane (**1a**), have been conducted[2a,2b,2c] and simple alkyl derivatives including methylcyclopropane (**1b**)[2d] and *cis*- and *trans*-1,2-dimethylcyclopropane (**1c**) have been photolyzed as well.[2c] The relative yield of *iso*- to *n*-butane derived by insertion of methylene (**2**) obtained by photolysis of **1a** at 165 nm in propane is identical to that obtained from other methylene sources such as propane, ketene, and diazomethane. Furthermore it is chemically indistinguishable from that from **2** derived by photolysis of ketene or diazomethane at longer wavelength. These observations then form a link between the species formed by vacuum ultraviolet and conventional photolytic methods and thus provide chemical evidence that the singlet state of **2** is involved in the vacuum ultraviolet studies with **1a**.[2c]

Products derived from methylene insertion in methylcyclopropane (**1b**) also have been detected upon photolysis of the latter at 147 nm.[2d] It is noteworthy that *cis*- and *trans*-1,2-dimethylcyclopropane (**2c**) undergo [3 → 2 + 1] cycloelimination and in the former case only *cis*-2-butene is formed as a cofragment, whereas only *trans*-2-butene is formed in the latter case.[2c] It is concluded using the familiar arguments advanced to diagnose the multiplicity of **2** for the reaction in the reverse direction that no triplet methylene is formed from the cis and trans isomers of **1c** although, of course, such arguments based on spin-imposed restrictions to cyclization and cleavage are subject to debate revolving around the question of the lifetime of potential intermediates and relative spin inversion rates. While trimethylene diradical has been postulated as an intermediate in the photolysis of cyclopropane,[2b] more recent studies[2c] indicate that such an intermediate is not required to accommodate the data available, at least for the formation of methylene and ethylene, and in fact if formed it must be very short lived.

Leermakers and Vesley[3a] report that the gas phase photolysis (254 nm) of benzylcyclopropane (**3**) leads to extensive fragmentation to produce a number of hydrocarbons including ethylene, which they suggest is a primary product formed in addition to benzylcarbene (**4**).

$$\underset{\mathbf{3}}{\text{cyclopropyl-}CH_2C_6H_5} \xrightarrow[254\text{ nm}]{h\nu} \underset{\mathbf{4}}{C_6H_5CH_2\ddot{C}H} + CH_2{=}CH_2$$

It is to be expected that aryl substitution of the cyclopropane ring should be accompanied by concomitant C—C bond weakening and more facile homolytic cleavage. The aryl chromophores also permit photoexcitation to be achieved readily. The cis-trans thermo- and photoisomerization of aryl- and aroyl-substituted cyclopropanes reflect the decreased bond strength.[4] Several reports of photolytic cycloeliminations involving aryl cyclopropanes are listed in Table 1.

Irradiation of phenylcyclopropane (**7a**) in solution in cyclohexene gives rise to methylene addition products, the isomeric methylcyclohexenes and norcarane. The methylene generated from **7a** in hydrocarbon solutions exhibits the same indiscriminate insertion into aliphatic carbon-hydrogen bonds as does methylene generated from diazomethane or ketene.[6a] [3 → 2 + 1] Cycloelimination upon photolysis of **7a** in the vapor phase is postulated to explain the formation of ethylene as a primary product.[3b]

TABLE 1

Cyclopropane	Carbene	Yield (%)	Reference(s)
 **5**	$:CH_2$ (**2**) $[H_2C{=}C:]$ (**6**)	Not reported	5
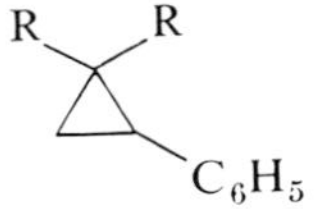 **7a**, R = H **7b**, R = Cl	$:CH_2$ (**2**) $Cl_2C:$ (**8**)	Not reported 9–15[a]	36, 6, 7
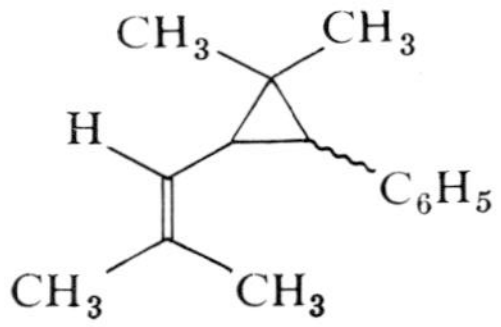 **9**	$(CH_3)_2C{=}CH\ddot{C}H$ (**10**)	~10	8
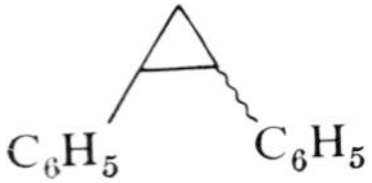 **11**	$C_6H_5\ddot{C}H$ (**12**)	7–8	9

13a, $R_1 = R_2 = R_3 = H; R_4 = C_6H_5$	$[H_2C{=}C{:}]$ (**6**)	~33	10a
13b, $R_1 = R_3 = R_4 = H; R_2 = C_6H_5$	$[C_6H_5CH{=}C{:}]$ (**14**)	~4	10a
13c, $R_1 = R_2 = H; R_3 = R_4 = C_6H_5$	$[H_2C{=}C{:}]$ (**6**)	0.5	10a
13d, $R_1 = R_2 = CH_3; R_3 = R_4 = C_6H_5$	$[(CH_3)_2C{=}C{:}]$ (**15**)	3[a]	10b
16	$(C_6H_5)_2C{:}$ (**17**)	38[c]	11
	$(CH_3)_2C{=}CH\ddot{C}H$ (**10**)	45	
18	$C_6H_5\ddot{C}H$ (**12**)	5[b]	9, 12

TABLE 1 (*Continued*)

Cyclopropane	Carbene	Yield (%)	Reference(s)
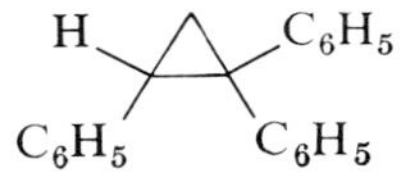 **19**	$C_6H_5\ddot{C}H$ (**12**) $(C_6H_5)_2C:$ (**17**)	Trace[c] 7–10[c]	12
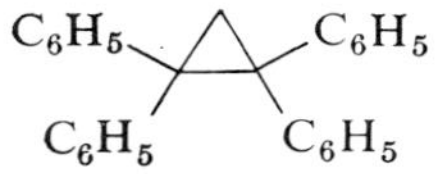 **20**	$(C_6H_5)_2C:$ (**17**)	50	12
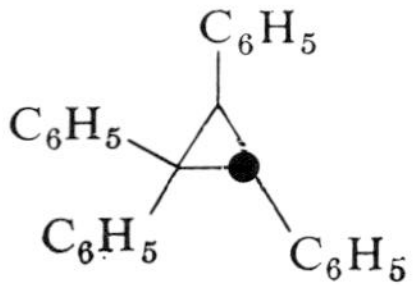 **21**	$C_6H_5\ddot{C}H$ (**12**) $(C_6H_5)_2C:$ (**17**)	Not detected[c] 40	12
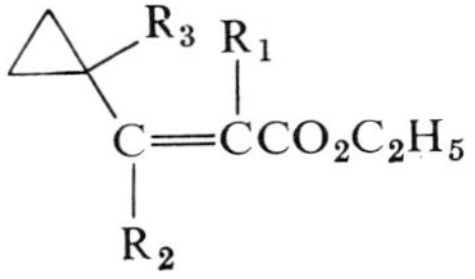 **22**	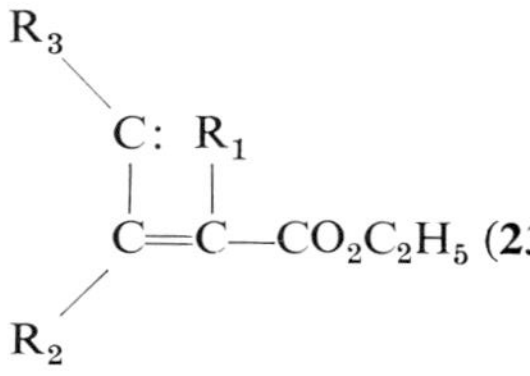 $R_3C{:}(R_2)C{=}C(R_1){-}CO_2C_2H_5$ (**23**)	Not reported	13

[a] Yield based on addition to alkenes to give cyclopropanes.
[b] Yield determined by the extent of C—H insertion.
[c] Yield of carbene determined from extent of ether formation in alcohols.

Competitive photolytic reactions are observed for alkyl-substituted cyclopropanes such as **24**.[14] A major photoproduct is in general a terminal olefin of the type **25**, which appears to be formed by initial homolysis of the more

CH_3 CH_3 C_6H_5 $\xrightarrow{h\nu}$ CH_2 C_6H_5 CH_3

24 **25**

highly substituted benzyl bond of the ring. It has been established by deuterium-labeling studies that subsequent H-transfer occurs from the methyl group directly to the benzyl position in a five-membered transition state rather than by H-transfer to the ring.[14,15]

It is noteworthy that 2,2-dichlorophenylcyclopropane (**7b**) undergoes photochemical cycloelimination to give primarily dichlorocarbene **8**.[7] The addition of **8** to olefins appears to occur in a cis fashion in a manner similar to that observed for other halocarbenes or carbenoids produced in the base-catalyzed decomposition of haloforms.[16,17] Fragmentation in the alternate sense to give phenylcarbene was not excluded and by-products incorporating solvent are formed.

The isomeric phenylvinylcyclopropanes **9** are photolabile and when irradiated with a mercury resonance lamp give the vinylcarbene **10** and 1-phenyl-2-methylpropene among other rearrangement products. No additional details are described.[8]

cis- and *trans*-1,2-Diphenylcyclopropane (**11**) undergo photointerconversion[4a] and the kinetics have now been studied in cyclohexane at 25 and 65° using a 254 nm light source. Geometrical isomerization predominates over branching to side products by a ratio of 4:1, which is sufficient, however, to prevent attainment of a "true" photostationary state.[9b] Upon prolonged irradiation of **11** in solution in *n*-pentane (24–48 hours) it is possible to achieve fragmentation to the extent of 7–8%.[9] The selectivity factor for secondary versus primary C—H insertion for phenylcarbene generated from **11** (8.3) is essentially the same as that observed when conventional phenylcarbene precursors such as phenyldiazomethane are used.[18] A trimethylene diradical accounts for the geometrical (as well as accompanying structural) isomerization.[4a,4b,4c,9b] In fact, a comprehensive study of the absorption and emission spectroscopy and photochemistry of six phenyl-substituted cyclopropanes recently has been carried out at 77° K. Long wavelength phosphorescences (maxima ~ 575 nm, τ_p ~ 7 millisec) are observable for *trans*-diphenylcyclopropane (**11**) as well as for the tri- and tetraphenylcyclopropanes studied (**19** and **20**, respectively). It is suggested that these phosphorescences may arise from diradicals of the trimethylene type **26** formed by photocleavage of the cyclopropane ring, presumably in the allowed disrotatory

$$\mathbf{11}\ (C_6H_5,\ C_6H_5\text{-cyclopropane}) \xrightarrow[\text{disrotatory opening}]{h\nu} C_6H_5\text{–}\dot{C}H\text{–}CH_2\text{–}\dot{C}H\text{–}C_6H_5\ (\mathbf{26}) \longrightarrow C_6H_5CH{=}CH_2 + C_6H_5\ddot{C}H$$

manner.[19,20] The diradical character of ring open intermediates derived from cyclopropanes is supported by theoretical studies employing nonempirical self-consistent field and configurational interaction methods.[21a] It is inviting to propose that arylcarbenes in general and phenylcarbene in particular arise in turn by subsequent thermal homolysis of trimethylene diradicals of which **26** is representative. Hoffmann contends that the "easiest passage" thermally from the trimethylene energy valley to the much deeper valley for cyclopropane is *via* conrotatory motion of both methylene groups,[20] which would account for the observed concomitant cis-trans isomerization; that is, conrotatory recyclization of **26** formed photochemically from the trans isomer to *cis*-1,2-diphenylcyclopropane. Furthermore, reasonable mechanistic routes may be imagined for conversion of **26** to the other by-products observed, 1-phenylindane and *cis*- and *trans*-1,3-diphenylpropene. In the cleavage of the cyclopropane ring, the question of whether bond breaking may lead to a singlet diradical, potentially zwitterionic in character, also may be entertained since the sites of charge and spin presumably may be sufficiently close to accommodate orbital overlap. This problem recently has been faced experimentally by Yankee and Cram[4f] for thermal transformations of certain cyclopropyl systems.

The methylene carbenes **6**, **14**, and **15** have been generated photolytically from methylenecyclopropanes.[5,10] Irradiation of 2,2-diphenyl-1-methylenecyclopropane (**13c**) results in extrusion of methylenecarbene **6**, which subsequently rearranges to acetylene faster than it can be trapped by the solvent (pentane). In contrast isopropylidenecarbene **15** may be trapped by photolysis of **13d** in cyclohexene and cyclohexane. The anticipated addition and insertion products, 7-isopropylidenebicyclo[4.1.0]heptane and 1-cyclohexyl-2-methylpropene, are obtained in yields of 3 and 1%, respectively.

In connection with their studies of the di-π-methane rearrangement, Zimmerman and Pratt[11] noted that 1,1-dimethyl-2,2-diphenyl-3-(2-methylpropenyl)cyclopropane (**16**) undergoes two types of carbene scission in *t*-butanol. Upon irradiation, 1,1-diphenyl-2-methylpropene and presumably the vinylcarbene **10** are formed. Diphenylcarbene **17**, which is also produced, reacts with the solvent to give benzhydryl-*t*-butyl ether (38%).

It is apparent from the yield data presented in Table 1 that if the structure

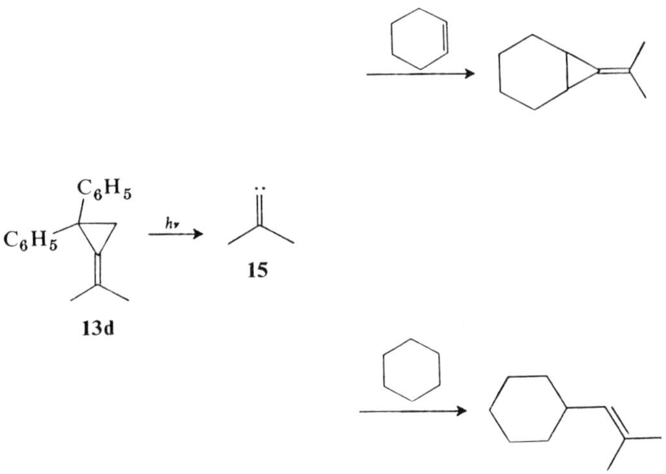

13d

15

of the cyclopropane is such that two options for cycloelimination exist—formation of phenyl- and diphenylcarbene as in the case of the tri- and tetraphenylcyclopropanes **19** and **21**—then fragmentation to give diphenylcarbene predominates. For example, no detectable benzylmethyl ether (which would be formed from phenylcarbene) is observed upon irradiation of *trans*-1,1,2,3-tetraphenylcyclopropane (**21**) in methanol-benzene under conditions where a substantial amount (40%) of benzhydrylmethyl ether is produced.[12] Similarly, photolysis of 1,1,2-triphenylcyclopropane (**19**) gives traces of phenylcarbene **12**, while diphenylcarbene **17** is formed in approximately 10% yield.

Although no direct spectroscopic (optical or EPR) evidence for the formation of phenyl- or diphenylcarbene from the cyclopropanes **18** through **21** (or for that matter any of the cyclopropanes in Table 1) has yet been obtained, the chemical evidence for formation of these species is convincing. As with **11** (*vide supra*), a quantitative determination[9a] of the relative reactivities of the C—H bonds in *n*-pentane toward insertion reactions of phenylcarbene **12** derived from **18** compares favorably with the values obtained using phenyldiazomethane as a carbene precursor.[9a,18] The formation of diphenylcarbene **17** from **19**, **20**, and **21** is adduced from the behavior observed upon photolysis of these substrates in 2-methyl-2-butene and/or methanol, whereupon according to expectations one obtains *sym*-tetraphenylethane and/or benzhydrylmethyl ether, respectively.[12]

The photochemical response of six α,β-unsaturated esters of the type **22** to ultraviolet radiation has been studied by Jorgenson[13] and fragmentation to carbenes **23** is a surprisingly general mode of reaction for such cyclopropylacrylic esters. It is worth noting that the cycloelimination reaction invariably takes place by that course which leads to ethylene and a resonance-stabilized carbene. No cycloelimination to methylene and a dienic ester, a pathway analogous to the formation of methylene from arylcyclopropanes **7a** and **7b**,[5,6,7] is observed with these esters.

B. Carbenes from Fused-Ring Cyclopropane Derivatives—Norcaradiene and Related Systems

A careful examination of the products obtained on photolysis of cycloheptatriene **27** in hydrocarbon solvents by methods which would have revealed the presence of methylene insertion products shows that at least under the conditions employed cycloelimination to this carbene does not occur,[6a] although a substantial driving force might be anticipated for fragmentation of norcaradiene **28**, a valence isomer of **27**.

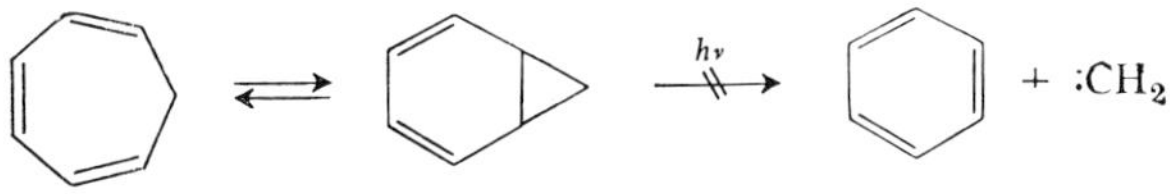

$h\nu$ + $:CH_2$

27 **28**

What may be regarded as a formal transfer of dimethylaminocarbene, however, is observed in low yield when 7-dimethylaminocycloheptatriene (**29**) is photolyzed.[6b]

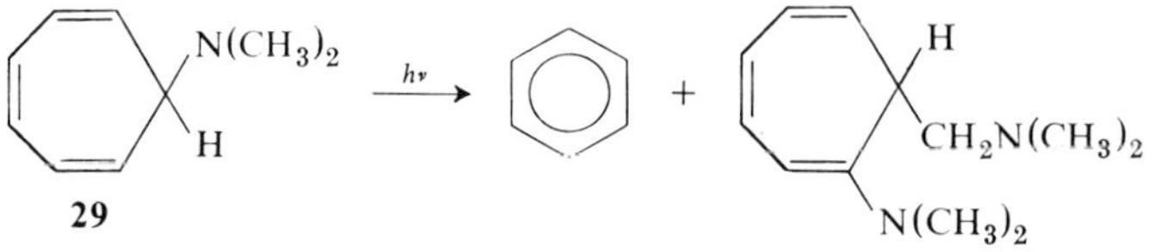

$N(CH_3)_2$ H $h\nu$ + H $CH_2N(CH_3)_2$ $N(CH_3)_2$

29

The spirocyclopropane **30**, accessible from acenaphthylene and diazofluorene, was found to decompose at 280° with formation of difluorenylidene **31** and acenaphthylene.[23] Although this conversion appears to be a reversal of a carbene addition reaction, the intermediacy of a carbene in the thermal process has not been substantiated.[24] In contrast, **30** is remarkably photostable, which is particularly surprising in view of the photolability of a variety of fused-ring norcaradiene derivatives (*vide infra*) and what would appear to be substantial cumulative driving force for formation of fluorenylidene, a diarylcarbene, and acenaphthylene.[12]

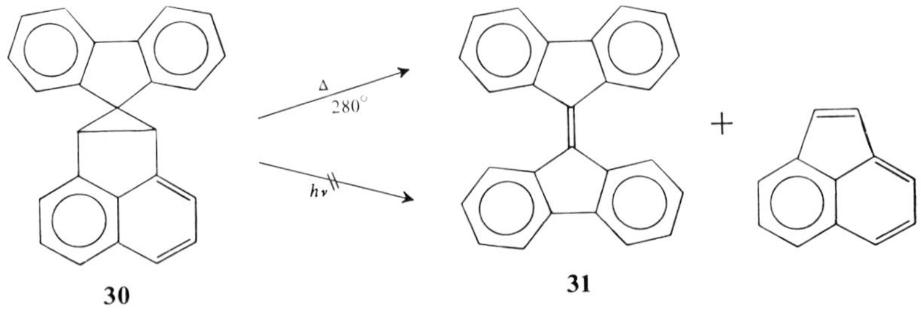

Δ 280° $h\nu$ +

30 **31**

TABLE 2

Precursor	Carbene	Yield (%)	Reference(s)
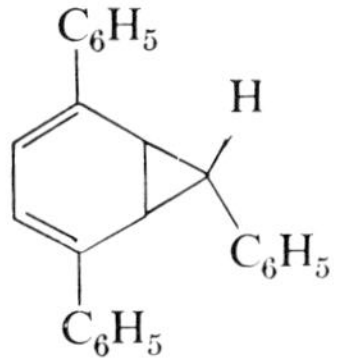 **32**	$C_6H_5\ddot{C}H$ (**12**)	Not reported	25
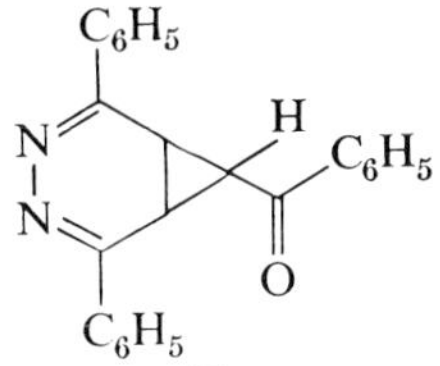 **33**	$[C_6H_5CO\ddot{C}H]$ (**34**)	Not reported	26
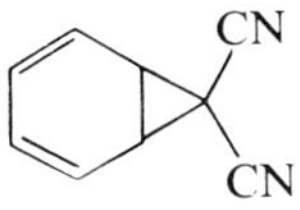 **35**	$:C(CN)_2$ (**36**)	60	27
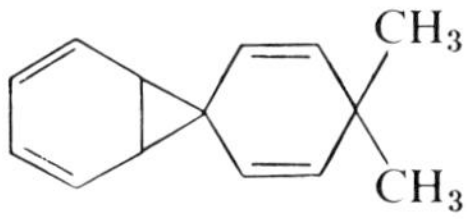 **37**	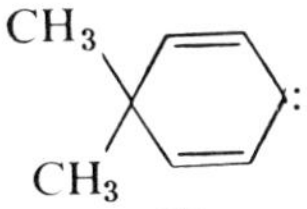 **38**	Not reported	28
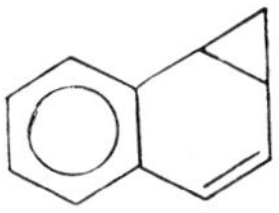 **39**	$:CH_2$ (**2**)	Not reported	28
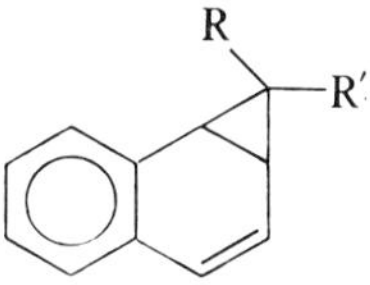			
40a, $R{=}R'{=}CN$	$:C(CN)_2$ (**36**)	22[a]	27
40b, $R{=}H;\ R'{=}CO_2C_2H_5$	$:CHCO_2C_2C_5$ (**42**)	8[a]	27
40c, $R{=}H;\ R'CO_2CH_3$	$:CHCO_2CH_3$ (**43**)	10[a]	29
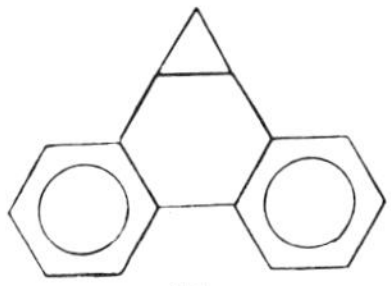 **41**	$:CH_2$ (**2**)	>90	5

[a] Yield determined by the extent of C—H insertion.

The photolysis of 2,5,7-triphenylnorcaradiene (**32**) in ether (Table 2) affords *p*-terphenyl (4%). Toda and co-workers[25] propose that phenylcarbene **12** is formed on the basis of analogous reactions of other stable norcaradienes (*vide infra*), although trapping experiments were unsuccessful. A similar fragmentation of the 7-benzoyl-3,4-diazonorcaradiene **33** to 3,6-diphenylpyridazine occurs and it is reasonable to assume, in the absence of data to the contrary, that a carbene such as **34** may be formed at least as a transient species.[26]

Photolysis of a cyclohexane solution of 7,7-dicyanonorcaradiene (**35**) gave cyclohexylmalononitrile in 60% yield.[27] Furthermore, the insertion products obtained from **35** using 2,3-dimethylbutane as a solvent substrate—2,3-dimethyl-1-butylmalononitrile (1° C—H insertion product) and 2,3-dimethyl-2-butylmalononitrile (3° C—H insertion product) were higher (27% in each case) than those obtained when dicyanodiazomethane was employed as a precursor for dicyanocarbene **36** (12 and 16%, respectively). The similar product ratios observed upon photolysis of **35** in 2,3-dimethylbutane and thermolysis of dicyanodiazomethane in the same substrate are cited as evidence that the species formed from these two sources have essentially the same relative reactivity toward 1 and 3° C—H bonds.[27]

The norcaradiene **37** upon photolysis through quartz undergoes cycloelimination with extrusion of 4,4-dimethylcyclohexadienylidene (**38**) in addition to isomerization. This carbene may be trapped with various alkenes and a comparison of the properties of **38** with those of the carbene generated from the corresponding diazo compound revealed them to be identical.[28]

Interest in the photochemical behavior of *o*-divinylbenzene led Pomerantz and Gruber to examine the photochemistry of 3,4-benzotropilidine (**44**).[29] Irradiation of dilute solutions of **44** in ether led to the benzonorcaradiene **39**

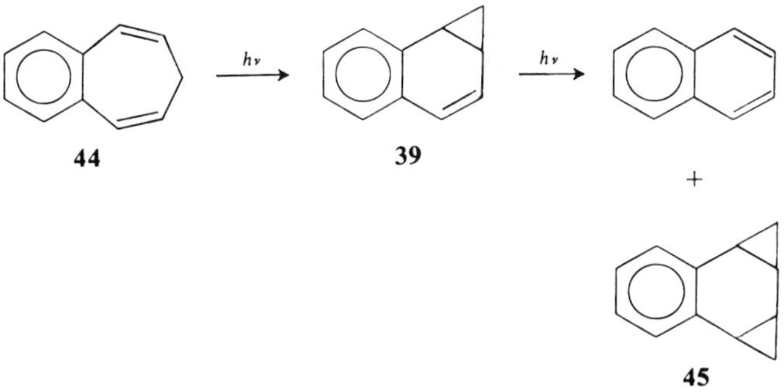

44 $\xrightarrow{h\nu}$ **39** $\xrightarrow{h\nu}$ naphthalene + **45**

(70–80%), which upon prolonged irradiation afforded naphthalene and the bismethanonaphthalene **45**. The hypothesis that methylene **2** is transferred in the course of the conversion of **39** into **45** and naphthalene was substantiated by trapping experiments utilizing cyclohexene.

Ciganek[27] studied the photolysis of 7,7-dicyano-2,3-benzonorcaradiene (**40a**) in cyclohexane under conditions similar to those employed with **35** and obtained cyclohexylmalononitrile (22%) in addition to naphthalene (22%) and a photoisomer of **40a**, assigned structure **46** (44%). By comparison, photolysis of 7-carboethoxy-2,3-benzonorcaradiene (**40b**) in cyclohexane

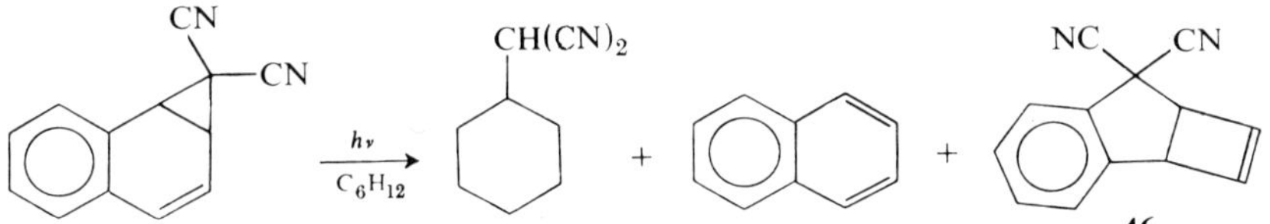

reportedly gives only a small amount of the carbene transfer reaction product ethylcyclohexylacetate (8%), in addition to naphthalene (11%) and intractable, probably polymeric, material.

Inexplicably a more complex spectrum of products was observed by Swenton and Krubsack[30] upon photolysis of 7-carbomethoxy-2,3-benzonorcaradiene (**40c**). The products include naphthalene (13%), the insertion product **47** (10%), methyl-1- and 2-naphthylacetate [**48** (14%) and **49** (5%), respectively] and **50** (11%).

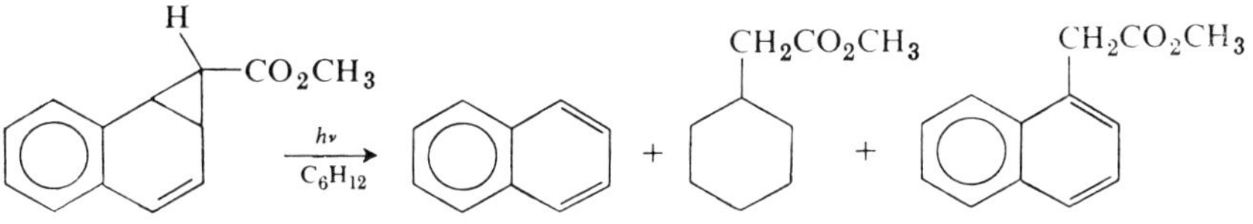

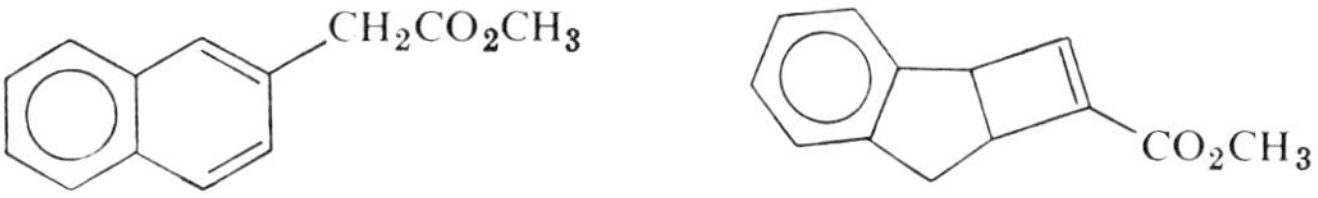

In an early cycloelimination study the dibenzonorcaradiene **41** is recommended as a shelf-stable high yield (>90%) source of active methylene to employ for small-scale synthesis and C^{14} labeling.[6a] The C—H insertion selectivity of methylene **2** formed photolytically from **41** was determined in a number of alkanes and compares favorably with the indiscriminate reactivity reported for methylene generated from diazomethane. Furthermore, photolysis of **41** in cyclohexene gives four C_7 hydrocarbons identified as the three methylcyclohexenes and norcarane. The relative ratios are the same as those produced by photolysis of diazomethane in cyclohexene. The stereochemical mode of addition of methylene formed from **41** also was examined by irradiation of this precursor in *cis*- and *trans*-4-methyl-2-pentene. In each case the products are identical to those obtained by photolysis of diazomethane and the stereochemical integrity of the alkene is maintained.

C. Carbenes from Oxiranes

vic-Diaryloxiranes have been shown to undergo cycloelimination on photolysis in solution to give arylcarbenes and carbonyl compounds. This mode of reaction has been studied extensively and shown to possess broad synthetic utility. The results of these investigations are summarized in Table 3.

TABLE 3

Oxirane	Carbene	Yield (%)	Reference(s)
C_6H_5, O, H; H, C_6H_5 **51**	$C_6H_5\ddot{C}H$ (**12**)	90[a], 65–70[a]	9a, 31, 32, 34
H, O, H; C_6H_5, C_6H_5 **52**	$C_6H_5\ddot{C}H$ (**12**)	Not reported	34
C_6H_5, O, C_6H_5; C_6H_5, H **53**	$(C_6H_5)_2C$: (**17**) $C_6H_5\ddot{C}H$ (**12**)	85[b] 15[b]	31, 33, 34
C_6H_5, O, C_6H_5; C_6H_5, C_6H_5 **54**	$(C_6H_5)_2C$: (**17**)	88–100[b]	31, 33, 34
C_6H_5, O, CH_3; H, C_6H_5 **55**	$C_6H_5\ddot{C}CH_3$ (**56**) $C_6H_5\ddot{C}H$ (**12**)	Not reported	36a
C_6H_5, O, CH_3; CH_3, C_6H_5 **57**	$C_6H_5\ddot{C}CH_3$ (**56**)	40–60[a]	34, 37

TABLE 3 (*Continued*)

Oxirane	Carbene	Yield (%)	Reference(s)
58	$[C_6H_5CO(CH_2)_4$-$CC_6H_5]$ **(59)**	Detected spectroscopically	34
60	$C_6H_5\ddot{C}$— **(61)**	75[b]	36b
62	$C_6H_5\ddot{C}CN$ **(63)** $C_6H_5\ddot{C}H$ **(12)**	62[b] ~5[b]	38, 39
64	$C_6H_5\ddot{C}CN$ **(63)** $(C_6H_5)_2C:$ **(17)**	71[b] ~5[b]	38, 40
65	$C_6H_5\ddot{C}CN$ **(63)**	60–65[a]	40, 41
66	$C_6H_5\ddot{C}CN$ **(63)**	15[b], 20[a]	38
67	$C_6H_5\ddot{C}CO_2CH_3$ **(69)**	20[a]	40, 43

TABLE 3 (*Continued*)

Oxirane	Carbene	Yield (%)	Reference(s)
C_6H_5, O, CO_2CH_3 / C_6H_5, C_6H_5 **68**	$C_6H_5\ddot{C}CO_2CH_3$ (**69**) $(C_6H_5)_2C$: (**17**)	40[a] Traces[a]	40
C_6H_5, O, C_6H_5 / H, OCH_3 **70**	$C_6H_5\ddot{C}H$ (**12**) $C_6H_5\ddot{C}OCH_3$ (**71**)	Principal product ~5[a]	44
C_6H_5, O, C_6H_5 / C_6H_5, OCH_3 **72**	$(C_6H_5)_2C$: (**17**)	75[b]	36b, 45
C_6H_5, O, C_6H_5 / C_6H_5, O_2CCH_3 **73**	$(C_6H_5)_2C$: (**17**)	Principal product	36b
CH_3, O, C_6H_5 / C_6H_5, O_2CCH_3 **74**	$C_6H_5\ddot{C}CH_3$ (**56**)	Principal product (~40)	36b
75	(**76**)	Proposed intermediate	47

[a] Yield based on addition to alkenes to give cyclopropanes.
[b] Yield of carbene determined from extent of ether formation in alcohols.

trans-Stilbene oxide (**51**) has been irradiated in methanol and alkenes.[31] Under these conditions phenylcarbene **12** is trapped as benzyl methyl ether and the appropriate cyclopropane. According to expectations two epimeric cyclopropanes are obtained from the addition of phenylcarbene derived from **51** to noncentrosymmetric olefins, the sterically less hindered, more stable anti isomer being formed preferentially. Similar results are obtained with phenylcarbene generated from phenyldiazomethane[35] and quantitative studies of epimer ratios attest to the similarity of the species derived from the two independent sources.[9a] It should be noted that phenylcarbene generated from **51** adds in a highly stereoselective manner to *cis*-2-butene, suggesting that a singlet mechanism is operative.[1f]

Insertion selectivity of phenylcarbene generated from *trans*-stilbene oxide (**51**) into aliphatic 1° and 2° C—H bonds of *n*-pentane also have been compared with the results obtained using the conventional phenylcarbene precursor phenyldiazomethane.[9a,32,36c] The results are presented in Table 4 and confirm

TABLE 4

Precursor	Insertion (%)	$\frac{\text{C-2} + \text{C-3}}{\text{C-1}}$	$\frac{\text{C-2}}{\text{C-3}}$
trans-Stilbene oxide (**51**)	45.4	8.33 ± 0.14	1.35 ± 0.04
Phenyldiazomethane	31.5	8.38 ± 0.14	1.33 ± 0.09

that the reactivity of the transient species formed photochemically from *trans*-stilbene oxide compares favorably with that obtained from phenyldiazomethane. Although no EPR signal corresponding to phenylcarbene could be detected,[33] the typical luminescence at 389 and 414 nm due to this species is observed at 77° K upon irradiation of both *cis*- and *trans*-stilbene oxide (**52** and **51**, respectively) in 3-methylpentane glasses.[34] In addition to phenylcarbene and benzaldehyde, *cis*- and *trans*-stilbene also are formed upon irradiation of these oxiranes in rigid glasses at 77° K. These alkenes are not observed when the photolyses are conducted at room temperature (*vide infra*).

The fragmentation of triphenyloxirane (**53**) potentially may occur by two routes: (*a*) to give benzophenone and phenylcarbene **12**, and (*b*) to give benzaldehyde and diphenylcarbene **17**. Comparison of the respective phosphorescences of the carbonyl fragments resulting from irradiation of triphenyloxirane (**53**) in 3-methylpentane glass at 77° K with those of a standard containing equimolar amounts of benzophenone and benzaldehyde indicates that the latter represents the preferred mode of fragmentation (~70%).[34b] The relative amounts of benzhydryl- and benzylmethyl ether formed from (**53**) (85 and 15%, respectively) when methanol is employed as a trapping agent supports the conclusion that cycloelimination occurs preferentially to give diphenylcarbene **17** and benzaldehyde.[34b] The emission characteristic of diphenylcarbene at 485 nm is observed upon irradiation of **53** in 3-methylpentane glass; however, no emission due to phenylcarbene was detected. Furthermore, whereas the characteristic EPR signal of ground state triplet diphenylcarbene **17** (zero field parameters $D/hc = 0.4053\ cm^{-1}$, $E/hc = 0.0190\ cm^{-1}$) is observable upon photolysis of triphenyloxirane (**53**) in a methylcyclohexane glass at 77° K, no EPR signals corresponding to phenylcarbene were detected.[33]

Photolysis of tetraphenyloxirane (**54**) in methanol/benzene gave benzhydrylmethyl ether in essentially quantitative yield.[31] However, benzophenone, tetraphenylethylene (a primary photoproduct), and diphenylcarbene **17** are observable upon irradiation of this substrate in a rigid glass at 77° K as described previously. The emission spectrum of diphenylcarbene derived from **54** is broad and structureless with an onset at 460 nm and a maximum at 485 nm.[34b] The EPR spectrum of triplet diphenylcarbene **17** generated photochemically from tetraphenyloxirane (**54**) also has been observed.[33]

Like triphenyloxirane (**53**), 2-methyl-2,3-diphenyloxirane (**55**) is unsymmetrical and might be expected to fragment in two ways to give phenylcarbene **12** and acetophenone or phenylmethylcarbene **56** and benzaldehyde. Fragmentation of **55** does indeed occur in these two ways with methylphenylcarbene being formed preferentially.[36a] *trans*-2,3-Dimethyl-2,3-diphenyloxirane (**57**) is a superior source of phenylmethylcarbene and undergoes photofragmentation to give **56** and acetophenone in methanol and several alkenes. The carbene **56** is trapped as α-phenethylmethyl ether when methanol is employed as the solvent.[37] The yields of phenylcyclopropanes which result upon photolysis in the alkene solvents range from 40–60%.[37] Irradiation of **57** in a 3-methylpentane glass at 77° K yields acetophenone and an alkene, presumably formed in low yield by oxygen extrusion, which has been characterized as *trans*-1,2-dimethyl-1,2-diphenylethylene.[34b]

The luminescence spectrum of an irradiated sample of 1,2-diphenyl-1,2-epoxycyclohexane (**58**) is composed of a phosphorescence similar to that of acetophenone and an emission similar to that of *cis*-stilbene.[34b] These spectroscopic data indicate that extrusion of oxygen from the oxirane ring does indeed occur to give an alkene concurrently with cycloelimination to form a ketone and arylcarbene. In the case of the bicyclic oxirane **58**, it also was established spectroscopically that opening leads to the expected ketocarbene **59**.[34b]

Irradiation of *trans*-2,3-diphenyl-2,3-dicyclopropyloxirane (**60**) in methanol affords phenylcyclopropylcarbene **61**, trapped as α-cyclopropylbenzylmethyl ether, in 75% yield.[36b] In contrast to expectations, little or no rearrangement of this carbene is observed when generated under these conditions.[36b]

trans-2-Cyano-2,3-diphenyloxirane (**62**) may fragment to give either phenylcyanocarbene **63** and benzaldehyde or alternatively phenylcarbene **12** and benzoylcyanide. When methanol is employed as the trapping agent α-cyanobenzylmethyl ether and benzylmethyl ether are formed in 62 and ~5% yields, respectively.[38] Photolysis of **62** in 2,3-dimethyl-2-butene affords

2,2,3,3-tetramethyl-1-cyano-1-phenylcyclopropane (28%) and 2-phenyl-3,3,4,4-tetramethyloxetane.[38] On this basis it is concluded that the preferred mode of fragmentation leads to phenylcyanocarbene and benzaldehyde. These results are consistent with those reported independently by Temnikova and co-workers.[39]

$$\xrightarrow{CH_3OH} C_6H_5CH(OCH_3)CN$$

$$\underset{\mathbf{62}}{C_6H_5(H)C\overset{O}{—}C(CN)C_6H_5} \xrightarrow{h\nu} \underset{\mathbf{63}}{C_6H_5\ddot{C}CN}$$

$$\xrightarrow{(CH_3)_2C=C(CH_3)_2} \text{2,2,3,3-tetramethyl-1-}C_6H_5\text{-1-}CN\text{-cyclopropane} + \text{2-}C_6H_5\text{-3,3,4,4-tetramethyloxetane}$$

A strong preference for phenylcyanocarbene **63** formation also is apparent when 2,3,3-triphenyl-2-cyanooxirane (**64**) is irradiated in methanol or unsaturated solvents. Photolysis of **64** in methanol affords α-cyanobenzylmethyl ether and benzhydrylmethyl ether in 71 and ~5% yields, respectively. Irradiation of **64** in 2,3-dimethyl-2-butene gives 2,2,3,3-tetramethyl-1-cyano-1-phenylcyclopropane in 70% yield. The addition of phenylcyanocarbene **63**, generated from **64**, to 2-methyl-2-butene also may be achieved and two epimeric 2,2,3-trimethyl-1-phenyl-1-cyanocyclopropanes are obtained in a ratio of 1.86:1.[38,40]

Irradiation of *cis*-2,3-dicyano-2,3-diphenyloxirane (**65**)[41] in 2-methyl-2-butene affords the isomeric 1-cyano-1-phenyl-2,2,3-trimethylcyclopropanes in 65–70% yield while in 2,3-dimethyl-2-butene 1-cyano-1-phenyl-2,2,3,3-tetramethylcyclopropane is obtained in 75% yield.[40] Temnikova and co-workers[42] have reported that photolysis of **65** in 2-methyl-2-butene and 2,3-dimethyl-2-butene gives the corresponding cyclopropanes, 1-cyano-1-phenyl-2,2,3-trimethylcyclopropane and 1-cyano-1-phenyl-2,2,3,3-tetramethylcyclopropane, in 80% yield. The expected fragment, benzoyl cyanide, was also detected; however, it is reported that considerable resinification occurs under the reaction conditions.[42]

Irradiation of the spiroöxirane **66** in methanol and 2,3-dimethyl-2-butene gives α-cyanobenzylmethyl ether and 1-cyano-1-phenyl-2,2,3,3-tetramethylcyclopropane (15 and 20%, respectively).[38] This represents one of the few

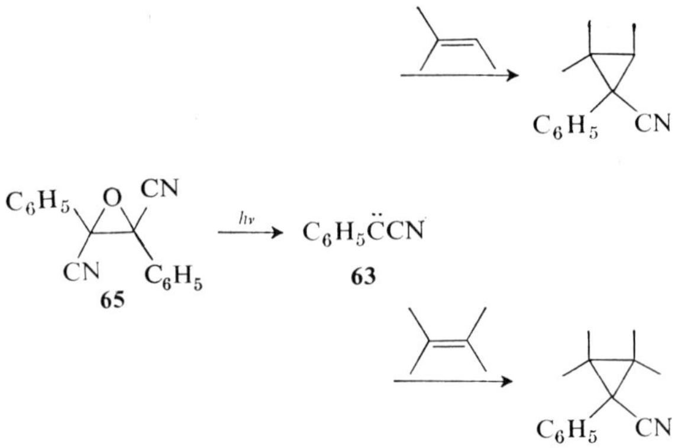

reported examples where cycloelimination of this type occurs in the absence of vicinal diaryl substitution on the oxirane ring.

Photolysis of the glycidic esters **67** and **68** in 2,3-dimethyl-2-butene affords 1-carbomethoxy-1-phenyl-2,2,3,3-tetramethylcyclopropane (20 and 40%, respectively).[40] The expected oxetanes formed from the alkene and benzaldehyde and benzophenone are also obtained upon photofragmentation of **67** and **68**, respectively. Irradiation of **67** in methanol gives α-carbomethoxybenzylmethyl ether in 70% yield and in 2-methyl-2-butene the epimeric 1-carbomethoxy-1-phenyl-2,2,3-trimethylcyclopropanes and 1-phenyl-2,2,3-trimethylcyclopropane (a minor product) in 70–75% yield.[43] Carbene **69** is clearly implicated.

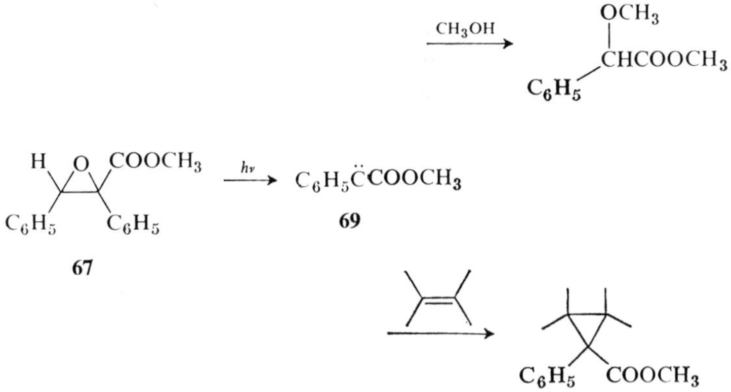

Irradiation of 2-ethyl-3-phenylglycidate in water saturated benzene has been reported to yield a single diastereoisomer of 4-carboethoxy-4-methyl-2,5-diphenyl-1,3-dioxolane. The product is believed to arise from photoaddition of benzaldehyde, formed along with methylcarboethoxycarbene by photofragmentation of the oxirane ring, to the glycidic ester. The carbene, however, could not be trapped.[40b]

2-Methoxy-2,3-diphenyloxirane (**70**) conceivably could undergo cycloelimination to methoxyphenylcarbene **71** and benzaldehyde and/or phenylcarbene and methyl benzoate. Irradiation of **70** in 2-methyl-2-butene and 2,3-dimethyl-2-butene indicates that the preferred mode of cleavage is by the second pathway to phenylcarbene and methyl benzoate, the principal products isolated being the isomeric 1-phenyl-2,2,3-trimethyl- as well as 1-phenyl-2,2,3,3-tetramethylcyclopropanes, respectively. The oxetane resulting from photocycloaddition of benzaldehyde to 2-methyl-2-butene is also observed (5%), indicating that reaction by the first pathway also occurs to a minor extent.[44]

Diphenylcarbene is formed on photolysis of 2-methoxy-2,3,3-triphenyloxirane (**72**)[36b] as established by trapping experiments with methanol in which the methyl ether of benzhydrol is formed (75%). The other fragment, methyl benzoate, was isolated by GLC. Luminescence and EPR studies confirm the intermediacy of diphenylcarbene.[45] It is noteworthy that there is no indication that **72** fragments in the alternate fashion to give benzophenone and methoxyphenylcarbene, which by analogy with other related systems should give the dimethylacetal of benzaldehyde upon reaction with methanol. These results are consistent with those reported independently by Temnikova and Stepanov.[46]

2,3,3-Triphenyl-2-acetoxyoxirane **73** behaves in a manner similar to **72**,[36b] and photocleavage occurs to give diphenylcarbene, which is conveniently trapped with methanol. No products derived from phenylacetoxycarbene, (i.e., formed by fragmentation in the alternate sense) were detected. This result might have been expected on the basis of the preferred direction of cleavage observed for the related methoxy derivative **72**. *trans*-2,3-Diphenyl-2-methyl-3-acetoxyoxirane (**74**) photofragments to methylphenylcarbene and the mixed anhydride of benzoic and acetic acids.[36b]

Irradiation of a benzene solution of the oxaspiropentane derivative **75** results in formation of the cumulene **77**, an allenic alcohol **78** and the allenic oxetane **79**. A cyclopropyl carbene **76** is invoked to rationalize the formation

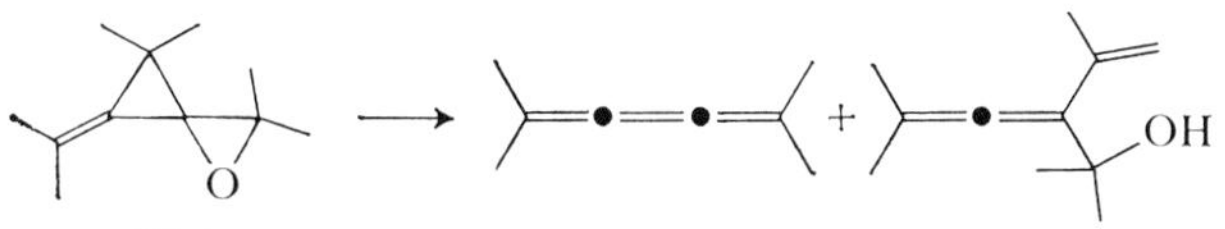

O OH

75 **77** **78**

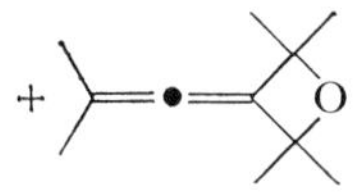

O

79

of 77,[47] as shown in the following scheme:

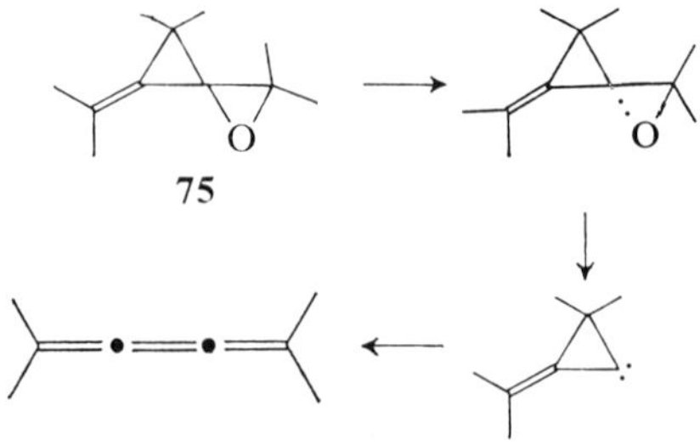

75

77 76

The oxetane **79** was shown to be a secondary photoproduct arising from acetone addition to 77.

Irradiation of the six isomeric 1,4-bis(2,3-diphenyloxiranyl)benzenes (**80**) in methanol afforded 1,4-bis(α-methoxybenzyl)benzene (69–75%) and 1-benzoyl-4-(α-methoxybenzyl)benzene (25–31%), the products resulting from the trapping of the carbenes *p*-phenylenebis(phenylmethylene) (**81**) and 1-benzoyl-4-phenylmethylenebenzene (**82**), respectively. *p*-Dibenzoylbenzene, the third potential large fragment, was not detected. It is quite probable that

C_6H_5 C_6H_5

81

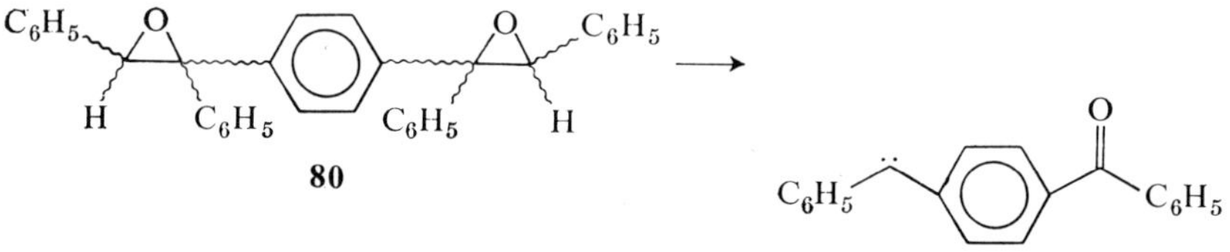

C_6H_5 O O C_6H_5

H C_6H_5 C_6H_5 H

80

O

C_6H_5 C_6H_5

82

the dicarbene **81** is never formed per se in solution but that the oxiranyl rings fragment in sequence, the second not reacting until the first has been converted to the methyl ether. These methanol trapping experiments indicate that it is highly probable that the dicarbene is formed upon irradiation of the bisoxiranes **80** in rigid glasses at 77° K, since the lifetime of the initially formed carbene should be sufficiently long to permit the two-stage process to occur.[48,36c]

Conjugation of the oxirane ring with a carbonyl substituent radically alters the photochemistry of aryloxiranes. The photochemical reactions of these epoxyketones have been reviewed extensively.[49,50] It is generally accepted that the n,π^* excited state is the chemically significant state involved in the photochemical transformations of this class of ketones. The two basic reaction types distinguished are (*a*) reversible photochromic valence isomerization of arylcyclopentenone oxides such as **83** to pyrylium oxides such as **84**, and (*b*) photoisomerization of α,β-epoxyketones, of which **85** is representative, to β-diketones, in this case **86**.

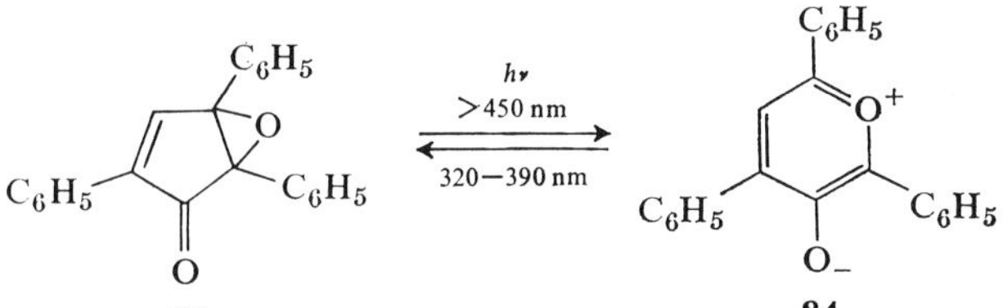

C_6H_5 C_6H_5 C_6H_5 O O — $h\nu$ >450 nm / 320–390 nm — C_6H_5 O^+ C_6H_5 C_6H_5 O_-

83 **84**

C_6H_5 O O C_6H_5 H C_6H_5 $\xrightarrow{h\nu}$ C_6H_5 O O C_6H_5 C_6H_5

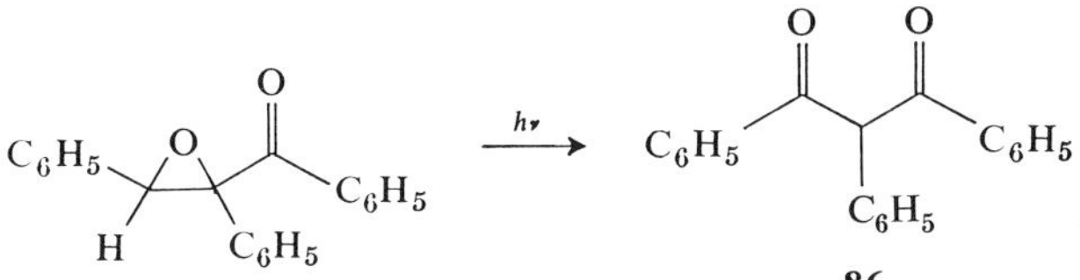

85 **86**

The preferred modes of fragmentation for the unsymmetrical polar-substituted aryloxiranes may be predicted if a zwitterionic intermediate such as **87** is assumed to be implicated. The overall cycloelimination reaction is

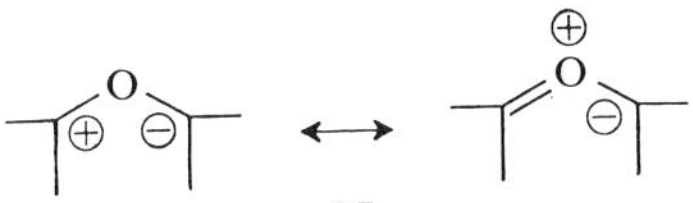

87

thus viewed as a photoinduced or thermal α-elimination of the carbonyl fragment from the more stable possible zwitterion formed by initial C—C oxirane bond cleavage. Some typical examples follow:

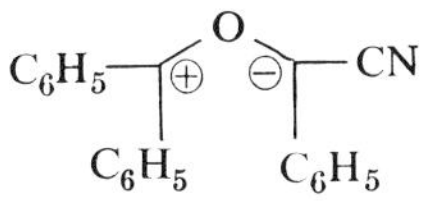

$$\longrightarrow C_6H_5\ddot{C}CN + (C_6H_5)_2CO$$
63

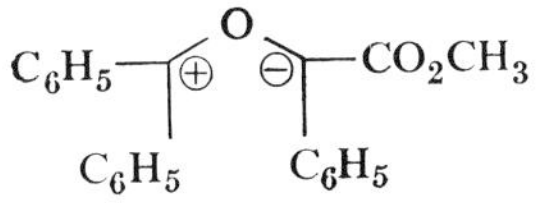

$$\longrightarrow C_6H_5\ddot{C}CO_2CH_3 + (C_6H_5)_2CO$$
69

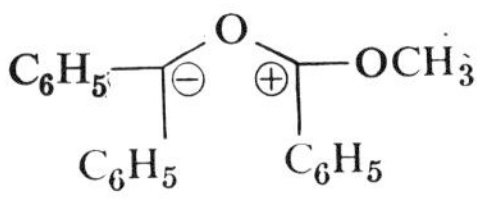

$$\longrightarrow (C_6H_5)_2C: + C_6H_5CO_2CH_3$$
17

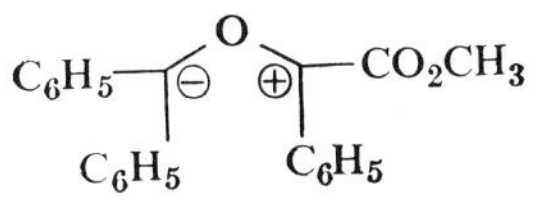

$$\longrightarrow (C_6H_5)_2C: + C_6H_5CO_2COCH_3$$
17

The mechanistic pathways delineated below[34b] are consistent with the original EPR and optical spectroscopic as well as chemical experimental observations made on oxirane substrates. The asterisks were employed at that stage to avoid any specific inference with respect to the electronic nature (and/or multiplicity) of the intermediates. Several questions pertaining to the nature of the energy required to achieve the proposed interconversions remained unanswered, however, recent theoretical studies using the non-empirical self-consistent field and configurational interaction methods predict comparable energetics for the formation of the C—C and C—O bond cleaved intermediates from oxiranes.[21b]

R O R″ / R′ R‴ ⇄ (hν; hν or Δ, path *a*) R O* R″ / R′ R‴ → (hν, path *a*) R R″ / R′ R‴ + [O]

hν or Δ ⇅ hν path *b*

path *a′* hν or Δ → RR′CO + R″R‴C:

R O R″ / * * / R′ R‴ → (hν or Δ, path *b′*) RR′CO + R″R‴C:

Although alkenes are not detected among the solution photoproducts of oxiranes at ambient temperatures, they are found to be produced at 77° K in rigid glasses by an intramolecular oxygen extrusion process as outlined in path *a*. It was demonstrated conclusively that the alkenes are not formed by bimolecular dimerization, which is not unexpected. It was proposed[34b] that at ambient temperatures the reaction rate via path *a′* may be so fast that it completely dominates the photochemistry of the oxirane; however, as the temperature is lowered, the rates of reaction by paths *a* and *a′* become comparable and the products of both reactions are observed.

Although the absorption and fluorescence emission spectra of phenyloxirane (styrene oxide) (**88**) are similar to the spectra of the phenyl-substituted oxiranes, the photochemistry of this oxirane observed in both rigid matrices at 77° K and 40° in fluid solution is atypical.[34b] The mixture of photoproducts detected spectroscopically upon irradiation of **88** in 3-methylpentane at 77° K includes benzyl radical, phenylacetaldehyde, and a small amount of styrene.[34b] Phenylacetaldehyde is also produced upon photolysis of **88** in fluid solution.[31b] When benzene is used as the solvent and the irradiation conducted at 40°, carbon monoxide and bibenzyl are produced.[31] Padwa and co-workers independently observed that bibenzyl is formed in 30–40% yield upon irradiation of phenylacetaldehyde.[51a] Acetophenone, 2-phenylethanol, and 1-phenylethanol are reportedly found among the photolysis products of styrene oxide in the liquid phase.[51b]

D. Photochromic Behavior of Vicinal Diaryloxiranes—The Mechanism of Photocycloelimination

The low-temperature photochromic behavior observed for vicinally substituted diaryloxiranes appears to be related to the cycloelimination of these aryloxiranes to carbenes and carbonyl compounds. Colors are generated upon irradiation of 2,3-diaryloxiranes in rigid glasses at 77° K, and their spectral characteristics are recorded.[34b] Furthermore, a single transient of 14 μsec lifetime has been observed upon flash-photolysis of tetraphenyloxirane (**54**), which has spectral properties in common with the colored species generated from **54** at 77° K.[36d] In all cases the colors are stable at 77° K; however, they can be bleached either by warming the sample above 140° K or by irradiation into the visible absorption bands.[34b] Arylcarbenes and carbonyl compounds are also observable simultaneously in the presence of the colored materials using optical and EPR spectroscopic techniques. The extent of cycloelimination observed during photolysis warm-up cycles, however, is estimated to be 20–25 times more than originally produced in the initial photolysis.[52] Although the thermal production of additional carbene and the bleaching of the color upon warming to near 140° K occur essentially simultaneously, it has yet to be demonstrated that they are related, concerted rather than independent events, that is, that the colored material is that species from which the additional carbene is derived by thermally induced fragmentation. In fact, recent studies by Huisgen on the thermal properties of oxiranes (*vide infra*) indicate that cleavage of the colored species is a photolytic process.

Although the mechanism of carbene formation has been discussed,[34b] only recently has information bearing on the electronic nature of the species involved in the photochromic process been published. DoMinh, Trozzolo, and Griffin[52] have presented evidence that the colored intermediates are most probably carbonyl ylides and that the opening and recyclization occur in a concerted disrotatory manner with conservation of orbital symmetry. The allowed transformations are exemplified using *cis*- and *trans*-stilbene oxide **51** and **52**.

Electronic spectral data support the contention that thermodynamic considerations dictate the mode of disrotation in those cases where "inward and outward" alternatives exist. For example, the colored species obtained from *cis*-stilbene oxide (**52**) could have structure **90** or **91**. The exo ylide **90** is more stable (less hindered) than the other possible structural isomer having endo stereochemistry **91** and that obtained from the trans isomer **89**. Indeed the visible spectral absorption maximum of the colored material obtained from **52** is red-shifted relative to that formed from **51**, which presumably reflects the predominate formation of **90** rather than **91**. Furthermore, these conclusions are supported by the work of Huisgen and co-workers,[53] who have trapped an azomethine ylide structurally related to **90** and derived from a *cis*-aziridine via the "outward" disrotatory mode.

Related evidence for the formation of a carbonyl ylide from an oxirane was obtained by Arnold and Kanrischky.[54a] Photolysis and pyrolysis (100° C) of the 5-oxabicyclo[2.1.0]pentane **92** gave purple species whose visible absorption spectra were essentially identical. It is proposed that the carbonyl ylide **93** is responsible for the purple color. The three-membered oxiranyl ring of **92** is incorporated in a fused system which is geometrically constrained to

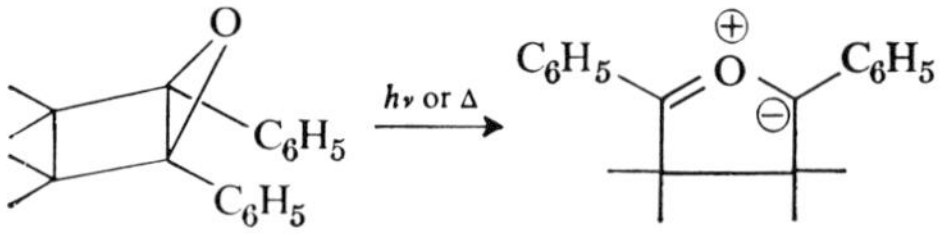

O C_6H_5 C_6H_5 $\xrightarrow{h\nu \text{ or } \Delta}$ C_6H_5 $\overset{\oplus}{O}$ C_6H_5 $\ominus$

92 **93**

open in a disrotatory manner. The colored intermediate, stable at high temperature and produced upon thermolysis, may be trapped using conventional dipolarophiles. Considerable strain is relieved upon bond cleavage which may account for the facile thermal conversion of **92** to **93** and the stability of the latter. In a subsequent study Arnold and Chang[54b] have shown that the carbonyl ylide derived from 1-carbomethoxy-4-phenyl-2,2,3,3-tetramethyl-5-oxabicyclo[2.1.0]pentane is significantly longer lived than **93** indicating that the carbomethoxy group can stabilize the negative charge.

Recently Huisgen and co-workers trapped the carbonyl ylides formed thermally from several aryloxiranes, including **62**, **65**, **94**, and **95**, with active dipolarophiles.[55] The stereochemistry in each case is explicable within the framework of orbital symmetry theory.

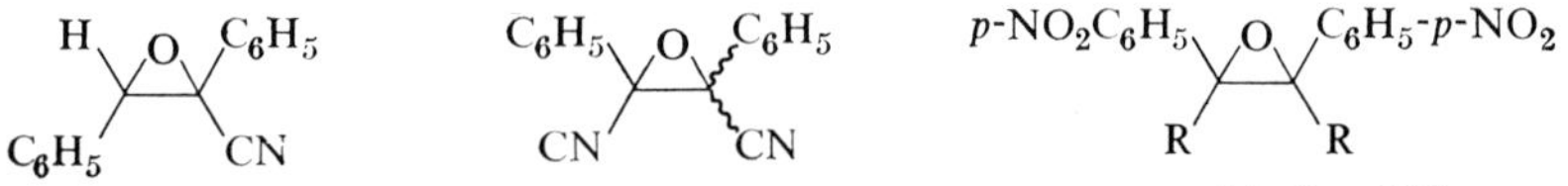

H O C_6H_5 / C_6H_5 CN — **62**

C_6H_5 O C_6H_5 / CN CN — **65**

p-$NO_2C_6H_5$ O C_6H_5-p-NO_2 / R R

94, R = CN
95, R = H

Selected examples of the adducts, isolated in high yield, are shown below; the temperature at which oxirane ring cleavage occurs is remarkably low.

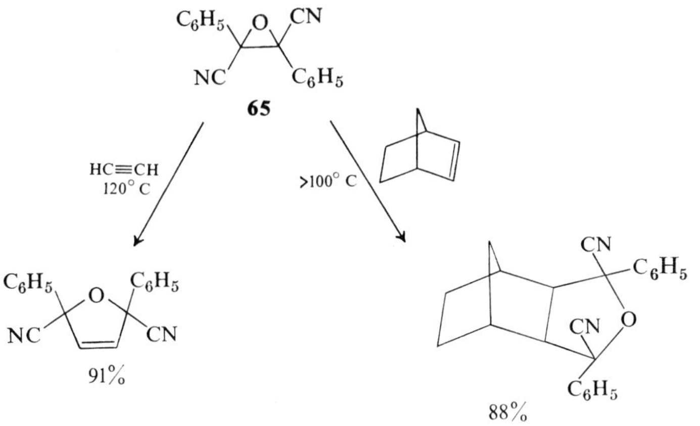

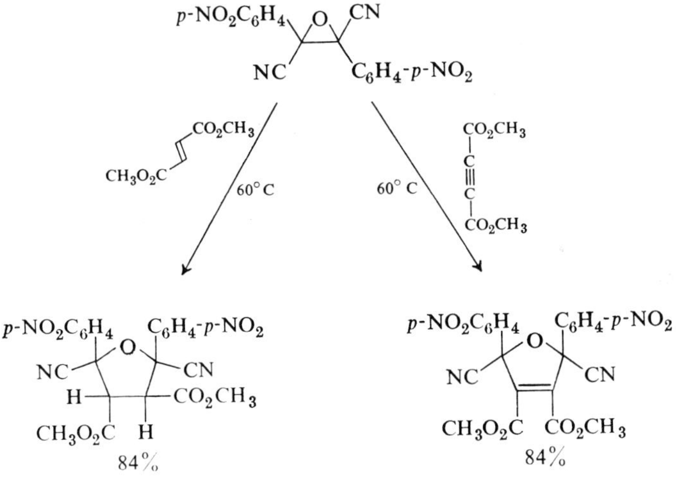

C_6H_5 O CN
NC C_6H_5
65

$HC{\equiv}CH$
120° C

>100° C

C_6H_5 O C_6H_5
NC CN
91%

CN
C_6H_5
CN O
C_6H_5
88%

p-$NO_2C_6H_4$ O CN
NC C_6H_4-p-NO_2

CO_2CH_3
CH_3O_2C
60° C

60° C
CO_2CH_3
C
|||
C
CO_2CH_3

p-$NO_2C_6H_4$ C_6H_4-p-NO_2
O
NC CN
H CO_2CH_3
CH_3O_2C H
84%

p-$NO_2C_6H_4$ C_6H_4-p-NO_2
O
NC CN
CH_3O_2C CO_2CH_3
84%

The failure of Huisgen and co-workers to detect phenylcyanocarbene **63** or its reaction products upon thermolysis of the oxiranes **62** and **65**, despite convincing evidence for ylide intermediates, is in accord with our observation that 2,2,4,4-tetraphenyloxetanone (**96**), a known precursor for the ylide **97** in rigid matrices at 77° K, by proper choice of conditions may be decarbonylated photochemically in fluid solution to give the oxirane **54** without formation of diphenylcarbene **17**. Clearly these results indicate that structurally identical ylides generated photochemically from oxiranes require a second photon for cleavage to carbenes and revert primarily to the oxirane on warming. Orbital symmetry constants, however, require that cis-trans isomerization accompany such a process, that is, opening via the excited state and thermal recyclization. Our inability to detect substantial cis-trans photointerconversion of oxiranes and photoracemization of active *trans*-stilbene oxide at ambient temperatures in solution may be due to intervention of an efficient cheletropic mechanism (concerted extrusion) under these conditions and/or the fact that the extinction coefficient of the ylide may be sufficiently greater than that of the residual oxirane to render subsequent photocleavage of the ylide by a second photon to carbene photochemically highly probable. Another alternative is that the intermediate formed by C—O bond cleavage may be the carbene precursor. Clearly further work is required to completely define the overall mechanistic path of oxirane photocycloelimination and such is in progress.

C_6H_5, C_6H_5, C_6H_5, C_6H_5 (oxetanone) $\xrightarrow{h\nu}$ C_6H_5—C(=$\overset{\oplus}{O}$)—$\overset{\ominus}{C}$—C_6H_5 with C_6H_5, C_6H_5 $\longrightarrow$ C_6H_5, C_6H_5, C_6H_5, C_6H_5 (oxirane)

96 **97** **54**

E. Carbenes from Aziridines

The formation of phenylcarbene upon photolysis of 1,2,3-triphenylaziridine (**98**) (stereochemistry not specified) has been reported by Nozaki and co-workers.[56] Irradiation of **98** in alcohols results in cleavage of the aziridine ring to give benzaldehyde acetals **99** and *N*-benzylaniline (**100**). Compctitive cycloelimination of phenylcarbene **12**, trapped as the alkyl benzyl ether **102**, appears to occur and the residual anil fragment **101** was also detected. This mode of cleavage is formally analogous to the photofragmentation of aryloxiranes to arylcarbenes and carbonyl compounds.

$C_6H_5CH(OR)_2 \xleftarrow[h\nu]{ROH}$ **98** (N-C_6H_5; C_6H_5, C_6H_5) $\xrightarrow[h\nu]{ROH} C_6H_5CH{=}NC_6H_5$

99 **101**

\+ **98** +

$C_6H_5CH_2NHC_6H_5$ **100** $C_6H_5CH_2OR$ **102**

$R = CH_3, C_2H_5, 2\text{-}C_3H_7, t\text{-}C_4H_9$

A photogenerated intermediate assumed to be implicated in these reactions may be intercepted with cyclohexene. A pair of stereoisomeric 1,3-cycloaddition products having the 1,2,3-triphenyloctahydroisoindole skeletal structure **103** are formed. The complementary thermal studies conducted by

Huisgen and collaborators on **92**[55d] confirm that whereas the azomethine ylide is formed at 150° and may be trapped,[55e] fragmentation to carbene is not observed, suggesting that if the ylide is involved in the photocycloelimination process, a second photon is required for cleavage. No evidence was found to indicate that phenylnitrene is generated under these conditions. Similarly, attempts to generate phenylnitrene from 1,2,2,3-tetraphenylaziridine also proved unsuccessful,[36e] although the amino nitrene **105** is liberated readily

C_6H_5 C_6H_5 C_6H_5 **98** $\xrightarrow{h\nu}$ C_6H_5 N—C_6H_5 C_6H_5 **103**

upon irradiation of the aziridine **104** and undergoes transfer to alkenes.[57]

104 $\underset{}{\overset{h\nu}{\rightleftharpoons}}$ **105**

F. Photocycloelimination Reactions of Thiiranes

Despite their structural resemblance to aryloxiranes, the photolytic cycloelimination reactions of arylthiiranes do not, at least in those cases studied, parallel the behavior of the former. A [3 → 2 + 1] photocycloelimination to sulfur and the parent alkene appears to be the preferred mode of cleavage as shown in examples cited for the thiiranes **106–110**.

Reference

106 $\xrightarrow{h\nu}$ [S] + [cis-stilbene] $\xrightarrow{h\nu}$ phenanthrene — 58

107 $\xrightarrow{h\nu}$ [S](^{3}P) + $(C_6H_5)_2C{=}C(C_6H_5)_2$ $\xrightarrow{h\nu}$ 9,10-diphenylphenanthrene — 36f, 36i

$\xrightarrow{h\nu}$ [S] + (36f)

108

$\xrightarrow{h\nu}$ [S] + (36f)

109

$\xrightarrow{h\nu}$ [S] + (59)

110

G. Photocycloelimination Reactions of Oxaziridines

Two [3 → 2 + 1] cycloelimination reactions might be anticipated for oxaziridines **111**, path *a* to give carbenes **112** and nitroso compounds **113** and/or path *b* to give nitrenes **114** and carbonyl compounds **115**. In those

$$R_3\text{—NO} + :C(R_1)(R_2) \xleftarrow{(a)} \mathbf{111} \xrightarrow{(b)} R_3\text{—}\ddot{N}: + O{=}C(R_1)(R_2)$$

113 112 111 114 115

cycloelimination reactions reported the reaction course invariably occurs by path *b*.[60,61] The arylnitrenes formed from the oxaziridines (or nitrones via prior cyclization) may be trapped chemically using substituted amines.[60]

Meyer and Griffin[60] originally reported that irradiation of the spiro oxaziridine **116** in diethylamine and cyclohexylamine affords diethyl- and cyclohexylamino-3H-azepines (**117** and **118**, respectively). Phenylnitrene was proposed as an intermediate in these transformations.

Splitter and Calvin[61] subsequently detected the EPR spectrum of triplet phenylnitrene (**119**) upon irradiation of 2-phenyl-3-(*p*-dimethylamino-phenyl)oxaziridine (**120**) or 2,3,3-triphenyloxaziridine (**121**) in either aceto-phenone or methylene chloride/fluorolube at 77° K. Similar independent

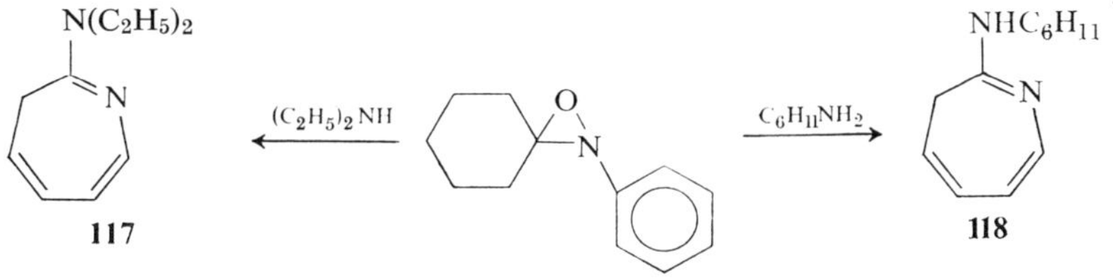
N(C2H5)2
N
(C2H5)2NH
O
N
C6H11NH2
NHC6H11
N

117 **116** **118**

results have been obtained by Griffin and Trozzolo using **116**, **121**, and the

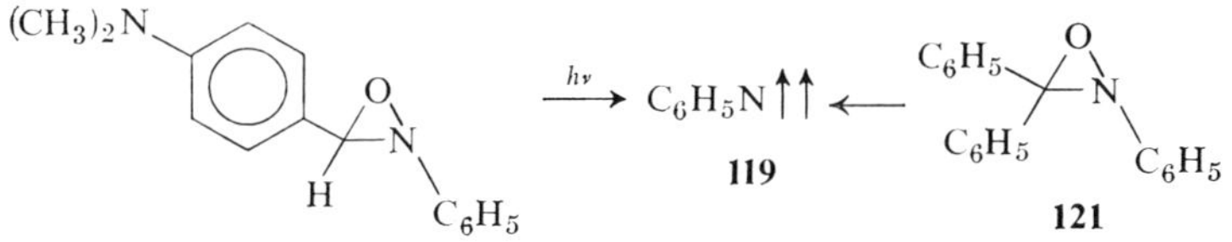
(CH3)2N
O
N
H
C6H5
hν
C6H5N↑↑
C6H5
O
N
C6H5
C6H5

120 **119** **121**

nitrones **122** and **123**, which presumably undergo photoisomerization to the corresponding oxaziridines under the reaction conditions.[62]

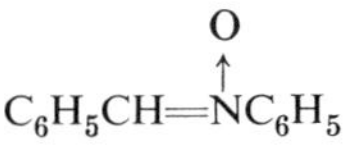
O
C6H5CH=NC6H5

122

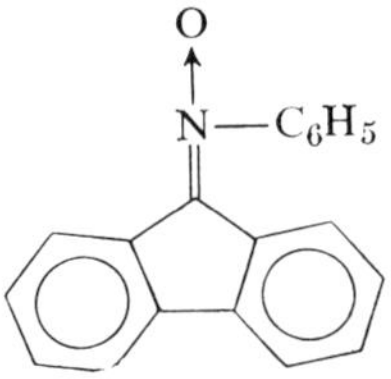
O
N—C6H5

123

H. Carbenes from Diazirines

The photochemistry of diazirines **124**, established carbene precursors, has been reviewed extensively elsewhere.[63,64] Irradiation of these cyclic isomers of diazo compounds results in loss of nitrogen and generation of the corresponding carbene **112**. The diazirines whose photodecompositions have

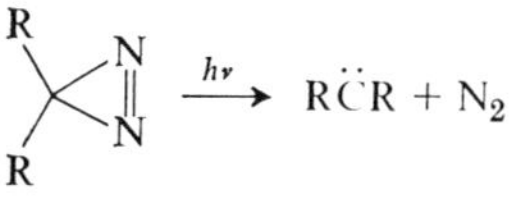
R
N
N
R
hν
RC̈R + N2

124 **112**

been described to date together with the carbene generated in the primary process are summarized in Table 5. The majority of these investigations were conducted in the gas phase and numerous secondary reactions occur.

TABLE 5

Diazirine	Carbene	Reference(s)
H, H (ring $N{=}N$)	$H\ddot{C}H$	65, 66
CH_3, H (ring $N{=}N$)	$CH_3\ddot{C}H$	67
CH_3, CH_3 (ring $N{=}N$)	$CH_3\ddot{C}CH_3$	68
CH_3, C_2H_5 (ring $N{=}N$)	$CH_3\ddot{C}C_2H_5$	69, 70
C_2H_5, C_2H_5 (ring $N{=}N$)	$C_2H_5\ddot{C}C_2H_5$	70
2-C_3H_7, H (ring $N{=}N$)	2-$C_3H_7\ddot{C}H$	70
2-C_3H_7, CH_3 (ring $N{=}N$)	2-$C_3H_7\ddot{C}CH_3$	70
t-C_4H_9, H (ring $N{=}N$)	*t*-$C_4H_9\ddot{C}H$	70, 71
t-C_4H_9, CH_3 (ring $N{=}N$)	*t*-$C_4H_9\ddot{C}CH_3$	70

TABLE 5 (*Continued*)

Diazirine	Carbene	Reference(s)
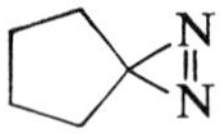 cyclopentane spiro-diazirine	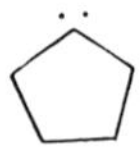 cyclopentylidene	64
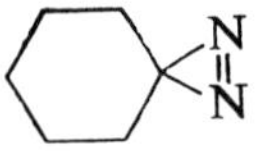 cyclohexane spiro-diazirine	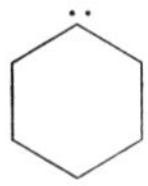 cyclohexylidene	72, 73
 F, F diazirine	$FC\ddot{F}$	64
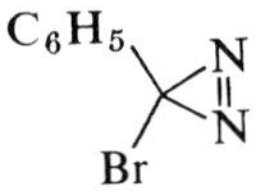 C_6H_5, Br diazirine	$C_6H_5C\ddot{B}r$	74
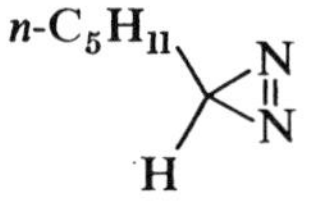 n-C_5H_{11}, H diazirine	n-$C_5H_{11}\ddot{C}H$	70

II. [5 → 3 + 2] PHOTOCYCLOELIMINATIONS

The photochemical conversion of substituted pyrazolenines (3H-pyrazoles) **125** to cyclopropenes **126**, which formally may be viewed as a [5 → 3 + 2] cycloelimination, is believed to occur by a two-photon process. The first step involves formation of the diazoalkane **127**, which subsequently loses nitrogen to give a vinylcarbene **128**, the immediate precursor of the cyclopropene.[75,76]

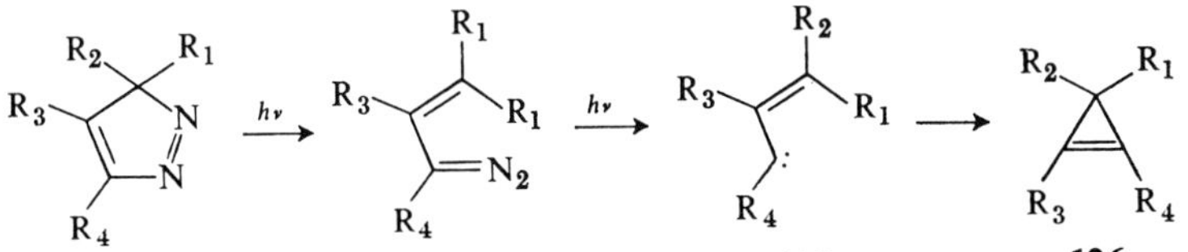

125 $\xrightarrow{h\nu}$ **127** $\xrightarrow{h\nu}$ **128** $\longrightarrow$ **126**

Those 3H-pyrazoles whose conversion to cyclopropenes has been investigated are tabulated in Table 6.

TABLE 6

3H-Pyrazole	Carbene intermediate	Cyclopropene	Reference(s)
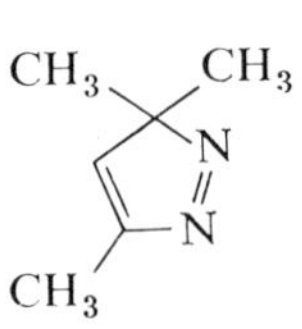 CH_3, CH_3, N, N, CH_3	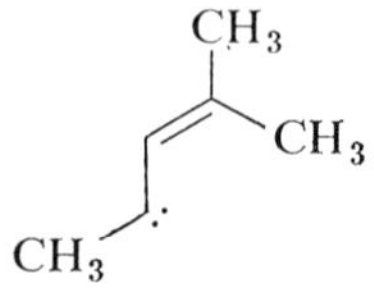 CH_3, CH_3, CH_3	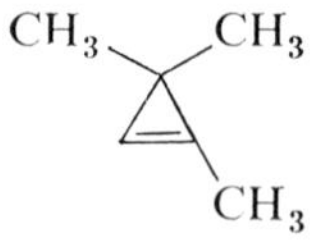 CH_3, CH_3, CH_3	77,78
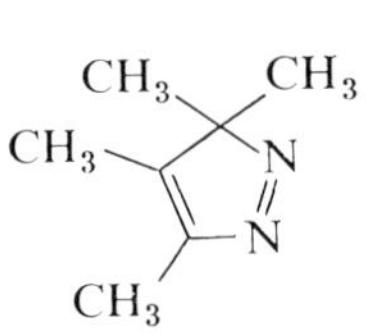 CH_3, CH_3, CH_3, N, N, CH_3	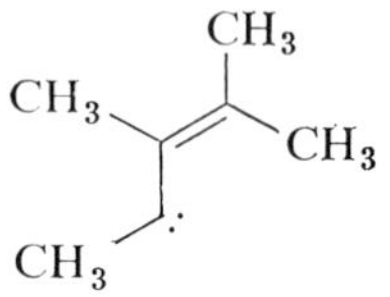 CH_3, CH_3, CH_3, CH_3	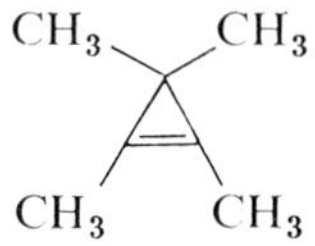 CH_3, CH_3, CH_3, CH_3	77
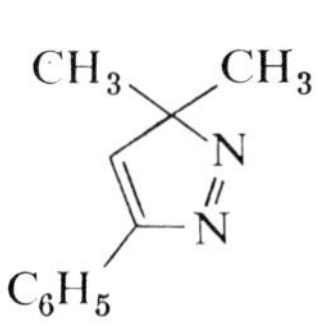 CH_3, CH_3, N, N, C_6H_5	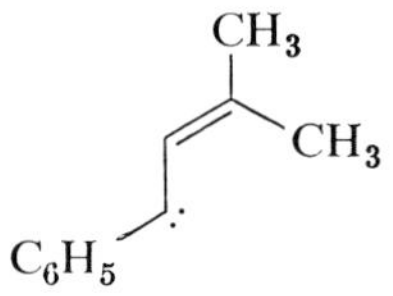 CH_3, CH_3, C_6H_5	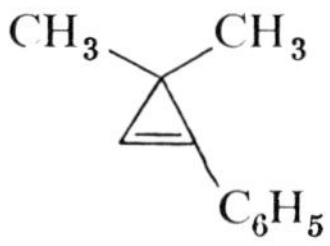 CH_3, CH_3, C_6H_5	77
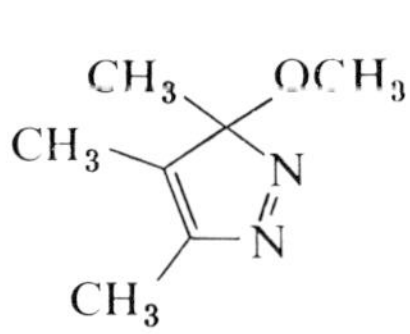 CH_3, OCH_3, CH_3, N, N, CH_3	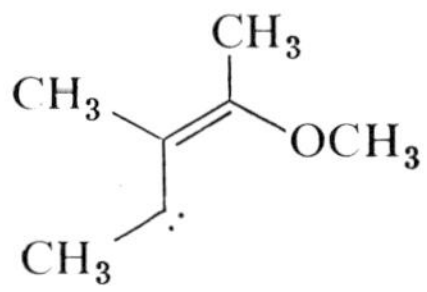 CH_3, CH_3, OCH_3, CH_3	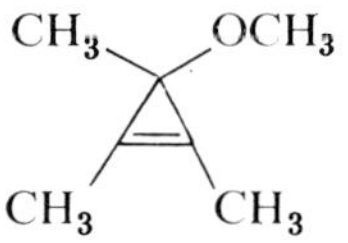 CH_3, OCH_3, CH_3, CH_3	77
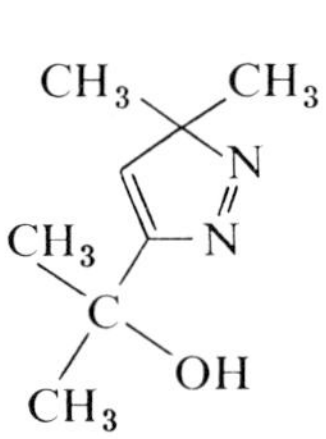 CH_3, CH_3, N, N, CH_3, C, OH, CH_3	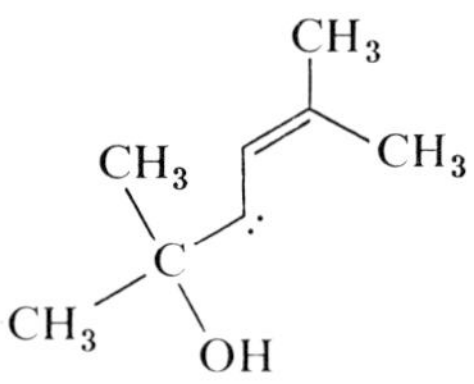 CH_3, CH_3, CH_3, C, CH_3, OH	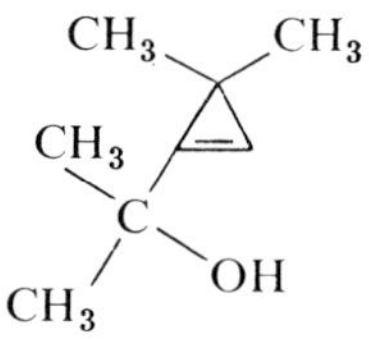 CH_3, CH_3, CH_3, C, OH, CH_3	76

TABLE 6 (*Continued*)

3H-Pyrazole	Carbene intermediate	Cyclopropene	References
CH_3, CH_3, N, N	CH_3, CH_3	CH_3, CH_3	75,77
CH_3, CH_3, N, N	CH_3, CH_3	CH_3, CH_3	75,77,78
CH_3, CH_3, N, N, CH_3O_2C	CH_3, CH_3, CH_3O_2C	CH_3, CH_3, CH_3O_2C	79
N, N, CH_3O_2C, CH_3O_2C	CH_3O_2C, CH_3O_2C	CH_3O_2C, CO_2CH_3	80

That the penultimate intermediate is the diazoalkane **127** is suggested by the development of a transient red color during the photolyses.[75,77] Direct evidence for the diazo intermediate was obtained by irradiation of the trimethyl-3H-pyrazole **129** using a filter which eliminates both visible light and ultraviolet light below 320 nm.[77] Under these conditions the expected diazoalkane **130** is formed in 50% yield.

CH_3 CH_3 N N CH_3 **129** —$h\nu$ (320–380 nm)→ CH_3 CH_3 CH_3 N_2 **130**

Bentrude[81] has reported that 1,3,2-dioxa-P^V-phospholes **131** undergo photocycloelimination of the [5 → 3 + 2] and [5 → 4 + 1] types. The former reaction is assumed to give the ketocarbenes **132**; however, these intermediates were not trapped.

OR_1, R_1O, OR_1, P, O, O, R_2, R_2 —$h\nu$→ $(R_1O)_3P + R_2—C(=O)—C(=O)—R_2$

→ $(R_1O)_3P + R_2—C(=O)—\ddot{C}—R_2$

131a, $R_1 = R_2 = CH_3$
131b, $R_1 = CH_3$; $R_2 = C_6H_5$

132a, $R_1 = R_2 = CH_3$
132b, $R_1 = CH_3$; $R_2 = C_6H_5$

IV. [5 → 4 + 1] PHOTOCYCLOELIMINATIONS

Generation of arylcarbenes from substituted 1,3-dioxolanes by [5 → 4 + 1] cycloelimination reactions of the type depicted in a general manner here have been reported by Griffin and co-workers.[82]

R_3 O R_1 R_4 O R_2 —$h\nu$→ $R_1R_2C\colon +$ R_3 O R_4 O

The results of their preliminary studies in this area are summarized in

TABLE 7

Dioxolane	Arylcarbene	Yield (%)[a]
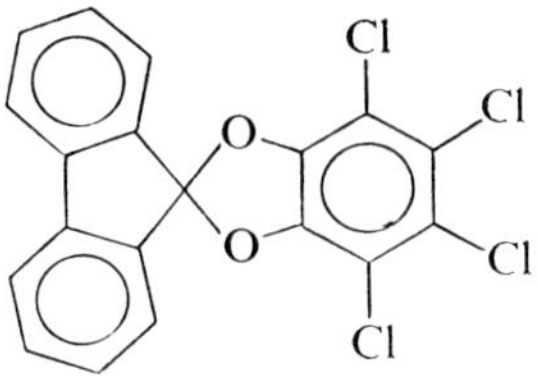 133	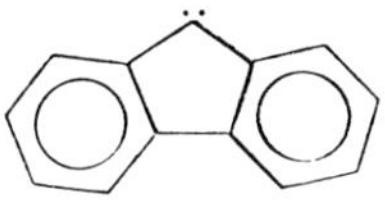 134	30
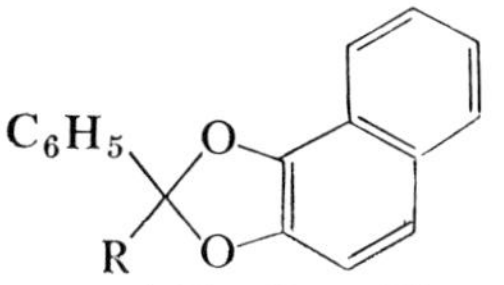 135a, R = H	$C_6H_5\ddot{C}H$ (12)	17
135b, R = C_6H_5	$(C_6H_5)_2\ddot{C}$ (17)	11
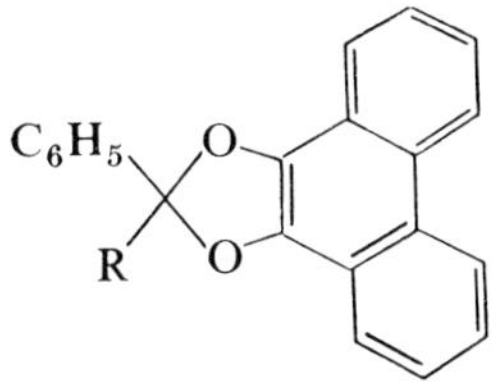 136a, R = H	$C_6H_5\ddot{C}H$ (12)	15
136b, R = C_6H_5	$(C_6H_5)C\colon$ (17)	5
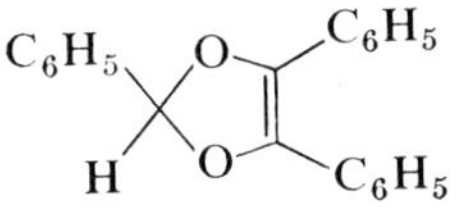 137	$C_6H_5\ddot{C}H$ (17)	16

[a] Yields based on ether formation in methanol.

Table 7. Photolysis of 4,5,6,7-tetrachlorospiro[1,3-benzodioxole-2,9′-fluorene] (**133**) in *cis*-2-butene leads to a mixture of *cis*- and *trans*-2,3-dimethylspiro[cyclopropane-1,9′-fluorene] (**138** and **139**), respectively). Since these cyclopropanes are interconvertible under the irradiation conditions, no significance may be attached to the apparent nonstereospecific addition of the fluorenylidene (**134**) to the alkene. Corroborative evidence for the formation of the carbene **134** was obtained by irradiation of **133** in methanol, which

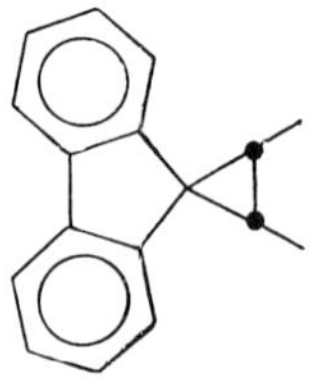

138 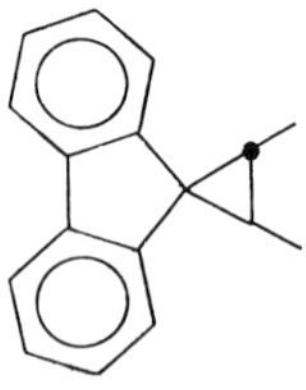 **139**

provides fluorenylmethyl ether in 30% yield. Photochemical reactions which preclude the isolation of the other fragment, 3,4,5,6-tetrachloro-*o*-benzoquinone (**140**), occur in *cis*-2-butene; however, when the irradiations are conducted in rigid matrices it is possible to observe characteristic oxidation products generated from **140**.

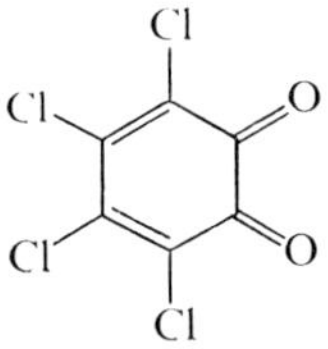

140

The remaining 1,3-dioxolanes (**135–137**) afford arylcarbenes in varying amounts, as evidenced by the formation of benzyl or benzhydryl methyl ether upon irradiation in methanol. Although it would appear from the data cited in Table 7 that aryl 1,3-dioxolanes afford arylcarbenes upon photolysis in lower yields than observed with aryloxiranes, the experimental conditions for such [5 → 4 + 1] cycloeliminations have not been optimized nor have possible structural modifications been fully explored.

V. [5 → 2 + 2 + 1] PHOTOCYCLOELIMINATIONS

A variety of 1,3-dioxolanes such as **141** undergo cycloelimination of the [5 → 2 + 2 + 1] rather than [5 → 4 + 1] type; that is, it is not invariably the ring carbon atom flanked by the oxygen atoms which is extruded in the divalent form. The course of the reaction may be altered by incorporating a trigonal center or heteroatom in the 2-position. This mode of cycloelimination is observed in the case of several 1,3-dioxolanes[32,83,84] (Table 8).

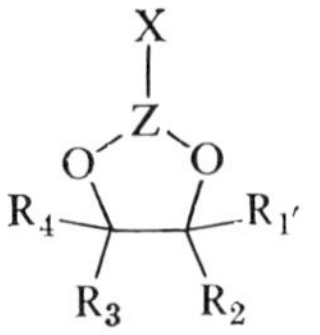

$$\mathbf{141} \xrightarrow{h\nu} R_1R_2C\colon + R_3R_4C{=}O + XZ{=}O$$

$Z = C, S, P;\ X{=}, O{=}, \quad (C_2H_5O)_3{-}$

141

TABLE 8

Dioxolane	Carbene	Yield (%)	Reference(s)
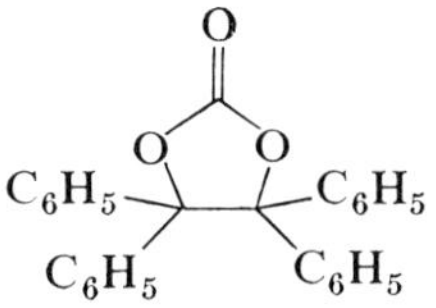 **142**	$(C_6H_5)_2\ddot{C}$ (**17**)	60[a]	83
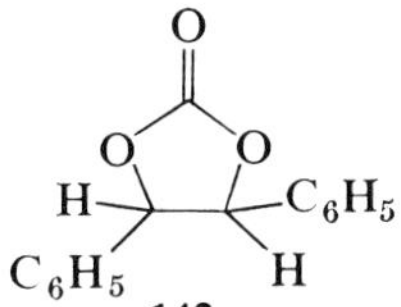 **143**	$C_6H_5\ddot{C}H$ (**12**)	5.2[b]	32
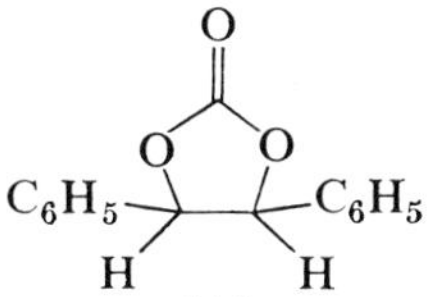 **144**	$C_6H_5\ddot{C}H$ (**12**)	3.8[b]	32
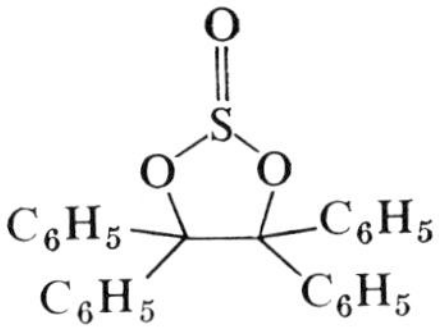 **145**	$(C_6H_5)_2\ddot{C}$ (**17**)	~40[a]	83
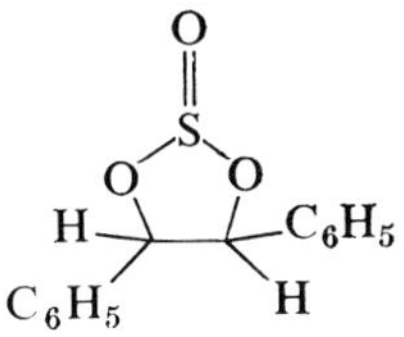 **146**	$C_6H_5\ddot{C}H$ (**12**)	6.6[b]	32

TABLE 8 (*Continued*)

Dioxolane	Carbene	Yield (%)	Reference(s)
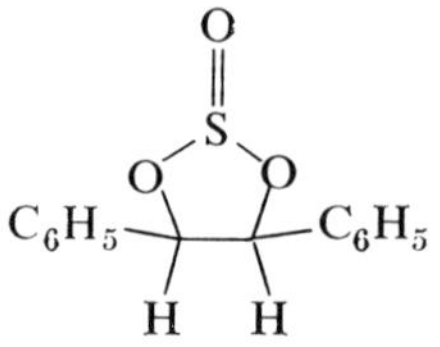 **147**	$C_6H_5\ddot{C}H$ (**12**)	5.5[b]	32
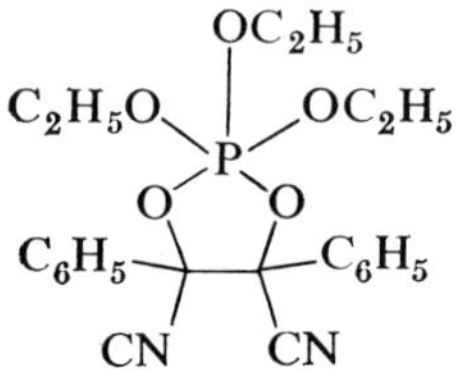 **148**	$C_6H_5\ddot{C}CN$ (**63**)	60–70[c]	84
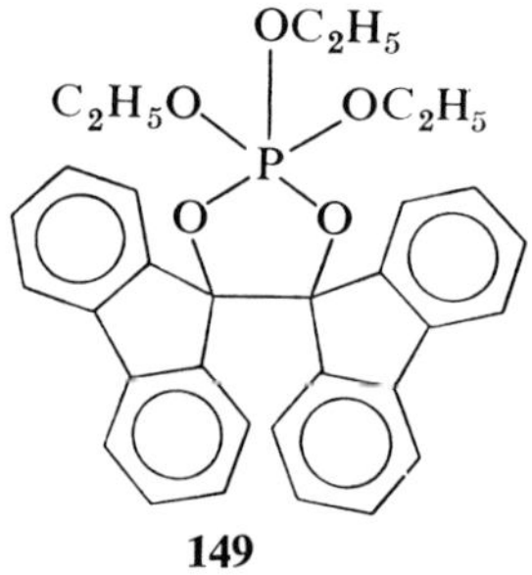 **149**	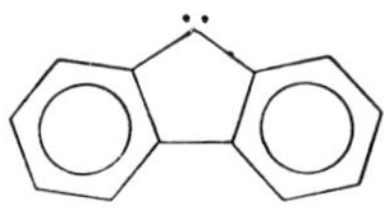	Not given	84

[a] Yield based on ether formation in methanol.
[b] Yield based on extent of aliphatic C—H insertion.
[c] Yield based on addition to alkenes to give cyclopropanes.

Irradiation of the cyclic arylcarbonates (**142–144**) results in photofragmentation to arylcarbenes, carbonyl compounds, and carbon dioxide.[32,83] Diphenylcarbene **17**, trapped as benzhydryl methyl ether, is formed in 60% yield upon photolysis of **142** in methanol. The benzpinacol sulfite **145** also affords **17** in high yield (~40%) upon irradiation in methanol.[83] The other fragments in this case are benzophenone and sulfur dioxide. Tetraphenyloxirane (**54**) could not be detected among the photolysis products of either **142** or **145**.

The insertion selectivity of phenylcarbene **12** generated from the *dl*- and *meso*-hydrobenzoin carbonates (**143** and **144**) and sulfites (**146** and **147**) into the aliphatic C—H bonds of *n*-pentane has been compared with those values obtained using the conventional phenylcarbene precursor phenyldiazomethane.[32] The results presented in Table 9 confirm that the reactivity of the

TABLE 9
Relative Insertion Ratios for Phenylcarbene in the C_1-, C_2-, and C_3-H Bonds of *n*-Pentane

Dioxolane	$\frac{C_2 + C_3}{C_1}$	$\frac{C_2}{C_3}$
143	8.27 ± 0.14	1.42 ± 0.03
144	8.47 ± 0.20	1.42 ± 0.03
146	8.00 ± 0.18	1.45 ± 0.02
147	8.48 ± 0.24	1.41 ± 0.05
$C_6H_5CHN_2$	7.14 ± 0.14	1.31 ± 0.09
	8.38 ± 0.19	1.33 ± 0.09

transient species **12** formed photochemically from **143**, **144**, **146**, and **147** compares favorably with that obtained from phenyldiazomethane. That phenylcarbene is generated in the singlet manifold from these precursors is implied by its stereospecific addition to *cis*-2-butene.[36b]

Irradiation of the 1,3,2-dioxo-P^V-phospholane **148** in 2-methyl-2-butene and 2,3-dimethyl-2-butene affords *syn*- and *anti*-1-cyano-1-phenyl-2,2,3-triphenylcyclopropane and 1-cyano-1-phenyl-2,2,3,3-tetramethylcyclopropane, respectively.[84] The cyclopropanes are obtained in 60–70% yields. Photolysis of the spiro phospholane **149** in *trans*-2-butene results in the formation of the expected cyclopropane **139** in low yield. The major product, however, is the spiro ketone **150**, which is identical to that obtained thermally from **149**.

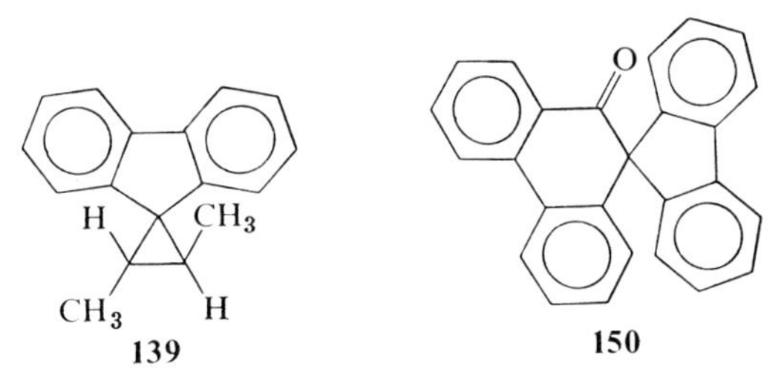

A [5 → 2 + 2 + 1] cycloelimination reaction yielding carbene and carbonyl compound has been reported by Bischoff and Brandstaedter.[85] Photolysis of 3,5-dimethyl-3,5-dihydroxy-1,2-dioxolane (**151a**) gives formic acid, which is formulated as arising from methylene. Irradiation of the related 4,4-dichloro compound **151b** in cyclohexene gives 7,7-dichloronorcarane in 0.4% yield which supports the contention regarding **151a**.

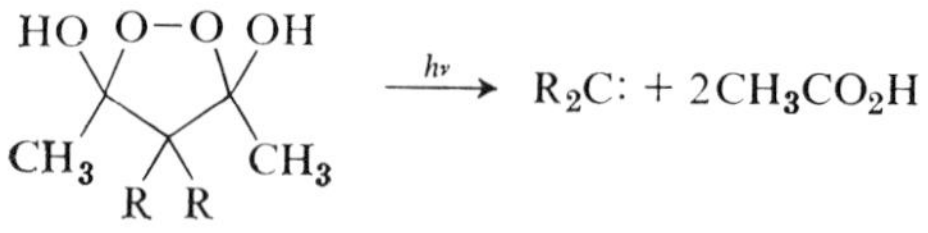

$$\text{151} \xrightarrow{h\nu} R_2C: + 2CH_3CO_2H$$

151a: R = H
151b: R = Cl

Slightly better yields of carbene (5–8%) were obtained upon irradiation of **152**; nevertheless, other photochemical processes dominate in this case too.[86]

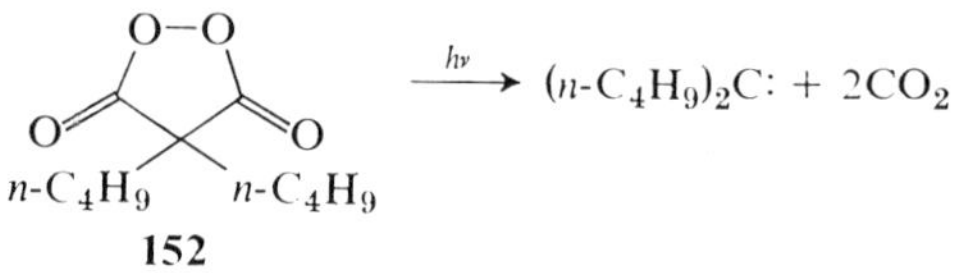

$$\text{152} \xrightarrow{h\nu} (n\text{-}C_4H_9)_2C: + 2CO_2$$

152

Acknowledgment

The authors would like to acknowledge The National Institutes of Health and the National Science Foundation for partial support of this work and also to thank Mrs. J. Thompson, Dr. A. Manmade, Dr. R. L. Smith, Dr. E. Elder and Dr. W. E. Franklin for their assistance in the preparation of this chapter.

References

1a. G. W. Griffin, *Angew. Chem.*, **83,** 604 (1971); *Angew. Chem. Int. Ed. Engl.*, **10,** 537 (1971).
1b. R. W. Hoffmann, *Angew. Chem.*, **83,** 595 (1971); *Angew. Chem. Int. Ed. Engl.*, **10,** 529 (1971).
1c. R. Huisgen, *Angew. Chem.*, **80,** 329 (1968); *Angew. Chem. Int. Ed. Engl.*, **7,** 321 (1968).
1d. R. B. Woodward and R. Hoffmann, *The Conservation of Orbital Symmetry*, Verlag Chemie GmbH, Academic Press, Germany, 1970.
1e. An alternate treatment of electrocyclic reactions based on PMO methods and stressing the significance of orbital topological control has recently been advanced by Dewar. See M. J. S. Dewar, Southeast-Southwest Regional ACS Meeting, New Orleans, Louisiana, 1970, Abstr. 381, p. 116; *Angew. Chem.*, **83,** 859 (1971); *Angew. Chem. Int. Ed. Engl.*, **10,** 761 (1971).
1f. W. Kirmse, *Carbene Chemistry*, 2nd ed., Academic Press, New York, 1971.
2a. C. L. Currie, S.J., H. Okabe, and J. R. McNesby, *J. Phys. Chem.*, **67,** 1494 (1963).
2b. A. A. Scala and P. Ausloos, *J. Chem. Phys.*, **49,** 2282 (1968).
2c. A. K. Dhingra and R. D. Koob, *J. Phys. Chem.*, **74,** 4490 (1970).
2d. R. D. Doepker, *J. Phys. Chem.*, **73,** 3219 (1969).
3a. P. A. Leermakers and G. F. Vesley, *J. Org. Chem.*, **30,** 539 (1965).
3b. P. A. Leermakers and M. E. Ross, *J. Org. Chem.*, **31,** 301 (1966).
4a. G. W. Griffin, E. J. O'Connell, and H. A. Hammond, *J. Am. Chem. Soc.*, **85,** 1001 (1963); see also G. W. Griffin, J. Covell, R. C. Petterson, R. M. Dodson, and G. Klose, *J. Am. Chem. Soc.*, **87,** 1410 (1965).
4b. W. von E. Doering and M. Jones, Jr., *Tetrahedron Lett.*, **1963,** 791.

4c. R. C. Cookson, M. J. Nye, and G. Subrahmanyam, *Proc. Roy. Chem. Soc. (London)*, **1964,** 144.
4d. G. S. Hammond, P. Wyatt, C. D. DeBoer, and N. J. Turro, *J. Am. Chem. Soc.*, **86,** 2532 (1964).
4e. L. B. Rodewald and C. H. DePuy, *Tetrahedron Lett.*, **1964,** 2951.
4f. It has recently been suggested that cyclopropanes with strongly electron-withdrawing substituents may open via zwitterionic intermediates; see E. W. Yankee and D. J. Cram, *J. Am. Chem. Soc.*, **92,** 6328, 6329, 6331 (1970).
5. R. K. Brinton, *J. Phys. Chem.*, **72,** 321 (1968).
6a. D. B. Richardson, L. R. Durrett, J. M. Martin, Jr., W. E. Putnam, S. C. Slaymaker, and I. Dvoretsky, *J. Am. Chem. Soc.*, **87,** 2763 (1965).
6b. A. P. ter Borg, E. Razenberg, and H. Kloosterziel, *Rec. Trav. Chim. Pays-Bas*, **85,** 774 (1966).
7. M. Jones, Jr., W. H. Sachs, A. Kulczycki, Jr., and F. J. Waller, *J. Am. Chem. Soc.*, **88,** 3167 (1966).
8. H. Kristinsson and G. S. Hammond, *J. Am. Chem. Soc.*, **89,** 5970 (1967).
9a. H. Dietrich, G. W. Griffin, and R. C. Petterson, *Tetrahedron Lett.*, **1968,** 153.
9b. E. Valyocsik and P. Sigal, *J. Org. Chem.*, **36,** 66 (1971).
10a. A. S. Kende, Z. Goldschmidt, and R. F. Smith, *J. Am. Chem. Soc.*, **92,** 7606 (1970).
10b. J. C. Gilbert and J. R. Butler, *J. Am. Chem. Soc.*, **92,** 7493 (1970).
11. H. E. Zimmerman and A. C. Pratt, *J. Am. Chem. Soc.*, **92,** 6259 (1970).
12. H. Kristinsson, K. N. Mehrotra, G. W. Griffin, R. C. Petterson, and C. S. Irving, *Chem. Ind. (London)*, **1966,** 1562.
13. M. J. Jorgenson, *J. Am. Chem. Soc.*, **91,** 6432 (1969).
14. H. Kristinsson and G. W. Griffin, *Tetrahedron Lett.*, **1966,** 3259; G. W. Griffin and E. Waldau, unpublished results.
15. P. H. Mazzocchi, R. S. Lustig, and G. W. Craig, *J. Am. Chem. Soc.*, **92,** 2169 (1970).
16a. P. S. Skell and A. Y. Garner, *J. Am. Chem. Soc.*, **78,** 3409 (1956).
16b. W. von E. Doering and P. M. LaFlamme, *J. Am. Chem. Soc.*, **78,** 5447 (1956).
17. G. L. Closs in N. Allinger, and E. L. Eliel, *Topics in Stereochemistry*, Vol. 3, Wiley, New York, 1968, p. 193.
18. C. D. Gutsche, G. L. Bachman, and R. S. Coffey, *Tetrahedron*, **18,** 617 (1968).
19. L. Edwards, R. O. Bost, R. S. Becker, G. W. Griffin, and M. V. Elam, Southeast-Southwest Regional ACS Meeting, New Orleans, Louisiana, 1970, Abstr. 472, p. 144.
20. R. Hoffmann, *J. Am. Chem. Soc.*, **90,** 1475 (1968); *ibid.*, in press.
21a. E. F. Hayes, Manuscripts of Contributed Papers, 4th IUPAC Photochemical Symposium, Baden-Baden (Germany), July 16—22, 1972, p. 92.
21b. A. K. Q. Siu and E. F. Hayes, *J. Am. Chem. Soc.*, in press; *ibid.*, **93,** 2090 (1971).
22. G. Maier, *Angew. Chem. Int. Ed. Engl.*, **6,** 402 (1965).
23. A. Schönberg and N. Latif, *J. Am. Chem. Soc.*, **75,** 2267 (1953).
24. G. W. Griffin, unpublished results.
25. T. Toda, M. Nitta, and T. Mukai, *Tetrahedron Lett.*, **1969,** 4401.
26. G. Maier and U. Heep, personal communication.
27. E. Ciganek, *J. Am. Chem. Soc.*, **89,** 1458 (1967).
28. R. H. Levin and M. Jones, Jr., *Tetrahedron*, **27,** 2031 (1971).
29. M. Pomerantz and G. W. Gruber, *J. Am. Chem. Soc.*, **89,** 6798, 6799 (1967).
30. J. S. Swenton and A. J. Krubsack, *J. Am. Chem. Soc.*, **91,** 786 (1969).
31a. H. Kristinsson and G. W. Griffin, *Angew. Chem. Int. Ed. Engl.*, **4,** 868 (1965).
31b. H. Kristinsson and G. W. Griffin, *J. Am. Chem. Soc.*, **88,** 1579 (1966).
32. R. L. Smith, A. Manmade, and G. W. Griffin, *Tetrahedron Lett.*, **1970,** 663.

33. A. M. Trozzolo, W. A. Yager, G. W. Griffin, H. Kristinsson, and I. Sarkar, *J. Am. Chem. Soc.*, **89,** 3357 (1967).
34a. R. S. Becker, J. Kolc, R. O. Bost, H. Dietrich, P. Petrellis, and G. W. Griffin, *J. Am. Chem. Soc.*, **90,** 3292 (1968).
34b. R. S. Becker, R. O. Bost, J. Kolc, N. R. Bertoniere, R. L. Smith, and G. W. Griffin, *J. Am. Chem. Soc.*, **92,** 1302 (1970).
35. G. L. Closs and R. A. Moss, *J. Am. Chem. Soc.*, **86,** 4042 (1964).
36a. H. Kristinsson and G. W. Griffin, unpublished results.
36b. P. Petrellis and G. W. Griffin, unpublished results.
36c. G. W. Griffin and N. R. Bertoniere, unpublished results.
36d. D. Whitten and G. W. Griffin, unpublished results.
36e. G. W. Griffin and A. Reine, unpublished results.
36f. I. Sarkar and G. W. Griffin, unpublished results.
36g. U. Mende and G. W. Griffin, unpublished results.
36h. A. Manmade, R. L. Smith, and G. W. Griffin, unpublished results.
36i. J. Font, O. P. Strausz, G. W. Griffin, I. Sarkar, R. C. Petterson, and A. L. Hebert, unpublished results.
37. H. Kristinsson, *Tetrahedron Lett.*, **1966,** 2343.
38. P. Petrellis, H. Dietrich, E. Meyer, and G. W. Griffin, *J. Am. Chem. Soc.*, **89,** 1967 (1967).
39. T. J. Temnikova, I. P. Stepanov, and O. A. Ikonopistseva, *Zh. Org. Khim.*, **2,** 2259 (1966).
40a. P. Petrellis and G. W. Griffin, *Chem. Commun.*, **1967,** 691.
40b. J. T. Przybytek, Ph.D. Thesis, University of Illinois at Chicago Circle, Chicago, Ill., 1972.
41. The stereochemistry of **65** was originally designated *trans* [T. Mukaiyaka, I. Kuwajima, and K. Ohno, *Bull. Chem. Soc. Jap.*, **38,** 1954 (1965)] and later questioned [J. H. Boyer and R. Selvarajan, *J. Org. Chem.*, **35,** 1229 (1970)]. More recently the former stereochemical assignment has been proven correct [See ref. 55a].
42. T. I. Temnikova, I. P. Stepanov, and L. O. Semenova, *Zh. Org. Khim.*, **3,** 1666 (1967).
43. T. I. Temnikova and I. P. Stepanov, *Zh. Org. Khim.*, **3,** 2203 (1967).
44. T. I. Temnikova, I. P. Stepanov, and L. A. Dotsenko, *Zh. Org. Khim.*, **3,** 2203 (1967).
45. A. M. Trozzolo, A. Manmade, and G. W. Griffin, unpublished results.
46. T. I. Temnikova and I. P. Stepanov, *Zh. Org. Khim.*, **2,** 1525 (1966).
47. J. K. Crandall and D. R. Paulson, *Tetrahedron Lett.*, **1969,** 2751.
48. N. R. Bertoniere, S. P. Rowland, and G. W. Griffin, *J. Org. Chem.*, **36,** 2956 (1971).
49a. A. Padwa, in O. L. Chapman, *Organic Photochemistry*, Vol. I, Marcel Dekker, New York, 1967, p. 91.
49b. N. R. Bertoniere and G. W. Griffin, in O. L. Chapman, *Organic Photochemistry*, Vol. III, Marcel Dekker, New York, in press.
50. O. Jeger, K. Schaffner, and H. Wehrli, *Pure Appl. Chem.*, **9,** 555 (1964).
51a. A. Padwa, D. Crumrine, R. Hartman, and R. Layton, *J. Am. Chem. Soc.*, **89,** 4435 (1967); R. B. Hartman, *Diss. Abstr. B.*, **27,** 3036 (1967), Univ. Microfilms Order No. 67-2455.
51b. R. J. Gritter and E. C. Sabatino, *J. Org. Chem.*, **29,** 1965 (1964).
52. T. Do-Minh, A. M. Trozzolo, and G. W. Griffin, *J. Am. Chem. Soc.*, **92,** 1402 (1970).
53. R. Huisgen, W. Sheer, H. Mader, and E. Brunn, *Angew. Chem. Int. Ed. Engl.*, **8,** 604 (1969).
54a. D. R. Arnold and L. A. Karnischky, *J. Am. Chem. Soc.*, **92,** 1404 (1970).
54b. D. R. Arnold and Y. C. Chang, *J. Heterocyclic Chem.*, **8,** 1097 (1971).
55a. H. Hamberger and R. Huisgen, *Chem. Commun.*, **1971,** 1190.

55b. A. Dahman, H. Hamberger, R. Huisgen, and V. Markowski, *Chem. Commun.*, **1971,** 1192.
55c. R. Huisgen, personal communication.
55d. J. H. Hall, R. Huisgen, C. H. Ross, and W. Scheer, *Chem. Commun.*, **1971,** 1188.
55e. J. H. Hall and R. Huisgen, *Chem. Commun.*, **1971,** 1187.
56. H. Nozaki, S. Fujita, and R. Noyori, *Tetrahedron*, **24,** 2193 (1968).
57. T. L. Gilchrist, C. W. Rees, and E. Stanton, *J. Chem. Soc. (C)*, **1971,** 988.
58. T. Sato, Y. Goto, T. Tohyama, S. Hayashi, and K. Hata, *Bull. Chem. Soc. Japan*, **40,** 2975 (1967).
59a. A. Padwa and D. Crumrine, *Chem. Commun.*, **1965,** 506.
59b. A. Padwa, D. Crumrine, and A. Shubber, *J. Am. Chem. Soc.*, **88,** 3064 (1966).
60. E. Meyer and G. W. Griffin, *Angew. Chem. Int. Ed. Engl.*, **6,** 634 (1967).
61. J. S. Splitter and M. Calvin, *Tetrahedron Lett.*, **1968,** 1445.
62. A. M. Trozzolo, A. Manmade, and G. W. Griffin, unpublished results.
63. H. M. Frey, *Pure Appl. Chem.*, **9,** 527 (1964).
64. H. M. Frey, *Advances in Photochemistry*, Vol. 4, W. A. Noyes, Jr., G. S. Hammond, and J. N. Pitts, Jr., Eds., Interscience, New York, 1966, p. 225.
65. M. J. Amrich and J. A. Bell, *J. Am. Chem. Soc.*, **86,** 292 (1964).
66. H. M. Frey and I. D. R. Stevens, *Proc. Chem. Soc.*, **1962,** 79.
67. H. M. Frey and I. D. R. Stevens, *J. Chem. Soc.*, **1965,** 1700.
68. H. M. Frey and I. D. R. Stevens, *J. Chem. Soc.*, **1963,** 3514.
69. H. M. Frey and I. D. R. Stevens, *J. Am. Chem. Soc.*, **84,** 2647 (1962).
70. A. M. Mansoor and I. D. R. Stevens, *Tetrahedron Lett.*, **1966,** 1733.
71. H. M. Frey and I. D. R. Stevens, *J. Chem. Soc.*, **1965,** 3101.
72. H. M. Frey and I. D. R. Stevens, *J. Chem. Soc.*, **1964,** 4700.
73. G. F. Bradley, Ph.D. thesis, University of Southampton, U.K., 1967.
74. R. A. Moss, *Tetrahedron Lett.*, **1967,** 4905.
75. G. L. Closs and W. A. Böll, *J. Am. Chem. Soc.*, **85,** 3904 (1963).
76. A. C. Day and M. C. Whiting, *J. Chem. Soc. (C)*, **1966,** 1719.
77. G. L. Closs, W. A. Böll, H. Heyn, and V. Dev, *J. Am. Chem. Soc.*, **90,** 173 (1968).
78. G. L. Closs and W. Böll, *Angew. Chem.*, **75,** 640 (1963); *Angew. Chem. Int. Ed. Engl.*, **2,** 399 (1963).
79. R. Anet and F. A. L. Anet, *J. Am. Chem. Soc.*, **86,** 525 (1964).
80. G. Ege, *Tetrahedron Lett.*, **1963,** 1665.
81. W. G. Bentrude, *Chem. Commun.*, **1967,** 174.
82. R. M. G. Nair, E. Meyer, and G. W. Griffin, *Angew. Chem.*, **80,** 442 (1968); *Angew. Chem. Int. Ed. Engl.*, **7,** 462 (1968).
83. R. L. Smith, A. Manmade, and G. W. Griffin, *J. Heterocycl. Chem.*, **6,** 443 (1969); G. W. Griffin and A. Manmade, *J. Org. Chem.*, **37,** in press (1972).
84. P. Petrellis and G. W. Griffin, *Chem. Commun.*, **1968,** 1099.
85. C. Bischoff and H. Brandstaedter, *Monastber. Deut. Akad. Wiss. Berlin*, **8,** 888 (1966); *Chem. Abstr.*, **68,** 68222h (1968).
86. W. Adam and R. Rucktäschel, *J. Am. Chem. Soc.*, **93,** 557 (1971).

Index